T0260981

Beyond Blue Skies

Outward Odyssey
A People's History of Spaceflight

Series editor
Colin Burgess

Beyond Blue Skies

The Rocket Plane Programs That Led to the Space Age

Chris Petty

Foreword by Dennis R. Jenkins

© 2020 by the Board of Regents
of the University of Nebraska

All rights reserved
Manufactured in the
United States of America

Library of Congress
Cataloging-in-Publication Data
Names: Petty, Chris, author.
Title: Beyond blue skies: the rocket plane
programs that led to the Space Age /
Chris Petty; foreword by Dennis R. Jenkins.
Description: Lincoln: University of Nebraska Press, [2020] |
Series: Outward odyssey: a people's history of spaceflight |
Includes bibliographical references and index.
Identifiers: LCCN 2020004246
ISBN 9781496218766 (hardcover)
ISBN 9781496223555 (PDF)
ISBN 9781496223548 (Mobi)
ISBN 9781496223531 (epub)
Subjects: LCSH: Research aircraft—United States—
History—20th century. | Rocket planes—
United States—History—20th century. |
Space vehicles—United States—History—20th century.
Classification: LCC TL567.R47 P48 2020 |
DDC 629.133/380973—dc23
LC record available at
https://lccn.loc.gov/2020004246

Set in Adobe Garamond Pro by Laura Buis.
Designed by R. Buchholz.

In Memory of Jerry S. Reedy

Now pilots are highly trained people, and wings are not easily won. But without the work of the maintenance man, our pilots would march with a gun. So, when you see the mighty jet aircraft as they mark their way through the air, remember the grease-stained man with the wrench in his hand; he is the man who put them there.

—From *The Forgotten Man*, author unknown

Contents

Part 3. The Lifting Bodies

Illustrations

Foreword

The high-speed X planes that once blazed across the desert skies over Edwards AFB have been featured in many books. So the obvious question is: Why another?

Despite the coverage they have received, there is a different perspective to be discussed. Unlike most writers, myself included, who focused on the technical aspects of the aircraft, Chris Petty brings the human side into the equation, providing additional details on the main characters and the motivation behind many of the decisions that affected the efforts. Fortunately, Chris has an approachable and readable style that lends itself to telling this more personal side of the story.

For the most part, few of the pilots or engineers who first arrived in the High Desert to test the x-1 knew they were the beginning of an effort that would ultimately take America into space. That realization came later; the interest started to show in the years leading up to Sputnik, which itself acted as a catalyst to solidify those desires into action. For a while, it appeared that the x-15 might somehow lead to men flying into space from Edwards, but in retrospect, the capsules being launched from Cape Canaveral had an unbeatable advantage, especially after John Kennedy challenged the Soviets to the space race.

In a world, aerospace and otherwise, that has become increasingly polarized, with governments seemingly unable to act, commercial partners suing each other, and most headlines seemingly searching to embarrass everybody concerned, it is interesting to remember how things unfolded during the 1950s.

The U.S. Air Force and the NACA, in particular, with the U.S. Navy largely staying out of the fray, were both searching for missions that would increase their importance and funding. At the time, the high-speed airplanes were highly visible symbols of this soft power. The NACA, which became NASA toward the end of the period of interest to this book, was founded largely to support American military airpower, correcting a bitter lesson learned from a lack of technological prowess during World War I. The air force, riding high on the successes of its predecessors in World War II, had money but found that it still needed the researchers from the NACA/NASA to make the breakthroughs required to go higher and faster. Neither organization really wanted to share the potential glory with the other; both realized they had to. Compromise was still alive and well during the 1950s.

Nevertheless, it was the personalities of a very special set of managers, engineers, and technicians in the High Desert that made it all work. Together, the two agencies, along with the navy and the contractors that actually built the machinery, greatly expanded the frontiers of flight, ultimately increasing the envelope from less than Mach 1 to almost Mach 7 and reaching the edges of space. Along the way, there was posturing and consternation, fights and disagreements, tragedy and glory. Despite all of this, somehow, the mission always took precedence. Chris takes this opportunity to examine the people and the compromises, bringing enough technical detail to the story to set the stage and put the broader effort into context.

The pilots and engineers at Edwards lived in relative obscurity. In fact, in the beginning, they seemed to barely live since the desert was a barren, inhospitable place. They were there because they wanted to be where the future was being born. Flying faster, and eventually higher, than ever before required a lot of room, although the relative security from prying eyes was also an important consideration in the burgeoning Cold War. They could always go to Pancho's to celebrate, relax, and torment each other with the latest accomplishment or failure. It was a dangerous game, and several extremely talented men did not make it out alive. Eventually, they gave up their temporary quarters for the relative comforts provided by the desert towns of Rosamond, Lancaster, and Palmdale. As the aircraft gained performance, Edwards, and the NACA facility in its corner, became more sophisticated to support them.

It was all in stark contrast to what became the show at Cape Canaveral. There were no hotels full of journalists watching their every move as would become commonplace during Mercury. Instead, there were long hours of carpooling to the base, followed by even longer hours in the simulators.

All for an eight-minute flight that would likely not find its way into any publication save maybe *Aviation Week*. Magazines were not staging events and documenting the pilots' every move, and car companies were not giving the test pilots, let alone the engineers, Corvettes to drive around in. Nobody minded.

As Chris will show, this unique set of people set the stage for the future space race. When it came time to go into space, NASA elected to emulate much of the operational model being used in the High Desert. Conceptually, it was a very similar task: hundreds of hours of research and simulation for a very short flight. Indeed, many of the people from Edwards played major roles in sending men to the moon.

It had taken forty-four years from the first flight of the Wright brothers until Chuck Yeager broke the sound barrier for the first time in the x-1. It would take only fourteen more years for Bob White to reach Mach 6 in the x-15. A mere twenty years later, John Young and Bob Crippen would pilot *Columbia* at Mach 25 as it flew home from orbit.

It was not a direct or linear path from Walt Williams and the establishment of the Muroc Flight Test Unit to *Columbia* arriving at the Dryden Flight Research Center, but without the efforts of the air force and NASA in the High Desert, it is unlikely that the space shuttle could have been developed and flown.

Engineering achievements do not happen in a vacuum, and the personal interactions among the players are an important, but oft overlooked, component in the success, or failure, of any effort. Enthusiast histories tend to focus on the technology; and social histories, on the people. The works are often sandboxed by whatever organization paid to have them written or by the author's particular interest, focusing on the air force or NASA side, as appropriate. All of these often miss the interactions between the organizations, between the people, and between the people and the hardware. Chris has bridged this gap, showing that neither the air force nor NASA could have accomplished the end result without the other. He also demonstrates what can be accomplished when reasonable people behave reasonably, acknowledging the strengths and weaknesses of the other, allowing each to share in the glory, and tragedy. It is a condition we should strive to return to in our reach toward the stars.

Dennis R. Jenkins

Acknowledgments

In the course of writing this book, I have often been told how much easier it would have been had I started two decades ago when many more of the protagonists were still able to tell their own stories. Unfortunately, that was not an option, but I have had the great privilege of speaking with at least some of the people who helped bring about this golden age of aviation, and my sincerest thanks go out to Johnny Armstrong, Charlie Baker, Vince Capasso, Bob Hoey, Bob Kempel, Wayne Ottinger, Jon Pyle, and Dave Stoddard. I am equally indebted to Michelle Evans, Al Hallonquist, and Carol Reukauf for opening the doors to so many wonderful characters. To all of those who feature in these pages but whom I can never meet, I thank you for your work and inspiration.

Every journey has its starting point, and for me that was a meeting with my now good friend Erik Reedy, during which he shared stories and artifacts from his grandfather Jerry's long career with the NACA and NASA at Edwards. Erik, I thank you and your family for the generosity and kindness shown over recent years. Another huge source of inspiration and advice has been Dennis Jenkins, whose tireless guidance has kept me on the right track and began with the words, "Don't worry, the first book is always the toughest!" With that in mind, I'd like to thank Robert Taylor, Sara Springsteen, Leif Milliken, and Courtney Ochsner at the University of Nebraska Press for making it less tough than it might have been. Thanks also to Jeremy Hall for helping me fashion a coherent whole from this book's many parts.

I'd also like to thank Christian Gelzer, Peter Merlin, Tony Moore, Mark Pestana, and George Welsh, who helped make my visits to the Antelope Valley so rewarding and fruitful. It is a special kind of place with a special kind of people, and I hope that never changes.

My search for suitable illustrations was greatly aided by the generosity of the following: Eleanor O'Rangers of the Southeastern Pennsylvania Cold War Historical Society, Carrie Rasberry of BellX2.com, and Ray Meissner of the Niagara Aerospace Museum. I must also thank the many people who have provided information, support, and encouragement at various stages, including Gabe Bennett, Colin Burgess, Jeanie Engle, Francis French, Fred Haise, Milton McKay, Kevin Rusnak, Charles Simpson, Rowland White, and Al Worden, and not forgetting Emily Carney and the Space Hipsters.

Final thanks go to my wife, Christine, and son, Jack, without whose support I could never have undertaken this journey.

Introduction

Between 1946 and 1975 an ancient dry lake bed in California's High Desert bore witness to an incredible aeronautical adventure. During those three decades, Rogers Dry Lake in the western Mojave Desert played host to a series of rocket-powered research aircraft built to investigate the outer reaches of flight or, in the words of Hugh L. Dryden, "to separate the real from the imagined."

Muroc Army Air Field (renamed Edwards AFB in 1949) became the focus of intense activity as these experimental aircraft probed the mysteries of supersonic and hypersonic flight before eventually breaking the boundaries of space. Along with the machines came the men and women who could make sense of the whole adventure: the aerodynamicists, engineers, human computers, flight planners, and countless other support staff needed to put the rocket planes in the air and digest the data they brought back. At the apex of this pyramid sat the test pilots, the highly trained individuals who would put their lives on the line, pitting themselves against both machine and nature. Theirs was the most visible role to the general public, often earning them the temporary adulation of the media during the early years, as they were bestowed such dramatic titles as "the fastest man alive" or "the first spaceman," but in reality, life in the High Desert proved far from glamourous.

The rocket planes that flew over this parched landscape did so under the combined flags of the air force, the navy, and the National Advisory Committee for Aeronautics or its successor, the National Aeronautics and Space Administration. The differing priorities and methodologies of these organizations occasionally saw strained relationships and rivalries, but the lessons learned through these joint ventures contributed directly to the success of America's space program. Even as the United States reeled from the shock of Sputnik and the "Flopnik" response of Vanguard TV3 in 1957, a revolutionary new aircraft was taking shape at North American Aviation's Inglewood plant in Los Angeles. As the media questioned America's technological capabilities and education system, few noted that the x-15 would be the first winged vehicle capable of flying beyond the atmosphere and across the boundary of space.

When attention turned to NASA's Mercury 7 astronauts and John F. Kennedy's moon-landing challenge, the rocket planes of Rogers Dry Lake continued to carve their mighty arcs across the blue Mojave skies, their pilots facing risks every bit as critical as their astronaut brethren. But for the most part, the newer generation of highly educated engineering test or research pilots did not become household names, feted by journalists, or idolized by the nation's youth. Theirs was a life of car pools, long hours in simulators, planning meetings, and postflight celebrations or wakes in the bars of Rosamond—all for a standard government salary with none of the perks afforded to their Houston counterparts. They were just one vital part of the small, close-knit teams that came to characterize these mighty endeavors, sharing a camaraderie forged through mutual respect and an indomitable can-do attitude.

Over the intervening years, there have been many technical histories concentrating on the what and how of the individual rocket plane programs. Some of the protagonists, especially the oldest and boldest of the test pilots, have recorded their own experiences in memoirs, but more often the stories of what was achieved during this intense period of progress have drifted away from the public consciousness like dust on the ever-present Mojave winds.

In this book, I aim to shine new light on some of the unsung heroes who helped advance our understanding of flight during those three decades. Behind the dry technical language of the flight reports lies a very human story of determination and perseverance, of great achievements and tragic loss. The legacy of these pioneers, all too often reduced to a few brief lines in the record books, deserves deeper examination so that it may inspire new generations to push on with the same spirit of adventure and curiosity.

I hope this book proves a worthy testament to the many men and women who toiled in the desert during those years in order to push our boundaries ever upward and outward, *Beyond Blue Skies.*

Abbreviations and Definitions

AAF	Army Air Field
AFB	Air Force Base
AFFDL	Air Force Flight Dynamics Laboratory
AFFTC	Air Force Flight Test Center
AFSC	Air Force Systems Command
AMC	Air Materiel Command
Angle of attack	The angle between the oncoming airstream and a reference line on the vehicle (often between the leading and trailing edge of the wing)
APU	Auxiliary power unit
ARDC	Air Research and Development Command
ARPS	Aerospace Research Pilot School
ATSC	Air Technical Services Command
FRC	Flight Research Center
GEDA	Goodyear L3 Electronic Differential Analyzer
HRE	Hypersonic research engine
HSFRS	High-Speed Flight Research Station
HSFS	High-Speed Flight Station
Hypersonic	A condition in which the airflow across the object exceeds Mach 5

ICBM	Intercontinental ballistic missile
JATO	Jet-assisted takeoff
L/D	The ratio of an aircraft's lift to its drag
LO2	Liquid oxygen
Mach number	The ratio of an object's true airspeed to the speed of sound in the surrounding air
MFTU	Muroc Flight Test Unit
NACA	National Advisory Committee for Aeronautics
NASA	National Aeronautics and Space Administration
OART	Office of Advanced Research and Technology
PARD	Pilotless Aircraft Research Division
PIO	Pilot-induced oscillation
RAPP	Research Airplane Panel
RCS	Reaction control system
RMI	Reaction Motors Inc.
SAG	Scientific Advisory Group
Scramjet	Supersonic combustion ramjet
Subsonic	A condition in which the airflow across an object is less than Mach 1 at all points
Supersonic	A condition in which the airflow across an object is greater than Mach 1 at all points
t/c	The ratio of a wing's thickness to its chord
Transonic	A condition in which the airflow across an object is less than Mach 1 at some points but exceeds Mach 1 at others
USAAC	U.S. Army Air Corps

USAAF	U.S. Army Air Forces
USAF	U.S. Air Force
USAF TPS	U.S. Air Force Test Pilot School
USNTPS	U.S. Naval Test Pilot School
WADC	Wright Air Development Center
WADD	Wright Air Development Division

Beyond Blue Skies

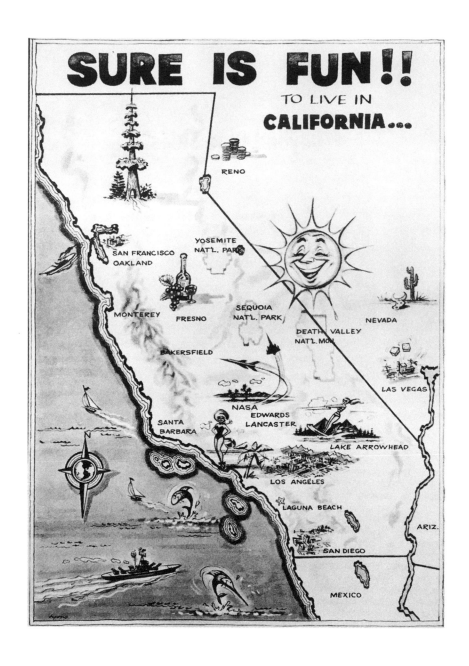

1. 1958 NASA map of California showing the location of Edwards AFB and the High-Speed Flight Station. Courtesy NASA.

Part 1

Breaking Barriers

1. A New Breed of Aircraft

At first glance, the many rocket planes that flew over Rogers Dry Lake between 1946 and 1975 seem an odd assortment. From the bullet-like XS-1 to the wingless flatiron-shaped X-24B, each was a pioneer of its time with form closely following function. But beyond the multitude of aviation firsts these aircraft achieved, all shared one common purpose; they were flying laboratories, built and flown in the pursuit of aeronautical knowledge.

Their story begins with the pioneering work carried out by Austrian physicist Ernst Mach in the late nineteenth century, which first revealed the detailed relationship between projectiles moving at high speed and the fluid, in this case air, through which they were traveling. As an object moves through air, it generates a series of pressure waves that radiate away in all directions at the speed of sound. If the object is moving at a subsonic speed (where none of the airflow around it exceeds the speed of sound), then pressure waves radiating ahead of it will inform air molecules in the object's path of its imminent arrival, allowing them to move smoothly out of the way; consequently, the air does not become compressed by the object. However, as the object accelerates, the distance between it and the pressure waves radiating ahead of it is reduced, allowing air molecules less opportunity to move away from the object's path. As the object approaches the speed of sound, the pressure waves and air molecules ahead are unable to move from the object's path before it arrives, and consequently they become compressed into a shockwave. For many years, scientists had known that projectiles could move faster than the speed of sound, but using the recently developed technique of Schlieren photography, Mach was able record the previously invisible patterns of shockwaves generated by fast-moving projectiles in detail for the first time.

The speed of sound itself is not a constant value; rather, it varies with the temperature and pressure of air. Within our atmosphere the local speed of sound generally decreases as altitude increases. For example, at sea level, under average conditions, the local speed of sound is in the region of 760 mph, but at an altitude of 35,000 ft., where both air pressure and temperature are considerably lower, this decreases to approximately 660 mph. In honor of his groundbreaking contributions to the field, the ratio of an object's speed with regard to the local speed of sound is known as its Mach number (unusually for a unit of measurement, "Mach" appears before a value, rather than after it). Mach 1 is achieved when airflow over the entire object exceeds the speed of sound, at which point the object is said to be supersonic. While aerodynamicists recognized that simple symmetrical shapes like bullets could readily exceed Mach 1, the prospects seemed less certain for more complex objects, such as an aircraft.

In the wake of the First World War, improvements in aerodynamics and propulsion led to a steady increase in the speed and altitude at which aircraft could fly. But although piston engines were becoming increasingly powerful and therefore capable of turning propellers at higher speeds, the propellers themselves were not generating the expected increases in thrust. Puzzled by this apparent shortfall, aerodynamicists, including Dr. Hugh Dryden at the U.S. Bureau of Standards, began to investigate the problem.

As propellers spin, their outer sections travel farther than their inner sections and therefore move at higher speeds. A propeller's airfoil generates lift (and therefore thrust) by forcing air to flow faster over its upper surface than its lower surface just like an aircraft wing, but as propellers began to spin faster, airflow over the outer sections of the blades was now exceeding 600 mph—a high percentage of the speed of sound. Wind tunnel tests using instrumented airfoil sections revealed that as airspeed approached Mach 1, the air ahead of the surface was being compressed to a point where a shockwave was forming, just as Mach's work had suggested. This shockwave caused a significant increase in drag but also led the airflow behind it to separate from the airfoil's surface leading to a loss of thrust. These detrimental aerodynamic effects became collectively known as compressibility and would pose a major challenge to high-speed flight during coming decades. Aerodynamicists defined the velocity above which an airfoil would begin to encounter compressibility as its critical Mach number. When this velocity is exceeded, the airflow over the airfoil becomes supersonic in some places but remains subsonic in others—a condition known as transonic flow.

Further wind tunnel research revealed that the airfoil's shape had an important influence on its critical Mach number. By measuring the maximum thickness between the upper and lower surfaces of an airfoil and its chord—the distance from the airfoil's leading to the trailing edge—a thickness/chord ratio (t/c) can be calculated. Aerodynamicists observed that airfoils with a lower t/c reached their critical Mach numbers at higher local speeds than those with a higher t/c, and these findings allowed the National Advisory Committee for Aeronautics (NACA) to develop a series of airfoil sections allowing propeller designers to use thicker (high t/c) profile on the inner section of the blade and thinner (low t/c) profile on the faster outer section. The use of variable section airfoils, along with the introduction of variable-pitch propellers, helped mitigate the compressibility problem, but as aircraft speeds and operational altitudes continued to increase, it seemed only a matter of time before other parts of the aircraft would also begin to suffer similar problems.

The 1930s saw a new breed of all-metal monoplanes emerging from the aircraft factories of Europe and the United States. These generally featured unbraced, cantilever wings and used lightweight aluminum skins in place of canvas to create stronger, more streamlined surfaces. Retractable landing gear were also developed to further reduce drag in flight. The growing popularity and prestige of air races, together with an increasing demand for faster airmail deliveries, fueled the inexorable push toward ever-higher speeds, and when more powerful supercharged engines arrived, the fastest aircraft were soon exceeding 400 mph in level flight. The decade also saw early research into both turbojet and rocket propulsion for aircraft, strengthening the likelihood that transonic flight and its attendant problems of compressibility would soon become a reality. With this in mind, the Fifth Volta Congress on High Speeds in Aviation, held in Italy during 1935, chose to concentrate on the issues relating to transonic and supersonic flight.

As the international delegates exchanged research and ideas, two of the more notable themes to emerge were the potential of forward- or rearward-swept wings to delay the onset of compressibility at higher speeds and the need for newer, more capable supersonic wind tunnels to enable more detailed high-speed research. While the importance of Adolf Busemann's swept-wing research was not immediately appreciated by all, the call for improved wind tunnels received a more positive reception from delegates; indeed a number of tunnels capable of simulating flights well above Mach I were already operating in Europe. However, a curious phenomenon was being observed in these tunnels, hampering their ability to investigate the pressing problems of transonic flight.

Even as delegates at the 1935 Volta Congress discussed the next steps in high-speed research, the clouds of war were once again gathering over Europe. The airplane had assumed an increasingly important role in military doctrine following the First World War, with aerial bombing now viewed as an effective alternative to long-range artillery. In order to challenge the prevailing wisdom that the bomber would always get through, a new generation of fast, well-armed fighter aircraft were developed using existing cantilever monoplanes as their starting point. When the Second World War began in 1939, classic fighter designs such as the Supermarine Spitfire and the Messerschmitt Bf 109 quickly entered combat, and these types underwent constant development throughout the conflict in an attempt to retain an advantage over their adversaries. The United States, initially some way behind the European powers in fighter development, quickly made up ground with a new generation of advanced pursuit aircraft. By the early 1940s, the U.S. Army Air Forces (USAAF) were able to field the Lockheed P-38 Lightning, the North American Aviation P-51 Mustang, and the Republic P-47 Thunderbolt, among other types.

As the operational practicalities of war demanded fighter aircraft capable of reaching ever-higher speeds and carrying more powerful armaments, it was perhaps understandable that the looming challenges of transonic flight became a secondary consideration for designers. Many new fighters featured thick, high t/c wings in order to accommodate more weapons, ammunition, and fuel. As the push for higher performance continued, research facilities such as the NACA's Langley Memorial Aeronautical Laboratory were pressed into action to refine fighter aerodynamics. Established during 1917 near Hampton, Virginia, Langley had already made significant contributions to aerodynamic research in the United States, but now the laboratory's full-scale wind tunnel was used for a series of drag-cleanup studies on military aircraft in order to improve their top speed and operational range.

As fighter aircraft edged ever-closer to the 500 mph mark, their pilots began to experience worrying control problems during high-speed dives. Many reported severe buffeting, with their controls becoming less effective. In the worst cases, the aircraft's controls might become reversed, or the dive would steepen of its own accord. Many pilots found themselves unable to regain control of their runaway aircraft until they reached lower altitudes, while for an unlucky few there was no recovery as their aircraft struck the ground or disintegrated in midair. As predicted, aircraft wings and tails were now exceeding their critical Mach numbers while diving through the thinner high-altitude air and consequently were encountering compressibility. As airflow over these surfaces became transonic, strong shockwaves moved back across their upper surfaces, causing the buffeting and increased drag. As the airflow became detached from the airfoil behind these shockwaves, the movable control surfaces were rendered ineffective or in some cases experienced uncontrolled movements, causing control reversal or a downward pitching known as Mach-tuck. In extreme cases, ailerons or elevators would oscillate wildly in the disturbed airflow, and if this oscillation—or flutter—hit a resonant frequency, the surfaces could exceed their structural limits and break up, often resulting in the loss of the aircraft.

During 1935, British aerodynamicist William Hilton observed a new phenomenon during high-speed wind tunnel research. In explaining his discoveries to the press, Hilton pointed to the huge increase in drag that "shoots up like a barrier against higher speeds as we approach the speed of sound." To Hilton it seemed that the massive increase in power required to break through this resistance might render supersonic flight impossible. In the hands of journalists, Hilton's description was inaccurately portrayed as an unbreachable speed limit in the sky—a sonic wall or, as it became more popularly known, a "sound barrier." As reports of wartime pilots' terrifying encounters with compressibility reached the popular press, some now wondered if this sound barrier might be to blame. But rather than providing proof of an impenetrable obstacle, Hilton's results had actually hinted at an experimental problem that was now frustrating aerodynamicists in their efforts to understand the complexities of transonic flight.

Wind tunnel designs of the time generally featured solid-walled test sections within which a test model would be mounted on supports or stings. At subsonic or supersonic speeds, the structure of the tunnel did not disrupt

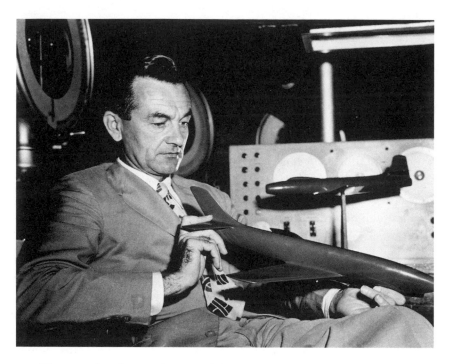

2. The NACA's John Stack, one of the original advocates for high-speed research aircraft. Courtesy NASA.

the airflow, but as airspeeds exceeded the model's critical Mach number and entered the transonic range, shockwaves emanating from the model or its stings were reaching the test section's walls and reflecting back inward. The chaotic airflow that resulted made any data gathered between Mach 0.7 and Mach 1.3—the exact range in which compressibility was occurring—extremely unreliable. As engineers sought a solution to this "choking" effect, some began to promote a different means for gathering the much-needed transonic data.

In July 1927 a headstrong young aeronautical engineer named John Stack had joined Langley, fresh from his studies at the Massachusetts Institute of Technology (MIT). Following his initial assignment to the laboratory's Variable Density Tunnel, a new facility allowing small-scale models to be tested more accurately through the use of variable air pressure, Stack became intimately involved in Langley's high-speed research programs. As choking began to affect the quality of data during these high-speed investigations, Stack began to consider other means for collecting transonic data.

Recognizing that choking would not be a problem for a free-flying laboratory, he began designing a hypothetical, highly streamlined piston-engine monoplane capable of performing sustained research at transonic speeds. In 1934 Stack presented his design study, and although the NACA chose not to pursue the project, the exercise proved useful in promoting Stack's belief that the apparent barriers to supersonic flight were not physical obstacles but rather shortcomings in experimental technique.

While Stack was promoting the idea of a dedicated high-speed research aircraft within the NACA, a young aeronautical engineer in the Midwest was thinking along similar lines. After graduating with a degree in mechanical engineering from the University of California in 1928, Ezra Kotcher had joined the U.S. Army Air Corps (USAAC) as a civilian instructor at the newly formed Air Corps Engineering School based at Wright Field near Dayton, Ohio. Like Stack, Kotcher was fascinated with the challenges of high-speed flight and shared the NACA engineer's conviction that a full-scale research aircraft would provide the best means to explore the transonic zone. In 1939 Kotcher was asked to prepare a report for the Kilner-Lindbergh board, established the previous year by USAAC major general Henry "Hap" Arnold in order to examine the future of aeronautical research in the United States.

In his submission, Kotcher proposed that a transonic flight research program should be initiated to verify and supplement the unreliable wind tunnel data. The far-sighted Kotcher suggested that either turbojet or rocket propulsion could be used to push the proposed research through the increased drag at transonic speeds. By 1939 Great Britain and Germany were both involved in turbojet propulsion research, but beyond limited studies by the NACA, little had been done in the United States. The prospects for rocket propulsion looked similarly distant, with existing efforts focusing on the development of rocket-assisted-takeoff systems for large aircraft rather than primary propulsion systems. Kotcher's proposal, like Stack's before it, was met with limited enthusiasm; however, events soon conspired to change American attitudes toward transonic research.

In February 1941 Arnold contacted NACA and National Defense Research Committee (NDRC) chairman Vannevar Bush to make him aware of the significant German advances in turbojet and rocket propulsion. Arnold was concerned by the lack of similar efforts in the United States and stated that in his view research into these new propulsion methods was a matter of national security and therefore fell under the remit of the NDRC rather than the NACA (who traditionally dealt more in pure research). His concerns were compounded during a visit to Great Britain some weeks later when he saw the Gloster E.28/39 test bed aircraft carrying a Frank Whittle–designed W.1 turbojet being prepared for its first flight. Arnold subsequently obtained an agreement from the British government to provide a full set of blueprints for Whittle's engine plus an example of the W.1 power plant itself. Once back in the United States, Arnold issued top-secret contracts to General Electric, who would re-create the Whittle engine, and the Bell Aircraft Company, who would design a suitable airframe to test the new power plants. By the time the USAAC became the USAAF in July 1941, Arnold had successfully laid the groundwork for America's early jet program.

In the meantime, Bush had responded to Arnold's earlier letter, agreeing on the need for increased urgency regarding turbojet and rocket propulsion research but insisting that the NACA rather than the NDRC would be best equipped to undertake this work. Some limited research into axial compressors had already been carried out at Langley, but Bush now formed a Special Committee on Jet Propulsion to be headed by venerable engineering professor and former NACA chairman William Durand.

Following Arnold's decision not to share news of the British W.1 engine's arrival in the United States with Bush or the NACA, initial work at Langley followed a different direction, focusing on a ducted fan system known as a Campini engine, after the Italian designer Secondo Campini. This system used a conventional reciprocating engine to drive a compressor, creating a high-pressure airflow into which gasoline was injected and ignited (essentially what we would now recognize as an afterburner). A test rig nicknamed the Jeep was constructed at Langley, and designs for a straight-wing, V-tailed research aircraft to test the system in flight were produced. Unfortunately, the fuel-hungry Jeep failed to live up to its early promise during ground runs in late 1942, leaving the Durand committee distinctly unimpressed. Work on the ducted fan approach was subsequently suspended before the project was finally canceled in 1943.

Unbeknownst to many within the Durand committee and the NACA, the United States had already flown its first turbojet-powered aircraft prior to the Jeep's cancellation. The rapid chain of events set in motion by Arnold following his 1941 trip to Great Britain had yielded the Bell xp-59a Aira-comet, powered by two General Electric type 1-a engines. The Airacomet was being tested by Bell and the usaaf at the secluded Muroc aaf test site in California, where it flew for the first time on 1 October 1942. Professor Durand himself had in fact attended initial flights of the prototype, but given the top-secret classification attached to the program, it seems unlikely he was able to share the news with the Durand committee as a whole.

Even as work on the ill-fated Jeep was underway at Langley, John Stack had continued his studies into compressibility, spending the early months of 1942 devising solutions to the Lockheed p-38's high-speed dive prob-lems. This led to the installation of underwing dive flaps, which allowed the pilot a means of recovery, but with several frontline aircraft running into similar control problems, it was clear that a more effective means to per-form compressibility research remained an urgent priority. In recognition of this, a new Compressibility Research Division was established at Lang-ley with Stack as its head, and from this position, he continued to promote the idea of a transonic research aircraft capable of gathering the data that wind tunnels could not. However, while the naca's remit covered funda-mental research into aeronautical issues, the agency had not generally ven-tured into applied research such as building and testing its own aircraft. If Stack wished to expedite a transonic flight research program, he would need to find another organization to sponsor the aircraft's construction. In preparation for this, he gathered a small team of engineers to work on potential configurations for the proposed aircraft, recruiting, among oth-ers, a determined young engineer named Walter C. Williams, who would later play a pivotal role in the research aircraft story.

Throughout 1943 the main focus for both the naca and the military remained the problems of compressibility on the current generation of war-planes, but in December of that year Durand organized an naca confer-ence to discuss British progress in turbojet propulsion. By now the close veil of secrecy surrounding the Bell xp-59a had lifted slightly, meaning many attendees were aware that turbojet propulsion had become a viable technol-ogy for the United States. Those with the appropriate security clearances would no doubt also have been informed of Britain's decision to pursue its own transonic and supersonic research aircraft program with the turbojet-powered Miles m.52, as well as persistent intelligence rumors of a 1,000 kmh aircraft program in Germany. When Durand asked delegates how the nascent turbojet technology should be utilized by the United States, Robert Wolf, an engineer for Bell Aircraft, responded that interested parties including the USAAF, the navy's Bureau of Aeronautics (BuAer), and the NACA should consider working with industry on specifications for a high-speed research aircraft. The resulting vehicle could be funded by the military but operated by the NACA, who would then distribute research results to all parties. The NACA responded positively, recognizing a close synergy between Wolf's suggestion and the proposals Stack had already put forward. In February 1944 the NACA's head of aeronautical research, Dr. George W. Lewis, called for the establishment of a High-Speed Panel to coordinate efforts between Langley and the NACA's other research facilities, the Ames Aeronautical Laboratory at Moffet Field, California, and the Aircraft Engine Research Laboratory near Cleveland, Ohio. The High-Speed Panel met for the first time the following month at Langley, with John Stack prominently among its members.

As an interim measure, engineers within the NACA were already working on more limited means to collect transonic data. Instrumented shapes known as drop bodies were released from high-flying aircraft and tracked by radar as they fell, but these relatively simple aerodynamic shapes couldn't give precise information on the interplay of shockwaves that the aerodynamicists so badly needed. In an attempt to obtain more detailed data, Langley engineer Robert Gilruth developed a wing-flow testing method in which small airfoil or aircraft models were mounted on the upper wing surfaces of an NACA P-51, attached to sensitive airspeed and pressure instrumentation within the Mustang's gun compartment. Once the pilot had reached high altitude, he could dive the P-51 to transonic speeds, creating a supersonic flow around the model. Although this method returned useful data, it was still not able offer a complete insight into the intricacies of transonic airflow, while the risks associated with repeated high-speed dives also made it an unappealing long-term solution for the conservative NACA.

As Stack continued to lobby for a dedicated research aircraft, Ezra Kotcher (who had by now been called up to active service with the USAAF) was working on a revolutionary project that seemed to support his earlier convictions that rocket propulsion might soon provide a gateway to high-speed flight. In 1943 Kotcher had been selected as the USAAF's project officer for the Northrop XP-79 rocket-powered, flying-wing interceptor, and it seemed as though the Aerojet XCALR-2000A-1 Rotojet rocket motor that was being developed for this aircraft might also prove suitable for a high-speed research aircraft. In order to assess the relative merits of turbojet and rocket propulsion, Kotcher initiated the wryly titled "Mach 0.999" study (the subject of supersonic flight was still regarded as far-fetched by some), which suggested that rocket propulsion would indeed offer better performance at higher altitudes, where Mach 1 could be achieved at lower speeds. Following this conclusion, Kotcher asked designers at Wright Field to draw up an initial configuration for a rocket-powered transonic and supersonic research aircraft. The design that emerged had a bullet-shaped fuselage with a thin midmounted straight wing and a flush canopy. Although the horizontal tail surfaces were mounted on the fuselage rather on than the vertical fin, this design essentially laid the blueprint for what was to follow in the coming years.

Another key milestone in the quest for a high-speed research aircraft occurred on 16 March 1944, when the NACA hosted two conferences on the issue of compressibility at Langley. Here, representatives from the USAAF and BuAer heard John Stack put forward the case for a research aircraft to investigate transonic flight while work continued on solutions to the wind tunnel choking problem. Subsequent discussions led to a broad consensus that joint military and NACA projects to develop a research aircraft were indeed desirable; however, there was less agreement regarding the overall goals for any such project. Stack's preferred approach was for a turbojet-powered aircraft capable of sustained transonic flight, allowing for a thorough examination of conditions approaching the speed of sound. Using such a vehicle, the NACA could perform transonic research with the aim of devising recommendations on how to mitigate the effects of compressibility in future aircraft designs. Supersonic flight, although interesting, was not the agency's primary goal at this stage. While the NACA's views were broadly shared by the BuAer representatives, the USAAF hoped to quickly push beyond Mach 1, opening up a new realm of flight for future turbojet or rocket-propelled operational aircraft.

The conferences concluded with a recommendation that the USAAF and BuAer should each create specifications for a research aircraft to satisfy their needs, at which point they should find suitable manufacturers and expedite the procurement processes. The NACA would provide technical support during the design and development phases, as well as research instrumentation and pilots as required once the resulting aircraft were ready to fly.

Although the 16 March conferences marked a true starting point for the research aircraft programs that Stack and Kotcher had been pushing for since the late 1930s, the decision to proceed was a somewhat mixed blessing for the NACA. Throughout the wartime years, leading figures within the military, most notably General Arnold of the USAAF, had increasingly felt the need to take matters into their own hands in order to keep their services abreast of the latest aeronautical technologies. Arnold's decision to keep the acquisition of the W.1 turbojet and the ensuing XP-59A project at a top-secret level was indicative of a growing perception that the NACA had failed in its remit to "supervise and direct the scientific study of the problems of flight." The proposed research aircraft now presented the civilian agency with an opportunity to conduct full-scale investigations into one of aviation's most pressing problems, but conversely, if the research aircraft proved to be an effective and affordable means for the various parties to advance their aeronautical knowledge, would there still be a need for large and expensive facilities such as those at Langley? During the springtime of 1944, with war still raging and intelligence reports of major German breakthroughs in high-speed aviation and rocketry slowly filtering through, there was little practical alternative for the NACA but to press on in search of valuable transonic data, no matter how it was obtained.

Within USAAF Materiel Command at Wright Field, however, the decision to sponsor a high-speed research aircraft was merely one step toward a larger aspiration. Traditionally, the air forces had relied on highly paid company pilots to carry out test programs on new aircraft, with any flight research generally falling to the NACA. The USAAF's own test pilots handled operational testing, verifying constructors' findings and ensuring that an aircraft was fit for squadron deployment. But by 1944 thoughts were turning to the USAAF's postwar role, and with aviation seemingly on the cusp of a technological revolution, General Arnold was keen to identify key goals for the coming years. He invited revered aerodynamicist Theodore von Kármán (director of the Guggenheim Aeronautical Laboratory at the California Institute of Technology) to head the new Scientific Advisory Group (SAG) with the aim of identifying how the USAAF could build and retain a position at the forefront of aviation technology during the coming decades. Von Kármán had been one of Arnold's trusted advisors for many years and had already verified Kotcher's approach regarding rocket-powered aircraft. When the SAG reported back to Arnold, they placed heavy emphasis on the need for supersonic aircraft and recommended that the USAAF pursue a more active role in research through further investment in its science and technology divisions. The first obvious step in this strategy was to push beyond Mach 1, but rather than relying on others to do this for them, Arnold was now eyeing an increased flight research role for the USAAF itself.

Against this background of shifting needs and priorities, the NACA, BuAer, and USAAF Materiel Command (which, along with the Air Services Command, became the Air Technical Services Command, or the ATSC, during the summer of 1944) continued discussions to refine their respective requirements. Stack and Kotcher were in broad agreement that the USAAF-sponsored aircraft should be of a generally conventional design, featuring straight wings and a tail comprised of vertical and horizontal surfaces. This would ensure that any data gathered could be directly applied to the first generation of jet fighters, expected to feature similar configurations. They also agreed that the aircraft should be built to tolerate loads as high as 15 g's, given uncertainties about the exact conditions it might encounter. Significantly, the NACA stipulated that the aircraft's horizontal stabilizer should be adjustable during flight, allowing the pilot better control under transonic conditions and that it should be mounted on the vertical fin rather than on the fuselage, to avoid downwash from the wing.

Two major points of contention remained at this stage: first, the thickness of the wing airfoil, with the ATSC proposing a thinner 10 percent t/c as opposed to the NACA's preferred 12 percent, and second, the means of propulsion. John Stack continued to favor turbojet power, arguing that this would be safer and more practical in the near term by removing the inherent risks involved in taking off with a full load of volatile rocket fuel, while also allowing for longer duration research flights. In saying this, he was also no doubt mindful that the NACA's chief research pilot Mel Gough had stated that "no NACA pilot will ever be permitted to fly an airplane powered by damned firecracker." However, Kotcher countered that rocket propulsion would allow the aircraft to take advantage of the lower local speed of sound at higher altitudes (where turbojets were less efficient) and guarantee the ability to perform research runs during level flight rather than dives. Although the NACA might be prepared to settle for a more conservative approach, Kotcher knew that a rocket motor could more quickly get the USAAF where it wanted to go—namely, 800 mph at an altitude of 35,000 ft. and Mach 1 and over. Given that the air forces would be paying for the research aircraft, Stack was forced to concede on this point, knowing that the USAAF-sponsored aircraft would not be the NACA's only option.

By 1944 the U.S. military was heavily committed in both the European and Pacific theaters. The country's considerable industrial base had been fully mobilized to produce wartime materials for both its own forces and those of allied powers. Within the aviation industry, unprecedented cooperation between companies meant that they were often building each other's aircraft as military needs demanded. Huge new factories were opened in the Midwest, and even automotive companies such as Ford were brought in to produce aircraft subassemblies. Many of the larger manufacturers had expressed a tentative interest in creating a high-speed research aircraft. But in reality, most were already at full capacity with wartime production, and the prospect of tying up top designers and engineers on a project to build a handful of aircraft with little likelihood of follow-on orders was simply not a practical proposition. Fortunately for Ezra Kotcher, the solution to this problem arrived on his doorstep in the form of Robert Woods of the Bell Aircraft Corporation.

As cofounder and chief engineer for the New York State–based company, Woods had helped build Bell's reputation for innovative design solutions. Having already created the XP-59A Airacomet, Woods and company president Larry Bell were receptive to comparatively small projects dealing with new technologies, and as none of the company's aircraft were then considered vital to the war effort, Bell was in a position to consider new opportunities. After hearing of Kotcher's requirements and his difficulties in securing a suitable manufacturer, Woods quickly pledged Bell Aircraft's support for the high-speed research aircraft program. Engineers at Bell had already been considering the problems of high-speed flight, and with Woods himself having earlier worked at Langley alongside John Stack, the company seemed ideally placed to serve the needs of all parties.

After Woods took Kotcher's requirements back to Bell's Wheatfield base near Buffalo, he was able to present the company's response during a meeting between ATSC and the NACA High-Speed Panel, held at Langley on 13 December 1944. With events now moving apace, Bell and the NACA engineers worked together to refine the proposed design, and by 21 December, Bell and the ATSC were able to agree on a set of characteristics for the aircraft. Although no formal contract had been signed at this stage, Bell began a detailed design study in advance of a follow-up conference between Bell Aircraft, the ATSC, and the NACA held in early March 1945, allowing all parties to review and approve the general configuration. On 16 March an official contract was signed between the ATSC and Bell Aircraft for the production of three experimental rocket-propelled research aircraft under the USAAF project designation MX-653.

As the USAAF's plans were finalized at Wright Field, the navy were also hard at work on their own research aircraft specifications. Following the March 1944 conferences, BuAer had continued to work with the NACA, obtaining whatever data the agency could provide to guide their discussions. Lieutenant Abraham Hyatt took on the role of coordinating the available information and advocating the need for accurate high-speed data within BuAer. In September 1944 Hyatt presented the case for a navy-sponsored research aircraft in a memorandum that called for a turbojet-powered vehicle capable of achieving 650 mph at sea level and withstanding loads up to 12 g's. Emphasis was placed on making the aircraft as streamlined as possible, with a wing t/c no greater than 10 percent seen as desirable. This more conservative approach was far closer to John Stack's preferred configuration, and on 19 December BuAer contacted the NACA to confirm that it intended to expedite procurement of a research aircraft matching Hyatt's guidelines. The next big question facing BuAer was who would build its transonic plane, and again, as with Woods's fortuitously timed visit to Kotcher's office, luck smiled on the navy research aircraft program as Eugene Root, head of the Aerodynamics Division at Douglas Aircraft's El Segundo facility in California, visited BuAer to seek out old college friend Commander Bill Sweeney.

During the course of their conversation, Sweeney passed Root a copy of BuAer's draft research aircraft specifications and asked his friend if he thought Douglas might be interested. Root responded positively, and on his return to California, he showed the specifications to the division's chief engineer, Ed Heinemann. By 1944 Douglas Aircraft had already given the subject of transonic flight a great deal of thought. As a major manufacturer of naval aircraft, the company had seen its SBD dive bombers encounter compressibility problems and were understandably eager to find solutions that could be implemented on future designs. In 1941 the USAAF had asked Douglas to investigate the aerodynamics of supersonic flight, and so when the BuAer requirements arrived, the company already had some expertise on which to draw. Although the El Segundo plant remained heavily committed to wartime production, Root and Heinemann saw merit in pursuing the project, and so in early 1945, Douglas Aircraft submitted an initial design study (carrying the company model number D-558) to BuAer. On 28 February, Douglas representatives discussed the proposal with both BuAer and the members of the NACA High-Speed Panel, including John Stack. Consequently, the NACA recommended that BuAer should move forward with the D-558, and on 9 May the proposal received formal approval.

And so, a mere day after the allied forces had declared victory in Europe, not one, but two high-speed research aircraft programs were finally underway in the United States. Although interservice rivalries seemed sure to play their part, regular communications between the NACA, ATSC, and BuAer project offices were established to help avoid unnecessary duplication of effort. Given the contrasting configurations chosen for the MX-653 and D-558 and the different goals of their sponsor organizations, it appeared from the outset that this would not be a major problem. With paperwork signed, the project teams at Bell and Douglas now began their work in earnest, but many challenges still lay ahead before either aircraft would take to the air to probe the mysteries of high-speed flight.

2. Like a Speeding Bullet

During 1944 a team headed by Stan Smith began to examine the challenges that lay ahead for Bell Aircraft. The engineers at Wheatfield had been tasked with designing a unique new aircraft that would not only offer performance beyond anything that had gone before it but also be strong enough to repeatedly endure whatever punishment might lie in wait as it approached Mach 1. Following their December meetings with the NACA and the ATSC, Robert Woods and chief engineer Bob Stanley had returned with a list of requirements for the research aircraft:

It should be practical, meaning the design should be as simple and straightforward as possible.

It should be able to carry between five hundred and one thousand pounds of research instrumentation, including all associated wiring and tubing.

It should be capable of transonic, and if possible supersonic, flight at an altitude of approximately 35,000 ft.

It should be able to take off from a seven-thousand-foot runway.

The pilot should be seated, not prone (there had been some discussion that a pilot lying prone in the cockpit may be better able to endure high g-forces).

It should have a flight duration of at least ten minutes.

It should be capable of enduring forces of 18 g's.

It should be completed within one year.

While these points offered valuable guidance for Smith's team, many fundamental issues still needed to be addressed, including the key question of fuselage shape. To gather relevant opinions and data on this matter, Smith dispatched engineer Benson Hamlin and aerodynamicist Paul Emmons on a fact-finding tour of the country's leading aeronautical institutions, but although the pair heard various opinions on the subject, none of the experts they met with were able to offer conclusive data to aid their design. Consequently, they fell back on one of the few certainties available to them—that bullets were known to be capable of stable supersonic flight. This led the Bell duo to the USAAF's Ballistics Laboratory at Wright Field, to learn more about the properties of high-speed projectiles. Based on the available research, the shape that appeared to offer the most promise was the .50 caliber bullet, as this demonstrated tighter grouping when fired from a machine gun, indicating good stability during flight. Armed with this insight, Hamlin and Emmons returned to Bell's Wheatfield plant to begin work on their design.

The thickness of the aircraft's wings had already been discussed at length by the ATSC and the NACA, with Kotcher and Langley's Gilruth favoring a thinner profile to raise the wing's critical Mach number. Stack, however, held out for a thicker profile, which he felt would be more representative of the wings used for the next generation of operational aircraft. Fortunately, a compromise was reached whereby the first aircraft would feature an 8 percent t/c wing, while the second would carry a thicker 10 percent version. In both cases, the horizontal tail surfaces would use a profile 2 percent thinner than the wing, ensuring that wing and tail would not encounter compressibility simultaneously, allowing the pilot to retain some control over the aircraft during transonic flight. Given that these thin wings were required to withstand a maximum force of 18 g's, it was no surprise that their design and manufacture proved challenging. After much trial and error, Bell devised new rolling techniques to create thick aluminum wing skins, tapering from a half-inch thickness at the wing's root to a one-thirty-second-inch thickness at the tip. Construction was further complicated by the need to drill 240 pressure-sensing orifices into the skin of the left wing, with the structure also needing to accommodate the associated piping plus a dozen strain gauges. During their design studies, Bell engineers had examined a variety of different tail configurations before reverting to a conventional vertical fin with high-mounted horizontal surfaces, but alongside the usual elevators, the tail also featured a movable horizontal stabilizer as requested by the NACA—an addition that would later prove key to the aircraft's success.

To preserve the fuselage's low-drag ogival lines, Hamlin and Emmons decided that the cockpit canopy should remain flush to the nose of the aircraft, but this raised concerns about the pilot's field of view during takeoff and landing. Consequently, a cockpit mock-up was constructed and reviewed by project pilot Jack Woolams, who decided that, although restricted, the visibility remained acceptable. Within the cockpit, an H-shaped control yoke was favored over a fighter-type stick, as it was thought that this would help pilots hold the aircraft steady during the anticipated transonic buffeting. Switches and levers for power settings, trim, and data recording were placed on the yoke, allowing pilots to make adjustments without releasing their grip on the controls. A small hatch on the right-hand side of the fuselage provided the only means of entry and exit from the cramped cockpit, but given its close proximity to the thin wing's leading edge, a successful bailout during flight looked an unlikely prospect for the research aircraft's pilots.

While design of the airframe progressed, the aircraft's propulsion and fuel systems were proving more of a challenge. Although Ezra Kotcher had clearly stated his preference for rocket propulsion, some Bell engineers were still unconvinced. However, when Hamlin examined both turbojet and combined turbojet and rocket configurations for the research aircraft and found both wanting, rocket power represented the only viable option. At Kotcher's behest, the 6,000 lbf. (pound-force) thrust Aerojet Rotojet motor intended for the Northrop XP-79 was initially considered for the project. Powered by a hypergolic (meaning they would ignite on contact) combination of red fuming nitric acid and aniline, the Rotojet did not require a separate ignition system; however, the prospect of making runway takeoffs laden with corrosive, highly reactive chemicals alarmed Bell engineers. By late 1944 the woeful safety record of Germany's Me-163 rocket-propelled interceptor was no doubt known to the USAAF, and when the Aerojet motor ran into development problems, the ATSC took the opportunity to look elsewhere rather than risk similar problems. Fortunately, an ideal replacement was already under development, albeit under the auspices of the navy.

Amateur rocketry enthusiasts Lovell Lawrence Jr., John Shesta, James Hart Wyld, and Hugh Franklin Pierce had begun working together on experimental rocket designs after meeting through the American Rocketry Society in the mid-1930s. Inspired by the work of German pioneer Herman Oberth and Austrian engineer Dr. Eugen Sänger, Wyld made a major breakthrough when he began to experiment with liquid cooling as a way to prevent combustion chambers from burning through. By 1938 he had constructed a small test motor in which propellant was circulated through tubes within a jacket around the combustion chamber, cooling the chamber walls while also being preheated, allowing for more efficient combustion—a technique known as regenerative cooling. Encouraged by the test motor's performance, Wyld and his colleagues decided to pursue the technology on a commercial basis, and with America's entry into the Second World War looking increasingly likely by late 1941, the group approached the military in search of development funding. When Wyld demonstrated an improved 100 lbf. thrust version of his motor to the navy just weeks before the Japanese attack on Pearl Harbor, BuAer offered the team a $5,000 development contract, and with this investment in place, Lawrence, Shesta, Wyld, and Pierce formed the country's first commercial rocket propulsion company, Reaction Motors Incorporated (RMI).

Based in the Pompton area of New Jersey, RMI had long working relationship with the navy, which began with the development of jet-assisted takeoff (JATO) motors for flying boats—the word *jet* still being preferred to *rocket* at this time due to the latter's science-fictional associations. Wyld's original design was scaled up into a 1,000 lbf. thrust unit using a gasoline propellant and liquid oxygen (LO2) as its oxidizer. BuAer then asked RMI to design a 3,400 lbf. thrust motor for use on a planned rocket-propelled interceptor, and this design gradually evolved into a larger, multichamber 6,000 lbf. thrust unit known as the 6000C (or by its naval designation, LR8). The new engine featured four regeneratively cooled 1,500 lbf. thrust chambers, each of which could be fired independently to give four graduated power settings, but unlike Aerojet's motor, the 6000C used a far less volatile water-alcohol mixture as its fuel, along with the LO2 oxidizer.

When Benson Hamlin learned of the 6000C through meetings with BuAer in early 1945, it seemed the RMI power plant might offer an ideal solution for the MX-653, and following navy consent, discussions between RMI and Bell led Kotcher to select a variant of the 6000C for the program in May 1945. The resulting motor became known as the 6000C-4 or more familiarly by its USAAF designation, XLR11. RMI were also contracted to supply a compact low-pressure turbopump to feed propellants to the motor.

Although both the NACA and the ATSC intended the aircraft to make conventional runway takeoffs, there was still some debate within the Bell team as to how safe and practical this would be. While Woods argued for the takeoff capability, viewing this as crucial to the prospects for a follow-on rocket-powered interceptor, Stanley favored air-launching, as it avoided the dangers associated with engine failure during takeoff, while preserving the aircraft's limited propellant supply for test runs at altitude. These discussions soon became moot, however, when delays to RMI's turbopump forced Bell to adopt an alternative system rather than face long delays. The turbopump was to have operated at low pressure, meaning the cylindrical fuel and oxidizer tanks could have relatively thin walls and occupy a large volume within the fuselage. Unfortunately, the only practical alternative was a pressure-feed system using high-pressure nitrogen gas to force propellants toward the rocket motor. To withstand the increased pressure, the fuel and oxidizer tanks would now need to be spherical and thick walled, meaning they would use less of the aircraft's internal volume while also being considerably heavier than the originals. In addition, the designers now needed to find room in the fuselage for twelve spherical nitrogen tanks and a system of regulators to control the pressure of the gas.

The overall effect of these changes was a near 50 percent reduction in the aircraft's propellant capacity along with a two-thousand-pound increase in its landing weight. Air-launch now offered the only practical means of getting the aircraft to its proposed 35,000 ft. operating altitude with enough propellant for a high-speed test run in level flight. Besides, the MX-653's tricycle landing gear was not strong enough to support the increased weight of a fully fueled aircraft, ruling out the possibility of runway takeoffs for high-performance flights. Although it viewed air-launching as cumbersome and complex, the NACA conceded that the technique offered a practical means to get flights underway until the turbopump was finally ready. Boeing's B-29 bomber was selected as a suitable launch aircraft, with surplus examples now becoming available following victory in the Pacific.

In October 1945, ATSC and NACA representatives visited Buffalo to inspect a full-scale MX-653 mock-up and review progress on the aircraft. Although work on the first airframe was well advanced, its thinner 8 percent t/c wing was still in production, meaning the aircraft would be fitted with the thicker 10 percent wing and corresponding 8 percent tail for its initial trials. Bell also reported that the XLR11 was still not ready, but both the ATSC and the NACA agreed that glide trials should be carried out as soon as possible. Accepting the design, the ATSC designated its new research aircraft the Experimental Sonic-1, or XS-1.

As the first aircraft neared completion during the final months of 1945, attention turned to the upcoming flight tests and the research program that would follow. Unfortunately, it appeared that the NACA, the ATSC, and Bell Aircraft had each reached slightly different conclusions regarding how these flights would be conducted and the roles the respective parties would play.

Following their earlier discussions, the working consensus had been that the USAAF would procure the research aircraft, allowing the NACA to perform flight research before distributing the findings. John Stack and his colleagues interpreted this to mean that NACA research pilots would perform all required research flights at Langley, but during the later months of 1945 it became increasingly clear that both the ATSC and Bell Aircraft had their own ideas regarding the program's execution. While all parties still agreed that the NACA should retain responsibility for research instrumentation, flight tracking, and data analysis, the question of who would actually make the flights and where became the subject of some debate.

This situation was perhaps not entirely surprising given the different expectations that each party brought to the program. The NACA's primary goal was the acquisition of data, leading to a comprehensive understanding of transonic flight. The XS-1 flights would supplement and verify existing wind tunnel data, while teams at Langley continued work on the wind tunnel choking problem. The USAAF, on the other hand, were looking squarely toward the future of military aviation. When Theodore von Kármán's SAG had delivered their report *Where We Stand* to General Arnold in August 1945, supersonic flight had been highlighted as an essential requirement to ensure future air superiority, and in order to make this a reality, an increase in high-speed research would be required. By 1945 Arnold was already keen to reduce the USAAF's dependence on the NACA, which he felt had lacked sufficient dynamism during the war years, and the XS-1 now offered a perfect opportunity to demonstrate that the ATSC could play an active role across all aspects of a flight research program. Bell Aircraft, meanwhile, had clear commercial incentives to prolong its involvement in the program. Having built America's first jet aircraft, the XP-59A, the company had failed to secure a major production contract for an operational type. Lawrence Bell and Robert Woods now hoped that the company might carve out a position at the forefront of supersonic flight. In line with standard USAAF practice, Bell pilots would fly the XS-1's initial demonstration phase, but the company held hopes that this arrangement might be extended all the way to Mach 1.

With flight tests slated to begin during January 1946, decisions were needed, and initially it was the NACA that was to be disappointed. The civilian agency's hopes that initial glide flights would take place at Langley were dashed as the ATSC suggested moving the operation to its test site at Muroc AAF. But although Muroc offered privacy as well as the preferred ten-thousand-foot runway, winter rains looked likely to interfere with the test schedule, and Bell began its own survey of suitable locations closer to home. While it was inconvenient for the company to ferry aircraft, equipment, and employees between Buffalo and the Mojave Desert, Bell's management also worried that Muroc's proximity to large Californian competitors, such as Lockheed, Douglas, and North American, might cost the company some of its best talent. In the end, Bell's suggestion to use Pinecastle Field near Orlando, Florida, proved acceptable to all. Pinecastle had been an auxiliary bomber base during the war, and although it had now been deemed surplus to USAAF requirements, it still offered a ten-thousand-foot runway; relative seclusion; access to technical assistance via nearby Orlando AAF; reasonable proximity to Buffalo, Langley, and Wright Field; and (hopefully) good weather conditions.

On 27 December 1945 the first XS-1 was officially rolled out from Bell's Wheatfield plant, allowing tests with the recently modified B-29 carrier aircraft to take place. NACA engineers journeyed to Buffalo to install much of the XS-1's instrumentation before their project team flew south to Florida during the early weeks of 1946. Although no powered flights would be possible until an airworthy XLR11 became available, glide flights could provide answers to many basic operational questions, and to build confidence in the air-launch technique, Bell performed captive flights of the mated aircraft ahead of the move to Pinecastle, using the XS-1's research instrumentation to record pressure data. These tests established that there was a downward pressure on the research plane, but to ensure a safe release, the

B-29's bomb shackles were modified, giving the XS-I a slight downward pitch, while guide rails were fitted on either side of the bomb bay to prevent the rocket plane from yawing once released. As a final measure, Bell installed a device to push down on the XS-I as it fell, but these modifications could only truly be tested during an actual drop. As well as investigating the XS-I's general handling characteristics, the Pinecastle flights would also indicate whether dead-stick landings on a fixed runway were practical going forward. For the NACA, the flights offered an opportunity to test research and tracking instrumentation and, if things went well, gather some initial data on the aircraft's stability and control ahead of powered flights. As Bell was responsible for conducting the demonstration phase, the ATSC chose to stay in the background, monitoring progress and offering assistance if required.

Bell's team in Florida would be led by Jack Woolams, a former USAAF fighter pilot who had made a name for himself as a daring but capable flyer. After joining Bell as a test pilot during the summer of 1941, Woolams had soon transferred across to the company's Experimental Research Division, participating in the XP-59A Airacomet program at the USAAF's Muroc test site. While flying the prototype jet fighter, he had distinguished himself by setting new altitude records and also earned a reputation as a practical joker. With the USAAF and Bell keen to prevent rumors of the classified jet spreading, following midair encounters with P-38s operating from Muroc's South Base, Woolams hit on a unique solution; rather than avoiding other aircraft, he took to donning a gorilla mask and derby hat in the cockpit. Pulling alongside a P-38, he would wave a cigar at the stunned pilot before tipping his hat and speeding away. Few hotshot fighter pilots cared to report being outflown by a courteous gorilla in his propellerless plane, and consequently tales of the prototype jet remained relatively contained.

The NACA's Pinecastle test team was headed by Walt Williams, who had previously worked on John Stack's transonic research aircraft studies. A forthright Louisianan, Williams had a reputation for scientific skill and a pragmatic, no-nonsense approach to testing. With Langley management recognizing that the NACA contingent needed a strong leader to stand up to Bell's equally forceful Bob Stanley, Williams was placed in charge, and with him he brought technicians from Langley's Instrument Research Division, including telemetry expert Gerald Truszynski. Joel Baker, one of Langley's junior research pilots, would ferry the team between Virginia and Florida while also observing the tests to glean valuable information on the XS-1's handling. By 19 January 1946 both teams had gathered at Pinecastle Field to begin their preparations, with the NACA technicians working doggedly to get the aircraft's remaining instrumentation installed and calibrated, while Truszynski obtained a surplus SCR-584 radar set from the USAAF, which he then modified to track the XS-1. This preflight period highlighted differences between the working methods of the government employees and the contractor team, with Williams's men often frustrated by the Bell crew's late starts and reluctance to work during weekends.

As the teams readied themselves for the XS-1's first flight, the Florida weather worsened, although the enforced delays allowed more last-minute fixes to be made. The B-29, piloted by Bell pilots Joe Cannon and Harold Dow, was finally able to carry the bright orange XS-1 aloft for its first flight on Friday, 25 January. As Bob Stanley flew chase in a P-51, a B-17 photo ship was also on hand to record the research aircraft's inaugural drop. Unfortunately, things did not go smoothly, as Woolams, having climbed down the small extendible ladder to enter the XS-1's cockpit, discovered that the aircraft's hatch would not seal securely. On the ground, meanwhile, Truszynski was rushing to repair a burned-out motor in the radar set. With Woolams suffering in the freezing temperatures as the B-29 reached its launch altitude of 25,000 ft., the pilot elected to make the drop as soon as possible with or without the ground radar, and as Cannon put the B-29 into a slight dive, Dow counted down from ten before releasing the rocket plane. To everyone's relief, the XS-1 dropped away cleanly without recontacting the bomber—earlier, Bell's ground crew had applied red paint to the bomb bay guide rails, but none of this found its way onto the research aircraft as Woolams fell away and began to feel out the controls. After several minutes of maneuvers, including stall tests to gauge its stability, the XS-1 approached the long Pinecastle runway, but as he lined up for landing, Woolams struggled with the poor visibility offered by the flush canopy he had earlier approved. Misjudging his descent rate and coming down short of the runway, Woolams landed heavily, causing minor damage to the aircraft. But first flights were all about learning, and the test pilot felt far from disheartened. As Woolams enthused about the XS-1's performance, a relieved Bob Stanley reported back to Larry Bell, "We've got a flying machine here." While the NACA team began examining data from the flight, Woolams and his colleagues headed out to celebrate, eventually winding up at a local rodeo, where the irrepressible test pilot disappeared briefly, only to reappear moments later on the back of a bull.

The XS-1 made ten glide flights at Pinecastle, with the series drawing to a close on 6 March. Air-launching the rocket plane had proven less troublesome than feared, allowing Bell to remove the pusher device from the B-29. The XS-1 had handled as well as anyone could have hoped and shown itself tough enough to withstand the stress of unplanned rough landings. Although landing gear problems had occurred during some flights—nose gear collapses would remain a persistent issue throughout the aircraft's life—the NACA, the ATSC, and Bell were generally satisfied, looking forward with optimism to the upcoming powered flights. What had been less successful, however, was the use of a fixed runway for dead-stick landings. When combined with the often-cloudy weather (which made visual tracking difficult), any hopes the NACA had still held for testing the XS-1 at Langley Field now evaporated; the rocket plane would now be heading for the Mojave Desert to show what it could really do.

Following the Pinecastle tests, Bell set to work preparing the first and second aircraft for their powered tests at Muroc. The number one aircraft had the 10 percent wing and 8 percent tail used at Pinecastle removed and replaced with the now-completed 8 percent wing and 6 percent tail. RMI had finally delivered a fully functioning XLR11, and this was installed in the second aircraft (featuring the thicker 10 percent wing and 8 percent tail) for use during the initial powered flights. Work on the number three aircraft was halted, awaiting progress on the troublesome turbopump. With Stan Smith transferring to the recently approved XS-2 rocket plane program, the role of project engineer was taken up Dick Frost, a former Bell test pilot who had switched back to engineering after sustaining serious burns in a flying accident. But even as the new man set about readying the XS-1 for its move west, tragedy struck the program.

Woolams and chief test pilot Tex Johnston had been preparing to take part in the Thompson Trophy air race in a pair of Bell Aircobras. In the final days before the race, Jack Woolams had taken his aircraft out over Lake Ontario for one final test run, but he never returned. The stunned

Johnston chose to race in his friend's memory and, triumphing over the competition, donated half the prize fund to Woolams's widow. Few at Bell Aircraft doubted that Jack Woolams could have pushed the xs-1 to its limits, but a new pilot was now needed in his place. That task fell to former navy test pilot, Chalmers "Slick" Goodlin. Although only twenty-three years old, Goodlin had packed a lot into his brief career, including a short spell flying Spitfires for the Royal Air Force, but this latest assignment presented a very different challenge. He would now be responsible for taking the xs-1 out to the contracted demonstration speed of Mach 0.8 and, Bell's management hoped, onward through Mach 1 and into the history books.

Some months earlier, the USAAF had officially named Muroc as a flight test center under the auspices of the newly formed Air Materiel Command (AMC), which assumed the responsibilities of its predecessor, the ATSC. Under the command of Colonel Signa Gilkey, the base had moved on from its role as a wartime training facility, becoming a key asset in General Arnold's plans for a more technologically independent air force. Although the xp-59A had flown from the secretive North Base, the South Base area with its permanent runway would now be used for testing new experimental and production aircraft, including the xs-1. Prior to his untimely death, Woolams had visited Muroc to initiate development of the new facilities needed to support the xs-1 program and to arrange suitable accommodation for the Bell team in the nearby town of Willow Springs. At Woolams's behest, large propellant tanks were constructed, and a sloping cruciform pit was dug so that the xs-1 could be loaded under the B-29. Preparations were also underway at Langley, with the chief of research airplane projects, Hartley Soulé, forming a temporary Muroc Test Unit to conduct the NACA research program on the xs-1. Reprising his Pinecastle role, Walt Williams would lead the group, and by September 1946 the first contingent left Virginia for their new posting in the High Desert.

While facilities at Muroc had gradually improved during the wartime years, conditions remained austere, with one NACA engineer later describing the accommodation as being "equivalent to high-type stables." Newcomers were greeted with the choice of temporary accommodation in the base's limited bachelor officer quarters or the tar-paper shacks of Kerosene Flats. For those who could afford it, accommodation in the Antelope Valley towns of Mojave, Rosamond, and Lancaster was available, as were limited barracks at the former marine air station at Mojave. The climate could scarcely have been more different to that of Virginia or upstate New York, with blazing hot days and freezing nights accompanied by the ever-present Mojave winds and all-pervading dust. Beyond the confines of the base, a fly-in motel and dude ranch run by pioneering female aviator Pancho Barnes offered the main social attractions. But however harsh the conditions may have seemed to the new arrivals, the huge lake bed and clear skies provided the ideal location for flight tests, and as the initial NACA and Bell Aircraft contingents arrived, the engineers and technicians were simply too busy to dwell on the desert's hardships.

The NACA detachment once again secured an SCR-584 radar set from the USAAF and began modifying this into a radar-phototheodolite in order to track the XS-1 and calibrate its altitude, but the required parts took months to reach the remote Muroc Field, frustrating Williams. As the NACA group awaited the XS-1's arrival, events half a world away offered a timely reminder of the risks involved in high-speed flight. On 27 September, British test pilot Geoffrey de Havilland was killed when his tailless DH 108 aircraft disintegrated during a transonic dive over the southeastern coast of England. Although the swept-wing DH 108 had little in common with the XS-1, the popular press seized on the incident as further evidence of the feared sound barrier.

The second XS-1 arrived at Muroc under its B-29 launch aircraft on 7 October, with the rest of the Bell Aircraft team following the next day, and as final preparations continued for the demonstration flights, Goodlin and Stanley met to discuss the test pilot's bonus. The original fee for the remaining flights had been set at $10,000, but given the perceived risks, Goodlin now sought more. Unable to alter the original sum, Stanley promised the test pilot a series of incremental bonuses totaling as much as $150,000 as he edged toward the goal of Mach 1. Satisfied with this verbal agreement and a handshake, Goodlin readied himself for his first flight in the rocket plane.

On 9 October, Bell technicians prepared the XS-1 for a glide flight, giving Goodlin an opportunity to familiarize himself with the aircraft's handling. This left the NACA team with little time to install and calibrate their instrumentation, but the headstrong Stanley was in no mood to wait for Walt Williams's men, reigniting a simmering animosity between the chief engineers. The roots of the problem lay in their differing objectives for the

3. Bell Aircraft test pilot Chalmers "Slick" Goodlin posing with the xs-1.
Courtesy Niagara Aerospace Museum.

demonstration phase. For Williams the key priority was the acquisition of
flight data to verify that the xs-1 was airworthy and fit for its forthcoming
research program. In Stanley's view, however, data collection came second
to proving that the xs-1 could actually fly. Matters were complicated by the
fact that only one set of technicians could work on the small research aircraft at any time, with Bell's team having priority. As soon as the xs-1 was
flight ready, Stanley's instinct was to get the aircraft in the air, even if this
meant that NACA instrumentation was not fully operational.

In order to prevent the situation from deteriorating further, AMC invoked
a clause in the original contract with Bell, stating that as well as demonstrating a speed of Mach 0.8 and performing a structural test involving an 8-g
maneuver, Bell would now be expected to perform twenty powered flights
before the aircraft could be accepted. It was hoped that these additional
flights might calm Stanley's haste to wrap up the demonstration phase,
while providing the NACA with sufficient opportunity to gather the data
it required. Both parties agreed to the proposal, but Williams made little
secret of the fact that he looked forward to Stanley's imminent departure

from Muroc, at which point the more amenable Dick Frost would assume control of the Bell operation.

The first attempt to make a glide flight, which occurred on 10 October, soon ran into problems. Having entered and pressurized the cockpit, Goodlin noticed that the pressure was continuing to rise. In an attempt to remedy this, he activated the cockpit pressure dump valve. But when this failed to open, the pressure continued upward, and by the time he reached an indicated altitude of 24,000 ft., Goodlin was experiencing a cabin pressure equivalent to that of 1,000 ft. below sea level. Fearing an explosion or permanent injury, Goodlin asked for chains to be reattached to the hatch before he opened it, hopefully equalizing the pressure. As he released the latch, the hatch exploded outward, striking the access ladder and becoming wedged between this and the xs-1. With his exit blocked, Goodlin faced a trip back to base beneath the bomber.

The damage was quickly repaired, and a second attempt was made the following day. Although cabin pressurization again proved troublesome, the flight proceeded as planned with Goodlin putting the xs-1 through a series of turns and stalls during his descent to the dry lake. Three more glide flights were made before the start of December, clearing the way for the xs-1's first powered flight; although the xlr11 rocket motor was still experiencing teething troubles on the test stand, the time had come to see how it and the high-pressure propellant system performed in flight. By the morning of the xs-1's planned powered debut on 6 December, Muroc was abuzz with senior AMC and Bell representatives who had descended on the remote base to witness the spectacle. Unfortunately, the Mojave weather cared little for human plans, and dawn brought only clouds and rain showers. Undaunted, the Bell and NACA teams kept working, hoping the skies might clear. By early afternoon, conditions appeared to be improving, and a decision was made to proceed. But even as the heavily laden B-29 strained for launch altitude, the weather closed in below it, resulting in an abort.

The teams reconvened for a second attempt under clearer skies on the morning of 9 December, and as the combined B-29 and xs-1 slowly circled up through 9,000 ft., Goodlin entered the xs-1 to begin his preflight preparations. Although he reported a problem with the tank pressures, the pilot felt that he could manage the situation using the cockpit regulators, and as the B-29 reached 27,000 ft., the xs-1 was dropped. Goodlin immediately noted just how rapidly the fully fueled rocket plane fell and wasted no time in lighting the XLR11's first chamber, quickly racing away from the lumbering bomber. Climbing now, he tested each of the four rocket chambers in turn and, once satisfied with their performance, lit two chambers simultaneously. Even using only half its available thrust, the XS-1 rapidly achieved Mach 0.79, just shy of the required demonstration speed. After cutting the motor and gliding back down to 15,000 ft., Goodlin lit all four chambers to experience the XLR11's maximum 6,000 lbf. thrust, but within seconds, the motor began to howl in protest as a fire warning light blinked into life on the instrument panel. Goodlin called on his chase pilot, Dick Frost, for advice. But Frost's P-51 had been left far behind, and the best the engineer could offer was an observation that the distant rocket plane appeared to be trailing smoke. On hearing this, Goodlin headed for home, and once the aircraft was safely on the ground, technicians confirmed that the XS-1 had indeed suffered a minor engine fire. Despite this, the first powered flight was hailed as a success, with Bell's publicity machine swinging into action to inform the press of Slick Goodlin's heroic first step toward the speed of sound.

The young test pilot now found himself in high demand, as reporters besieged him with questions and interview requests. In its eagerness to make what capital it could from the company's involvement with the program, Bell was only too happy to grant a seemingly endless stream of requests for Goodlin's time, much to the chagrin of the NACA and the AMC. As the press breathlessly reported how Goodlin would single-handedly take on the dreaded sound barrier, the other program partners and the scientific remit for the XS-1 received scant mention. Worse still, a worrying amount of detail about both the classified rocket plane and the test facilities at Muroc was finding its way into print. Finally, the AMC stepped in to halt further coverage. Goodlin himself had done little to court the attention, but the stream of inaccurate reporting and journalistic embellishments did little to endear the pilot to others within the program, creating a negative impression that would later cost him dearly.

Back at Muroc, the commencement of powered flights drew new arrivals to the NACA unit. Roxanah Yancey and Isabell Martin were among the newcomers, the first of many women to join the NACA detachment as computers. Initially these female pioneers had to make do with temporary accommodation in the makeshift shacks of Kerosene flats or even the base infirmary. The working conditions remained primitive, with the computers taking over part of a temporary administrative building alongside the NACA's shared hangar, and although the High Desert climate was very different to that of Virginia, the ladies were still expected to conform to Langley's strict dress code, with skirts and dresses mandatory even when their daily duties might include clambering over an aircraft to examine instrumentation. A computer's main role was the reduction of raw data from the filmstrips used by the aircraft's research instrumentation. Yancey and her colleagues would spend long hours straining over light boxes, using a film scale and pencil to mark various data points before turning to their noisy Friden mechanical calculators to tabulate the data for the program engineers. Although these ladies performed a seemingly thankless task, their efforts soon made them an indispensable part of the NACA team at Muroc.

As February 1949 drew to a close, Goodlin had a dozen powered flights in the second XS-1 behind him and had successfully demonstrated the required Mach 0.8 speed and 8-g structural maneuvers to validate the aircraft's design. Despite the earlier disagreements between Bob Stanley and Walt Williams, the AMC intervention had ensured that the NACA now possessed a wealth of data on the rocket plane's stability and handling, allowing engineers to plan the next phase of their research program. On 28 February, Bell's B-29 left Muroc carrying the number two aircraft back to Buffalo, where it would undergo modifications and have its research instrumentation removed for installation in its thinner-winged sibling. The number one XS-1 was soon due to make its Muroc debut, but as the end of Bell's contracted demonstration phase drew near, AMC and NACA representatives began considering arrangements for the follow-on research program. The eventual outcome of these conversations would bring major changes to the way in which the XS-1 and all subsequent flight research programs at Muroc would be conducted, while also dramatically curtailing Bell's role in the quest for Mach 1.

During the hiatus between the second XS-1's departure and the first aircraft's arrival at Muroc, the uneasy politics that had underpinned the program again took center stage. Although Bell still needed to complete its demonstration flights, the NACA had concluded that with some modification, the xs-1 would be suitable for its transonic research program. In discussions with the AMC, Hartley Soulé outlined NACA plans for flight research once both aircraft had been accepted. For their part, the postwar USAAF were experiencing an ever-tightening budget, and while AMC representatives expressed frustration at the length of time it was taking NACA research reports to find their way to Wright Field, both parties restated their commitment to cooperate on the xs-1 program. Although recognizing that the NACA would likely move at a slower pace than it would ideally like, the AMC's earlier plans for a quick assault on Mach 1 through a follow-on contract with Bell now appeared in jeopardy due to financial constraints.

In the course of these discussions, Bob Stanley was brought in to represent Bell, but on learning that the NACA would soon assume control of both aircraft, he raised serious reservations. In the NACA's cautious hands, Stanley argued, the xs-1 might find itself obsolete before ever achieving its true potential (privately, he referred to the NACA approach as "slow and tedious and fruitless"). Instead, he proposed a dual xs-1 test program, with the NACA performing research in the number two aircraft, while Bell undertook an accelerated program to push through Mach 1 in the first aircraft. When RMI's turbopump finally became available, the third xs-1, with its greater fuel capacity, could then carry Bell and the USAAF toward Mach 2 and into the upper reaches of the stratosphere. Stanley's efforts to secure an ongoing role for Bell in the high-speed program were entirely understandable. With few orders on its books, the company had more than money riding on the xs-1—Bell's reputation had become inextricably tied to the image it had created of Slick Goodlin shattering the sound barrier in the bullet-like xs-1. The determined Stanley began to apply pressure using Larry Bell's high-level USAAF contacts, and eventually, after visiting NACA Headquarters in Washington DC to seek approval, he won provisional agreement from both the NACA and the AMC for the accelerated program. Unfortunately for Stanley, his apparent victory was to be short lived.

Stanley's proposal had generated considerable interest at Wright Field, but not in the way the Bell engineer had intended. Details of his verbal agreement with Goodlin regarding the pilot's bonus schedule had now reached Colonel Smith, head of aircraft projects at the AMC. Why, Smith wondered, should the USAAF pay huge bonuses to a company test pilot when the AMC had test pilots of its own? Smith contacted Colonel Albert Boyd, chief of the Flight Test Division, to ask if Boyd felt that his own people could perform the program instead. Boyd, like General Arnold, believed strongly that the USAAF should be developing its own capabilities rather than relying on contractors to perform services. As an experienced test pilot in his own right, Boyd had earlier established the Accelerated Service Test Section to train handpicked test pilots and flight test engineers, and confident in the caliber of his people, he informed Smith that the Flight Test Division could (and indeed should) conduct the XS-1 flights.

When Bell presented additional details of its accelerated program, the NACA and the AMC were alarmed to learn this would now consist of up to sixty flights spanning more than a year, with costs to match. Stanley appeared to have misjudged the strength both of Bell's hand and of AMC's financial resources, and to compound matters, the company now began applying pressure on USAAF leaders to order an XS-1-derived rocket interceptor. In response to Bell's lengthy "accelerated plan," the AMC proposed a more limited fixed-price contract, but Stanley rejected this, believing that it would not cover the considerable cost of Bell's ongoing presence at Muroc (not to mention the controversial pilot bonuses).

As negotiations continued, XS-1 number one arrived at Muroc on 5 April, allowing the demonstration phase to resume. The first flight was to be unpowered, allowing Goodlin to judge what effect the thinner wings and tail might have on the aircraft's handling when compared to the second XS-1. More than a year had passed since the first aircraft's last free flight, but as XS-1 number one fell away from the B-29 on 10 April, Goodlin was pleased to report that it handled just as well as its sibling. Although he noted some minor differences in stability while putting the airplane through its paces, the test pilot felt happy to move straight to powered flights. After the XLR11 performed perfectly during its ground run the following morning, Goodlin was back in the air that afternoon as XS-1 number one made an almost flawless first powered flight, exhibiting none of the frustrating pressurization problems Goodlin had endured in the second aircraft. Unfortunately, as he descended toward the lake bed, Goodlin's oxygen supply began to run low, and the slightly disorientated pilot bounced the XS-1 on landing, leading to a heavy second contact, which broke the nosewheel support structure. Slick Goodlin had become the inaugural member of the Muroc nosewheel club, the first of many pilots to fall victim to the XS-1's tricky landing characteristics. The number one aircraft was back in the air by 29 April, and over the course of the next month, Goodlin made the required performance and structural demonstrations, clearing the way for the AMC to accept the first aircraft. On 22 May, Bell's Tex Johnston made a checkout flight in the second XS-1 before Goodlin used the same aircraft to make the final demonstration flight on 29 May, but before the aircraft were handed over, the Bell team were asked to perform one final duty.

On 5 June 1947 Muroc Field hosted an air display for the Aviation Writers Association, giving the USAAF an opportunity to showcase the latest advances in aviation, while also strengthening its case for an increased slice of the defense budget. The XS-1s were among the key attractions at the event, with Goodlin due to make a high-speed flyby in the number one aircraft while Dick Frost impressed the influential audience with a ground run of the second aircraft's rocket engine. As Frost ignited the XLR11, its third chamber exploded, but this went unnoticed amid the incredible noise generated by the remaining three chambers. Frost only became aware of the problem when a fire warning light blinked on in the cockpit, and rushing to the rear of the aircraft, the engineer was horrified to see the XS-1's orange paint blistering before his eyes. The burning water-alcohol mixture's invisible flames were engulfing the engine section, but thanks to the prompt actions of Bell technicians and the Muroc fire department, disaster was narrowly averted. The watching writers seemed oblivious to the drama and were soon distracted by Goodlin's spectacular flyby, but the second XS-1 would require major repairs before beginning its government career.

As the demonstration program wound up at Muroc and preparations were made to repair the damaged second aircraft, the battle for control of the program also reached its conclusion. Following Stanley's dismissal of its fixed-price contract offer, the AMC decided that Colonel Boyd's Flight Test Division would now perform the accelerated supersonic research program. This came as a bitter blow to Bell Aircraft, who had been trying to resolve the problem of Stanley's verbal agreement with Slick Goodlin over bonus payments. Bell's lawyers had been in talks with Goodlin's representative for many months and had concluded that the unwritten arrangement was not legally binding and therefore need not be honored. When Goodlin returned to Buffalo with the damaged XS-1, he found a letter on his desk informing him of the company's position. Following a brief exchange of views with Bob Stanley, Slick Goodlin resigned from Bell Aircraft with his dreams of supersonic flight, and a financially lucrative bonus, in tatters. Following his departure, Goodlin joined the Israeli Air Force, flying both Spitfires and Messerschmitt BF109s in combat before becoming the service's chief test pilot. During this period, he also flew DC-4 transports in an operation to ferry Jewish refugees from Germany to Israel, before moving into the aircraft resale industry.

Although Goodlin's financial demands are often cited as the reason that Bell Aircraft lost its role in the XS-1 program, he was largely the victim of events beyond his control. Goodlin had proven himself to be a highly competent test pilot, and later suggestions that he somehow delayed progress toward Mach 1 in an attempt to obtain more money ignore the fact that he flew the demonstration program exactly as requested by the NACA and the AMC. Publicity generated by Bell itself had placed Goodlin in a difficult position, damaging his reputation in the eyes of many at Muroc, while the company did little to shield him from the insatiable attentions of the press.

The task of breaching the mythical sound barrier in the XS-1 would now fall to one of the USAAF's own test pilots, and at Wright Field, Colonel Boyd had already begun a selection process to find the right man for the challenge.

Given the potential dangers involved, Boyd invited unmarried pilots from the Fighter Test Division to put their names forward for the XS-1 program on a purely voluntary basis. Although many of the candidates came with years of flight test experience, Boyd selected the relatively inexperienced Captain Charles Yeager as his prime pilot.

Having enlisted in 1941, Yeager flew with distinction in the skies above Europe, becoming an ace in his P-51 *Glamorous Glen* (named for his wartime sweetheart and eventual wife). Downed over France, Yeager escaped to Spain with the help of the Maquis, and following personal appeals to General Eisenhower, the West Virginian was allowed back on combat duties in the wake of the D-Day invasion. Following the war, Yeager served at Wright Field as an assistant maintenance officer, a role that took him out to Muroc with the Bell XP-59A test program. Impressed by his natural piloting ability and his eagerness to fly any aircraft available, Boyd brought Yeager into the Flight Test Division in 1945, where he attended USAAF test pilot school at Wright Field. Despite his junior status within the division, Yeager's ability and attitude earned Boyd's confidence (although Yeager later came close to losing his seat in the xs-1 when it came to Boyd's attention that he was married).

Yeager's backup would be Bob Hoover, an experienced combat pilot who had spent the later stages of the war in a German POW camp before escaping and flying to safety in a hastily requisitioned Focke-Wulf 190. His exceptional skills earned him a place as a junior test pilot at Wright Field, where he trained alongside Yeager. The two pilots would be accompanied by Captain Jack Ridley, a talented test pilot in his own right but also an intuitive engineer who had studied under Theodore von Kármán at CalTech. Boyd felt that Ridley's ability to explain complex engineering concepts in simple terms would make him an invaluable asset to Yeager and Hoover, while his thorough understanding of aerodynamics would earn the respect of Williams's NACA team.

On 30 June a meeting between AMC and NACA representatives formalized the two-track approach that the xs-1 program would now follow. Boyd's team would perform an accelerated research program up to and beyond Mach 1 using the first aircraft, while the NACA's own pilots would fly a thorough series of stability, drag, and load tests at transonic speeds in the more heavily instrumented second aircraft. The NACA would retain responsibility for instrumentation, tracking, and (initially) maintenance for both aircraft, with the AMC taking care of fuel, launch crews, chase duties, and other logistical requirements at Muroc. Bell would provide both teams with initial training, and at the AMC's request, Dick Frost would remain with the program for a six-month period. One other Bell employee would also stay at Muroc. Jack Russell had become an indispensable authority on the xs-1's often-temperamental systems, and in order to retain his knowledge, the AMC would employ Russell as a civilian contractor to continue as crew chief for the number one aircraft. The final member of the team would be Major Bob Cardenas, a senior multiengine pilot from the Flight Test Division who would serve as B-29 pilot and senior officer for the accelerated program.

Following a period of training in Buffalo, Yeager and his colleagues journeyed to Muroc during July 1947 to make final preparations before Yeager was to climb down from the B-29's bomb bay on 6 August and squeeze into the XS-1 to make his first glide flight. As the fourth man to fly the aircraft, Yeager was just as impressed as his predecessors, describing the XS-1 as the "best damn plane I ever flew." Two more glide flights followed in quick succession, with Yeager even engaging Bob Hoover's P-80 in a mock dogfight during the third flight, before the decision was made to move on to powered flights.

August also saw the arrival of NACA research pilots Herb Hoover and Howard Lilly, ahead of their familiarization flights in the second XS-1. The NACA had recently taken over Bell's facilities at Muroc, significantly increasing the agency's available workshop, office, and hangar space. New additions to the NACA Muroc group included Joe Vensel—a former research pilot from the Cleveland laboratory who became director of flight operations with responsibility for the unit's pilots, crew chiefs, and maintenance staff—and De Elroy Beeler from Langley, who became Williams's deputy, heading up load research on the XS-1 while taking charge of forthcoming research aircraft projects. Gerald Truszynski also rejoined the team to oversee instrumentation, telemetry, and the radar-based tracking of research flights.

Although the XS-1's instrument panel carried a Mach meter, there were doubts regarding just how reliable static pressure readings from the aircraft's air data probe might be during transonic flight. In order to gain accurate data, Langley engineers devised an ingenious experimental method to determine errors between the indicated Mach number shown in the cockpit and the actual Mach number. An SCR-584 radar set—modified with cameras, a VHF radio system, and phototheodolite—was used to track a support aircraft flying at a specified range and altitude. This aircraft trailed a pressure measurement device known as a bomb, to record very accurate surrounding atmospheric static pressure data, with the radar phototheodolite recording a precise geometric altitude for the aircraft. When the XS-1 flew its research run, the radar phototheodolite could guide and track the rocket plane to ensure that it passed through the same range and altitude. By using the support aircraft's pressure data as the reference, together with information obtained by the XS-1's instrumentation, Roxanah Yancey and her colleagues were able to calculate an accurate, or true, Mach number for the flight. As more data became available from flights, the NACA team were able to accurately determine errors in the Mach number indicated by the research aircraft's Mach meter, allowing more accurate instrumentation to be devised for future programs. Although complex and time consuming, this method would prove highly effective during the xs-1 program (the radar phototheodolite was not, however, intended to measure the aircraft's speed during flights, as has often been claimed).

By August a number of developments had lent an increased sense of urgency to the accelerated xs-1 program. In July 1947 the usaaf finally gained independence when President Truman recognized it as a separate service under the National Security Act. The new U.S. Air Force (usaf) was due to begin operations in September. As the fight for funding allocations with the other services intensified, a supersonic flight might prove a powerful bargaining chip for the air force, but the navy also had its eyes on this prize. BuAer's transonic research aircraft, the turbojet-powered d-558-1, had recently arrived at Muroc, and in the hands of Douglas Aircraft test pilot Gene May, it quickly exceeded the xs-1's best speed to date. A third potential candidate for supersonic flight, North American Aviation's new swept-wing xp-86 fighter prototype, was also due to make its debut at Muroc in coming months. Against this background, Yeager aimed to push forward as quickly as safety would allow, and on 29 August the air force team took a major step forward in the number one xs-1, now rechristened *Glamorous Glennis*.

As Cardenas coaxed the lumbering b-29 toward its 21,000 ft. launch altitude, Yeager began pressurizing the xs-1's systems and priming the xlr11 motor in preparation for his first powered flight. Readying for the drop, Yeager was shaken by Bob Hoover as the backup pilot buzzed the research aircraft in his p-80 while heading for a high-chase position some ten miles ahead. Dick Frost positioned himself off the b-29's starboard wing in a second p-80 to observe the drop, and when Yeager signaled that he was ready, Cardenas started to count down from ten (omitting one number, as always, to keep Yeager on his toes) before releasing the rocket plane at a speed of 225 mph. Falling away, Yeager lit the xlr11's first chamber before testing the other three chambers in quick succession and rolling the xs-1 for good measure. Shutting off the motor, Yeager made an unpowered dive reaching Mach 0.8, equaling the xs-1's demonstration speed. Leveling out at 5,000 ft., he relit one chamber and began climbing before firing the remaining three chambers to unleash the xlr11's full 6,000 lbf. of thrust. Although still climbing steeply, Yeager hit a maximum speed of Mach 0.85 before, propellants spent, he began his descent toward a smooth lake bed landing. The Muroc team were elated with the xs-1's performance, but Colonel Boyd was less satisfied. The planned maximum speed for the flight had been Mach 0.82, and now Boyd demanded a written explanation as to why this had been exceeded. When Yeager explained that his excitement at the aircraft's performance had gotten the better of him, Boyd advised him to pay more attention in the future but kept faith in his pilot; as Bob Cardenas would later note, "Boyd's genius was in getting the right people together, then leaving them alone until they got the job done."

With Yeager now fully checked out in the xs-1, the high-speed research could begin. During September he made four flights, with each one gradually pushing the orange rocket plane further into uncharted territory. As it approached Mach 1, the xs-1 began exhibiting stability and control issues, indicating the onset of compressibility and causing concern among the NACA team. In the absence of accurate wind tunnel data to suggest how the aircraft might behave beyond Mach 0.9, the pilots came up with an ingenious solution. While flying straight-winged jets such as the P-80 and P-84, Yeager and Hoover had noted that by rolling the aircraft and pulling back on the stick to generate 2–3 g's of force, the resulting conditions mimicked those usually found during level flight at marginally higher speeds. Recognizing that this technique could provide a glimpse into what lay ahead, Yeager took to performing the same maneuver once he had achieved his target speed during each xs-1 flight.

Following the sixth powered flight, Yeager and Ridley returned to Wright Field to present their progress report in person, but rather than offering his congratulations, Boyd instead expressed concern that the duo was becoming complacent and taking risks despite the recent evidence of high-speed instability. Chastened, the pair returned to Muroc with a fresh warning to proceed carefully as they edged closer to Mach 1. Boyd's caution was vindicated on 10 October; having reached an indicated Mach 0.94 during his eighth powered flight, Yeager rolled the xs-1 over, pulling back on the yoke only to find he got no response from the elevators. Yeager cut the motor and headed home, but his troubles weren't over, as the cockpit canopy now froze over. Fortunately, Hoover and Frost soon closed around the xs-1, talking Yeager down to a blind landing on the lake bed.

The loss of elevator effectiveness raised serious concerns on the ground,

4. The Bell XS-I in flight over Rogers Dry Lake. The runways of South Base can be seen at the top of the image. Courtesy USAF.

and once the NACA computers reduced the flight data, there was both good and bad news for the air force pilot. Pushing through Mach 0.9, the XS-I's shockwave had reached the aircraft's tail, moving rearward until it stood directly at the elevator hinge point, rendering the control surfaces ineffective. Unless a solution could be found, Yeager would be unable to control the aircraft's pitch as he advanced toward Mach 1. Better news came with his adjusted Mach number. High dynamic pressures on the XS-I's nose boom had caused the Mach meter to give a false reading during the flight. Thanks to data from the radar phototheodolite, the NACA's corrected figures showed that Yeager had actually reached Mach 0.997 at an altitude of 37,000 ft. Mach 1 was so close, but without a means to control the XS-I's pitch, it might remain just beyond Yeager's reach. During discussions with his NACA counterparts, Jack Ridley suggested that the solution lay in the all-moving horizontal stabilizer—the very feature that the NACA had insisted be included during the aircraft's design process. Rather than relying on his elevators beyond Mach 0.9, Ridley argued, Yeager could control the XS-I's pitch by moving the entire stabilizer. With no better options available, Ridley performed ground tests that appeared to validate his theory. The decision was made; the next flight would go for Mach 1.

Away from base, Yeager had more down-to-earth problems to deal with as the 14 October flight approached. Some days earlier he had been injured in a late-night incident, and fearing that the flight surgeon might ground him, he sought alternative medical advice. On learning that he had broken a couple of ribs, the heavily strapped Yeager returned to Muroc to ask Ridley's advice. Although he felt sure he could handle the flying, Yeager was less certain he had enough power in his arm to secure the XS-1's hatch before the drop. Ridley did a quick calculation before fashioning a makeshift lever from a broom handle, giving Yeager sufficient purchase to force the latch closed. Although he did his best to conceal his injury, Bob Cardenas and Walt Williams were both well aware of the pilot's predicament, but neither man felt the need to remove him from the Mach 1 attempt. As the aircraft were readied for flight, a thin coating of shampoo was applied to the inside of the XS-1's canopy to prevent a repeat of the previous flight's icing problems.

At 10:00 a.m. Cardenas powered the B-29 and its orange cargo down the Muroc runway and up into the clear desert sky. Reaching 5,000 ft., Ridley and Yeager moved back into the bomb bay, and after squeezing gingerly into the rocket plane's cockpit and using Ridley's improvised lever to secure the hatch, Yeager began preparations for his ninth powered flight in the XS-1. At 20,000 ft. Cardenas pushed the B-29 into a shallow dive before counting down from ten (omitting number four this time) and releasing the rocket plane at 250 mph. As he fell, Yeager kicked the XLR11 into life and aimed for Bob Hoover's P-80, flying at 42,000 ft. some ten miles ahead. Using all four rocket chambers, the XS-1 roared upward through 30,000 ft. while Yeager kept a wary eye on the aircraft's stability. Pushing over into level flight at 42,000 ft., Yeager hit Mach 0.94, again noting the loss of elevator effectiveness but placing his trust in the all-moving tail to make the required adjustments. The XS-1 continued to surge onward through moderate buffeting, past Mach 0.96 and 0.98 until, suddenly, the Mach meter's needle jumped clear of its final increment—Mach 1.

Yeager was now traveling faster than anyone had before him, faster than many had believed possible, and rather than the terrifying white-knuckle ride many had anticipated, the transition to supersonic flight was relatively smooth. All buffeting disappeared, and for a few brief moments Chuck Yeager was able to outrun sound itself before he decelerated back through Mach 1 with a slight bump as his shockwave caught up with him. The jubilant pilot performed victory rolls on his way down to the lake bed where his first duties included a call to Wright Field, to inform Colonel Boyd that the task had been accomplished. Once Walt Williams contacted Hartley Soulé with the news that Yeager had exceeded Mach 1, the air force clamped a lid of secrecy on Yeager's achievement. Supersonic flight was a matter of national security, and it would be some time before the public learned about the xs-1's historic flight through Mach 1.

As the air force team celebrated their supersonic breakthrough and Yeager enjoyed his free steak dinner at Pancho's, things were moving more slowly for the NACA. By late October the second xs-1 had only been airborne once since returning to Muroc, when Yeager had carried out an acceptance flight on the NACA's behalf. A series of modifications and technical issues, compounded by maddening delays as spare parts slowly meandered their way to the base, conspired to keep the rocket plane hangar bound, but finally the time came for an NACA pilot to take the controls.

Herbert H. Hoover had trained as a pilot in the USAAC before flying for Standard Oil in South America. On his return to the United States in 1940, he joined the NACA as a research pilot, logging countless hours flying weather research in the worst conditions he could find. Among his colleagues, Hoover had a reputation for being cool and reliable in a crisis, and he went on to become chief pilot at Langley, making him a natural choice to fly the xs-1. Although the program was based at Muroc, Hoover only traveled west when needed for flights (often a two-day trip each way), in order to continue his work at Langley. One week after Yeager's historic flight, Hoover became the first NACA pilot to fly the xs-1, making a glide flight to familiarize himself with the aircraft. Although impressed by the research plane's handling, Hoover misjudged his landing, collapsing the nosewheel. The resulting repairs and the onset of winter rains kept xs-1 number two grounded until 16 December, when Hoover made his first rocket flight, reaching a maximum speed of Mach 0.71. The aircraft was aloft again the very next day, as Hoover reached Mach 0.8 during a second powered checkout flight, bringing activities to a close for the year.

5. The entire staff of the NACA Muroc Flight Test Unit in 1947, including Roxanah Yancey (back row, sixth from left), De E. Beeler (back row, far right), Joe Vensel (front row, second from left), Herbert Hoover (front row, third from left), Walt Williams (front row, fourth from right), and Howard Lilly (front row, second from right). Courtesy NASA.

During September 1947 the NACA's presence at Muroc had been cemented when the Muroc Flight Test Unit (MFTU) was granted permanent status by the newly appointed director of research, Dr. Hugh Dryden. Although still managed from Langley, Williams's team was now twenty-seven strong and continuing to grow. While life in the High Desert remained spartan, some of the ex-Langley hands had grown accustomed to the conditions, even preferring the dry desert air to the stifling humidity of the Virginia coast. The new posting, with its pioneer feel, also helped break down the barriers that prevailed among the workforce at Langley. In such a small group, the MFTU employees mixed more freely, be it at work, socially, or during carpools to the towns of the Antelope Valley. Occasionally, there would be opportunities to go "down below" to Los Angeles for the weekend or hiking trips into the spectacular nearby mountain ranges, and as 1948 dawned, Williams's constant petitions to improve living conditions for NACA employees finally paid off as work on new accommodation began within the boundaries of Muroc Air Force Base (as Muroc AAF had been renamed in February 1948). While the new dormitories were far from luxurious, they did at least provide a convenient and reasonably weatherproof improvement over the temporary arrangements that had preceded them.

As the NACA XS-1 research program began in earnest during January 1948, the air force allowed Jack Russell to assist the MFTU as crew chief for their often-temperamental rocket plane. Under Russell's supervision, the schedule became more regular, with seven flights taking place in January alone. As Herb Hoover began a series of stability and control investigations, the MFTU's second pilot, Howard Lilly, made his first flights in the XS-1. Lilly, a former naval aviator, had joined the NACA in 1942, quickly transferring from Langley to the Lewis Flight Propulsion Laboratory in Ohio, where he had performed flight tests on a wide variety of aircraft. Having made the move to Muroc, he was expected to become project pilot for the XS-1 and D-558-1 programs, allowing Hoover to return to Langley.

On 10 March, Hoover took the second XS-1 through Mach 1, becoming the first civilian to fly faster than sound. The flight ended typically, with landing gear problems as the nosewheel failed to extend. But damage was minor, and the rocket plane was soon airworthy again. Lilly made his own supersonic flight before the end of the month, but the majority of NACA flights continued to concentrate on conditions just below Mach 1.

Away from the High Desert, administrative changes during 1948 recognized the growing importance of the research aircraft programs within the NACA. Hugh Dryden now brought Hartley Soulé onto his Headquarters staff as research airplane project leader, formalizing the role Soulé had been performing at Langley up to this point. Soulé's elevation, together with Dryden's earlier confirmation of the MFTU's permanent status, afforded the NACA's Muroc operation a growing sense of autonomy within the organization. This was further bolstered when Soulé established the Research Airplane Panel (RAPP) to coordinate flight research–related activities across all its facilities. The panel initially consisted of one representative from each of the NACA's three laboratories, as well as Walt Williams and Soulé himself.

With the NACA research program finally on track, the air force continued flights in the number one aircraft, with Yeager expanding the XS-1's performance envelope to Mach 1.45 by March 1948. Bob Hoover moved on from the program without ever flying the rocket plane, having broken both legs while ejecting from a burning F-84. Following his recuperation,

Hoover moved into civilian life, becoming a test pilot for North American Aviation, whose XP-86 prototypes were breaking Mach 1 over Muroc by early 1948, albeit in full-power dives rather than level flight. In Hoover's place, two new pilots, Captain James Fitzgerald and Major Gustav Lundquist, joined the program to perform stability tests beyond Mach 1. Sadly, Fitzgerald's time with the XS-1 proved brief, as he died in September 1948 following an F-80 crash.

News of Yeager's Mach-busting flight had begun to leak out toward the end of 1947, with *Aviation Week*'s detailed accounts of the flight drawing empty legal threats from the government. On 22 December the *Los Angeles Times* carried the banner headline "U.S. Mystery Plane Tops Speed of Sound," breaking the news to the general public, but it was June 1948 before the air force confirmed that the X-1 (as the XS-1 was now known) had laid the sound barrier myth to rest. With the veil of secrecy lifted, Chuck Yeager could finally receive belated plaudits, and in December 1948 he accompanied Larry Bell and John Stack to the White House, where President Truman presented the trio with one of aviation's most prestigious honors, the Collier Trophy. There was a certain irony in Stack receiving an award for the X-1 on the NACA's behalf, however, given his opposition to the rocket-powered research aircraft.

Following Fitzgerald's death, Colonel Boyd needed a new pilot to take over the later stages of the program, and his choice was another West Virginian fighter pilot whose name, like Yeager's, would become synonymous with rocket plane flights. Major Frank "Pete" Everest had served with the USAAF in Europe, the Middle East, and China during the war. While flying against Japanese supply lines in the Far East, he was downed and imprisoned, remaining a POW until the conflict ended. Once back in the United States, he transferred to Wright Field, where, having proven himself to Boyd, he underwent test pilot training before joining the fighter test section. Everest was assistant chief of the section by early 1949 when Boyd offered him the chance to fly the X-1 on a series of high-altitude flights. At the time, the world altitude record was still held by the *Explorer II* balloon, which had flown beyond 72,000 ft. in late 1935. Everest readily accepted the challenge, and by March he had made his first supersonic checkout flight in the X-1 at Muroc.

During his high-altitude flights, Everest would be exceeding the Armstrong limit—the point around 62,000 ft. at which the air pressure is so low, bodily fluids can begin to boil. This meant a trip to the David Clark Company to be fitted for a partial pressure suit, followed by training at Wright-Patterson AFB's Aero Medical Laboratory to experience the constriction of the suit's tightening capstans and learn the counterintuitive pressure-breathing technique. Having to concentrate on forcing the breath out of the lungs every five or so seconds took a lot of getting used to, especially when combined with the other demands associated with rocket flight, but on 25 March, Everest dropped away from the B-29 to make his first high-altitude attempt wearing the suit. Unfortunately, the flight was marred by the engine problems that dogged the air force X-1 program during its later years. After bringing Dick Frost back on board to troubleshoot the XLR11, the team were soon ready to try again, and after a number of attempts during the summer of 1949, Everest finally reached 71,902 ft. on 8 August, just short of the existing record. Permission was given for another attempt on 25 August, but this flight would bring Everest perilously close to disaster.

As he entered the rocket plane, Everest noticed a small crack in the canopy's inner layer of glass, but not wanting to scrub the flight for such a minor issue, he chose to proceed. All seemed well until an ominous whooshing sound caught his attention as he raced through 65,000 ft. Instantly, his partial pressure suit inflated, pressing in on Everest's body and head as he realized that the minor canopy crack was now a major fissure. With the X-1's cockpit now depressurized, all that lay between Everest and the thin, freezing high-altitude air was his suit, and while it appeared to be doing its job, he was not about to test how long his luck would hold. Shutting off the rocket engine, Everest pointed the aircraft's nose down toward the safer, life-sustaining altitudes that lay beneath him. Some miles below, Yeager watched as the X-1's vapor trail came to an abrupt halt, but although Everest could hear his chase pilot's desperate calls, he found himself unable to speak under the suit's tight grip, managing only a series of unintelligible grunts. Finally, Yeager heard Everest's voice as he leveled the X-1 off at 20,000 ft. and could finally release the suit's pressure. Although it was not the record he had been aiming for that morning, Pete Everest had just become the first pilot to have his life saved by a pressure suit. In the aftermath of this drama, plans for further altitude flights were shelved until the next generation of advanced X-1 rocket planes became available.

6. Chuck Yeager and Jack Ridley alongside Glamorous Glennis, the aircraft in which Yeager first exceeded Mach 1. Courtesy USAF.

The first x-1 continued to fly on into 1950, giving other air force pilots, including Colonel Boyd and Jack Ridley, their first taste of faster-than-sound flight. On one memorable occasion, Ridley experienced an engine fire and, having reported that he could smell smoke, was reassured by chase pilot Yeager that there was nothing in the cockpit that could burn (due to the nitrogen atmosphere). "Like hell there isn't," replied the alarmed engineer. "I'm in here!" Fittingly on 12 May 1950 Yeager himself made the final flight of the aircraft with which he would forever be associated, before personally delivering the first x-1 to the Smithsonian Institution in Washington DC. The NACA x-1 would continue its research career into 1951 before being withdrawn to undergo major modifications that would see it reach new heights in years to come.

In the years since Kotcher had first proposed his rocket research aircraft, the x-1 had done more than simply prove his belief that supersonic flight was possible. The program had laid a template for the research aircraft that would follow it. Working relationships (however uneasy) had been established between military, industrial, and government partners, building the foundation on which future programs would be built. The once controversial air-launch technique had proven successful and was subsequently adopted by all future rocket research aircraft, but perhaps most importantly, the x-1 program had brought a highly capable cadre of engineers, technicians, and test pilots out to the High Desert, where they would continue to push the boundaries of aviation for decades to come.

3. A New Pretender

Like so many of his Muroc contemporaries, Bill Bridgeman's aviation career began with military service, but unlike Yeager or Everest, Bridgeman was no fighter pilot. Having joined the navy in 1940, he was initially posted to a PBY flying-boat squadron stationed at Pearl Harbor, where in December 1941 he witnessed the infamous Japanese attack that brought the United States into the war. The young aviator spent the next two years flying long patrols over the vast waters of the Pacific, searching for the elusive Japanese fleet, but he eventually became frustrated by the lack of action. In 1943 Bridgeman left the lumbering PBY Catalina behind him, transferring to the newly formed Bombing Squadron 109, commanded by the hard-driving Captain "Buzz" Miller. VB-109 flew repurposed B-24 bombers on low-level dawn raids against Japanese shipping and coastal defenses, supporting America's advance through the South Pacific. Their exploits became celebrated back home, earning them the nickname the Reluctant Raiders, but Miller's uncompromising attitude and the unrelenting schedule took their toll on his aircrews. By the summer of 1944, the exhausted squadron transferred back to San Diego where it gradually disbanded.

Bridgeman transferred to the Navy Air Transport Service, where he served as a ferry pilot, flying a wide variety of aircraft types from fighters to bombers. Although this role broadened his piloting experience, he found it a nomadic existence, building an inner restlessness that seemed to stay with him for the rest of his career. Bridgeman finished the war as an instructor on the PB4Y-2 bomber (a naval derivative of the B-24s he had flown in VB-109), before seeing out his service years flying transport planes over the now-peaceful Pacific. Readjusting to civilian life, he initially became a pilot on the short, island-hopping routes of Hawaiian Airlines, but after two years, he grew tired of the same skies and decided to move on. Back on the mainland, a brief stint plying the Pacific coastal routes for Southwest Airlines followed, but again he soon became restless, yearning for a fresh challenge away from the numbing routine of airline flying. Unsure of quite what this would be, he headed back to Malibu, where he had spent many happy childhood summers with his grandmother. As an accomplished surfer, Bridgeman quickly settled back into the beach lifestyle, spending the summer months of 1948 relaxing while considering his options. During this period a chance conversation with an old friend who flew for Douglas Aircraft led Bridgeman to the company's El Segundo plant, located just south of Los Angeles International Airport.

The immediate postwar years were difficult times for the aviation industry, particularly the military sector. While the air force and navy both lobbied for the development of a new generation of aircraft to face the rising specter of communism, their budgets were under considerable pressure. Douglas Aircraft retained enough civilian and military orders to keep its Californian facilities operating in 1948, with the El Segundo plant handling navy contracts, including the powerful new AD Skyraider attack aircraft. As every aircraft rolled off the production line, it needed to be put through a series of checks by a Douglas pilot before it could be certified for delivery to the navy. Unfortunately, the pilots' office at El Segundo was chronically understaffed, meaning a long backlog of Skyraiders had built up. After meeting with senior pilot Laverne Brown, Bridgeman was taken on as a production test pilot and, following the briefest of orientations, immediately sent aloft in his first AD.

Work at El Segundo was hard but rewarding. Bridgeman was soon checking out up to six Skyraiders a day, sometimes having to nurse a stricken aircraft back to the field following serious mechanical failure before jumping straight into the next plane on the line. The workload often exceeded what Bridgeman and Brown could handle, and on these occasions, help would arrive in the shape of engineering pilots from Douglas's nearby Santa Monica plant. With no formal engineering education Bridgeman became fascinated by the technical, jargon-laden conversations these pilots would share. Eager to understand more, he began studying aeronautical engineering manuals he found around the office. Gradually, these conversations started to become clearer, and he was able to put forward his own ideas with a better grounding in the technical aspects of his work.

Bridgeman's ambition to learn appeared to have made a positive impression on his superiors when, nearing the end of his first year with Douglas in the spring of 1949, he was summoned to Brown's office. A request like this was highly unusual in the generally informal environment, so Bridgeman approached apprehensively, and he was surprised to find not only Brown but also Bert Foulds—Douglas's assistant chief pilot—waiting for him. Rather than delivering a reprimand, the two men offered Bridgeman the chance to transfer to Muroc AFB to test Douglas's hottest new ship—the jet- and rocket-powered D-558-II Skyrocket. Taken aback, Bridgeman wondered why he was being offered this opportunity when Douglas already had a cadre of highly qualified and experienced engineering pilots up at Santa Monica? Through office gossip, he knew enough of the navy-funded research program to understand its importance in gathering vital transonic and supersonic data (as well as providing a navy response to recent air force triumphs in the X-1). This was an incredible opportunity and exactly the sort of challenge the restless pilot reveled in, but was he really the right man to take it on? After further discussions with Foulds and Brown, Bridgeman decided to travel out to Douglas's test facility at Muroc, where he could meet the Skyrocket team, including current pilot Gene May, with whom he might soon share testing responsibilities.

Following BuAer's 1945 decision to award Douglas Aircraft the contract for the navy's transonic research aircraft, Ed Heinemann's team at El Segundo had developed the D-558 into a practical design capable of fulfilling the joint BuAer and NACA brief. The specifications had called for a straight-winged vehicle capable of taking off and landing under its own power and spending extended periods between Mach 0.75 and Mach 1.0. While supersonic performance might be possible under certain conditions, it wouldn't be a priority. The new aircraft needed to be extremely robust (it was required to fulfill the same maximum load criteria as the X-1, up to 18 g's) and capable of carrying a large instrumentation payload into the transonic zone. The original contract was for six aircraft, acknowledging that some airframes might be modified at a later date to accommodate different turbojets, airfoils, and even supplementary rocket propulsion.

The aircraft that Douglas designed to meet these specifications was outwardly very simple in configuration, but it featured a number of innovative ideas. At just over thirty-five feet long, the fuselage was tubular with a constant circular section and featured a single large nose air intake. As air entered, it was deflected around the narrow cockpit before reaching a single General Electric J35 turbojet. The straight wings had a span of twenty-five feet and were mounted low on the body, blending smoothly into the fuselage to reduce drag. Using a thin 10 percent t/c airfoil, there was little room in the wings for the main gear, but the Douglas team developed an innovative compact design that just fitted. The all-moving horizontal tail surfaces sat around a third of the way up the large unswept vertical fin, where it was hoped they would be free from any shockwaves produced by the wings. One of the more unusual features was Douglas's decision to eschew an ejector seat in favor of a novel detachable nose section. If the pilot needed to leave the aircraft in a hurry, a lever would release the forward section at a point just to the rear of the cockpit. When this section stabilized and slowed to an acceptable speed, the pilot could release the seat back and tumble away before descending to the ground under their own parachute. The first examples of the aircraft were due for delivery in 1947, but even as the D-558 design was undergoing review, new research was pointing to a very different configuration.

In May 1945 Eugene Root and Douglas colleague A. M. O. Smith traveled to Germany as part of a BuAer mission examining aeronautical facilities captured at the end of the war. In the wake of the 1935 Volta conference, German aerodynamicists and engineers had eagerly pursued both the advanced propulsion and aerodynamic configurations that could open the door to high-speed flight, with the results of this research appearing in European skies toward the end of the war. While other teams sought information on jet and rocket propulsion, the Douglas engineers were interested in work on swept-wing configurations. Originally theorized by Busemann in 1935, swept wings promised to alleviate some of the problems of high-speed flight by delaying the onset of compressibility, thus reducing the buffeting, high drag, and loss of lift that pilots had been experiencing in higher-performance fighters. Root and Smith wasted no time in passing what they found back to BuAer and the NACA, who analyzed it alongside independent studies carried out by Robert Jones at Langley earlier that year. As a result, in August 1945 they decided a new swept-wing research aircraft should be ordered alongside the D-558.

Initially, Heinemann's team looked at modifying the existing straight-wing turbojet design to accommodate swept wings, but BuAer and the NACA felt that the single J35 turbojet wouldn't be powerful enough to take the new aircraft into the zones of interest. Instead, they decided that it should adopt rocket propulsion as the main power source but still retain a compact turbojet for use during takeoff and landing (an approach Bell had rejected during the x-1's development). As it became clear that an entirely new design would be required, the original D-558 contract was reconfigured to allow for a two-phase program; the original straight-winged turbojet would now become phase one (the D-558-1), with the swept-wing rocket- and turbojet-powered aircraft as phase two (the D-558-11).

Douglas pushed ahead with the development of the phase one aircraft, and by February 1947 when the first examples rolled out of the El Segundo factory in gleaming red livery, the D-558-1 had gained a new name—the Skystreak. Flight-testing was to take place out at Muroc, and Gene May, one of Douglas's most experienced engineering pilots, would take the controls. Following the Skystreak's arrival out in the High Desert, taxi tests and instrument calibration were carried out before May took the new aircraft into the air for its first, albeit brief, flight on 15 April. Engine problems caused an early abort that day, but over the coming months the Crimson Test Tube, as the D-558-1 became known, proved its airworthiness and was ready to begin research flights.

Generally, the navy was content for pilots from Douglas or the NACA to carry out research on their behalf, but occasionally interservice rivalries intervened. During June 1947 Colonel Boyd flew a series of high-speed tests in a modified Lockheed p-80 Shooting Star at Muroc. In doing so, his streamlined xp-80r achieved a new closed-course airspeed record of 623.7 mph. Seeing an opportunity to quickly relieve the air force of this record, the navy dispatched Commander Turner Caldwell and Lieutenant Colonel Marion Carl of the usmc to Muroc to familiarize themselves with the D-558-1. After a few orientation flights, both test pilots quickly got to work, with Caldwell setting a new record of 640.6 mph on 20 August. Carl, not wishing to be outdone, worked out how to squeeze a fraction more performance from the J35 and consequently raised the mark to 650.6 mph a mere five days later.

As Gene May continued to expand the Skystreak's performance envelope in aircraft number one, the MFTU took possession of the second air-craft. An expanded instrument package, similar to the one used in the X-I, was installed, and in November 1947 NACA pilot Howard Lilly began a transonic flight research program. Unfortunately, progress was soon inter-rupted as winter rains flooded the lake bed, but MFTU technicians used the opportunity to make one additional modification to their Skystreak. During earlier flights of both the D-558-II and the X-I, observers had found it dif-ficult to visually track the brightly colored research planes. Walt Williams now decided that all NACA research aircraft at Muroc should be painted white to alleviate this problem. Consequently, as weather delays continued through the winter of 1947, Skystreak number two lost the majority of its crimson coat, although the control surfaces retained their original finish to avoid the introduction of any balance issues.

By February 1948 the lake bed had hardened, and NACA flights resumed with Lilly investigating the Skystreak's stability at various Mach numbers until tragedy struck on 3 May. Having aborted one flight that morning due to landing gear problems, Lilly began his nineteenth Skystreak flight that afternoon, but as he left the ground, the J35's compressor disintegrated. The aircraft's detachable nose escape system was of no use at such low altitude, and with the aircraft's control cables sheared by debris, Lilly was unable to save himself. As the Skystreak crashed back to the lake bed, Howard Lilly became the first NACA pilot to be lost in the line of duty.

A few months after Lilly's fatal crash and almost a year after Yeager's his-toric supersonic flight, Gene May took the number one Skystreak to Mach 1, but the D-558-I had never been intended as a supersonic research vehi-cle. That role would fall to the phase two aircraft—the D-558-II Skyrocket.

The swept-wing design that Heinemann and his team had created for the D-558-II was a classic of its time. The smooth, elegant lines of the Sky-rocket, with its long, pointed nose and rakishly swept surfaces, spoke of speed before it ever took to the air. The forty-two-foot-long cylindrical body housed both the Westinghouse J34 turbojet and the RMI LR8 rocket motor. The J34 received its air through two small flush inlets mounted low on the body ahead of the wings, and its exhaust exited through an unusual flush, downward-canted ventral tailpipe. The four-chamber LR8 motor was situated at the rear of the aircraft and fed by a water-alcohol mixture and $LO2$ tanks located above the turbojet. Unlike the first two X-Is with their heavier nitrogen feed system, the Skyrocket was able to take advantage of RMI's finally functioning turbopump, driven by steam created when concentrated hydrogen peroxide was passed over a catalyst screen.

The Skyrocket's thin 5 percent t/c wings were midmounted and swept back at 35°, giving the aircraft a span of twenty-five feet. As on the Skystreak, the horizontal tail surfaces were mounted midway up the fin to avoid the wake of the wings, but these were swept at 40° to prevent the wings and tail from experiencing buffeting simultaneously. The cockpit canopy was originally flush to the body, as on the X-1, but was quickly changed to a raised V canopy after it was found that it provided insufficient visibility for the pilot. Again, Douglas chose a detachable nose section rather than an ejector seat to provide the pilot with a means of escape. Finally, the already crowded fuselage carried a 625-pound NACA instrumentation package, broadly similar to that carried by the number two XS-1 and with the following features:

Strain gauges to measure the aerodynamic forces on the airframe and the force of control inputs made by the pilot during test runs.

Manometers to record air pressure over four hundred ports drilled in the wings and tail of the aircraft, requiring some four miles of tubing.

A cockpit camera to record the data displayed by the aircraft's instrument panel during test maneuvers.

As the first of the three new research aircraft rolled out of the El Segundo plant during November 1947, it was easy to see how its long nose-mounted data probe and sleek lines had earned it the nickname the Flyin' Swordfish among Douglas engineers, but one key component was missing. The number one ship still lacked its LR8 rocket motor due to delays in testing and certifying the power plant's installation. However, rather than delaying the project, Douglas, BuAer, and the NACA agreed that initial airworthiness tests could be conducted using turbojet power alone.

With Gene May committed to the D-558-1 program, the job of flying the D-558-II had been offered to his engineering pilot colleagues at Douglas's Santa Monica plant earlier in 1947, but unfortunately, it had not proven a popular prospect. Concerns around the as-yet-unbreached sound barrier and rocket propulsion meant that many saw the Skyrocket as a risky career move when compared to some of Douglas's other upcoming projects. As was usual practice, the pilots were each asked to submit bids for the contract to fly the new rocket plane, but a plot was hatched within the pilots' office to make these bids unrealistically high in an attempt to avoid selection. Unfortunately, for senior pilot Johnny Martin, none of his colleagues saw fit to share this plan with him, and as he was overseas on a delivery flight, he simply wired Santa Monica with what he felt to be a fair bid. Unsurprisingly, then, on his return, Martin was informed that he was to be the Skyrocket's test pilot.

Following the number one aircraft's delivery to Muroc AFB in the final weeks of 1947, Martin began ground tests and taxi runs in January 1948. Almost as soon as powered tests began, concerns were raised about the poor performance of the J34 turbojet, and the first flight on 4 February did little to improve matters. Martin found that the engine produced low RPM, feeling generally sluggish, and knew that this situation would deteriorate further once the aircraft was carrying its rocket motor and full propellant load. After more flights and much investigative work, Douglas and Westinghouse were forced to concede that the single J34 turbojet, poorly served by the small flush inlets and unusual tailpipe, was simply not powerful enough for the Skyrocket. With no immediate prospect of a higher-performance engine capable of replacing the J34 within the cramped confines of the D-558-II, Douglas engineers decided that JATO units could provide a temporary solution. These short-duration, solid-fueled rockets would give the Skyrocket the boost it needed to leave the lake bed in a reasonable distance, before being jettisoned and allowing the aircraft to proceed under turbojet power alone.

Martin made the first JATO flight on 13 July, and while the Skyrocket could finally take off with some safety margin, the new boosters brought their own problems. If a unit failed to jettison, as happened on Martin's thirteenth flight, the aircraft was left dangerously unbalanced, meaning an immediate emergency landing. However imperfect, the JATO solution at least allowed demonstration testing to continue. Early flights also revealed that the Skyrocket exhibited poor lateral stability at low speeds, resulting in a rolling or yawing movement known as dutch roll (apparently for the characteristic rolling gait of speed skaters on the Dutch canals). This was partially alleviated by adding additional height to the vertical fin, but it was never fully eliminated. In September 1948 with sixteen Skyrocket flights behind him, Johnny Martin was promoted to the role of chief pilot at Douglas Aircraft, meaning a recall from Muroc. With both contractor and NACA research programs well underway in the D-558-1, Gene May was able to take Martin's place in the Skyrocket, but with the veteran test pilot now in his late forties and a grandfather, Douglas recognized that a new pilot was needed to share the load and eventually take the program over.

Following familiarization and envelope-expansion flights in the turbojet-powered number one aircraft, May was soon able to test out the dual turbojet-rocket configuration when the number three Skyrocket—the first to have the LR8 installed—arrived at Muroc in December 1948. In an attempt to circumvent the ongoing problems with JATO jettison, May began to experiment with using two of the LR8's rocket chambers to assist on take-off instead. While this technique did work, it still didn't remove the fundamental concerns that many had regarding lake bed takeoffs. Now that the LR8 was being used, the aircraft was carrying a full propellant load, and the long takeoff runs gave ample opportunity for gear failure or tire blowouts—situations that could easily lead to the loss of both aircraft and pilot. Nonetheless, the team devised a plan to achieve a memorable first for the Skyrocket program and its navy sponsors. Using all four of the LR8's chambers, May hoped to make the first all-rocket takeoff from the lake bed. Unfortunately for Douglas, word of this plan leaked out around the Muroc flight line, and before May could make his attempt, he found himself upstaged by Yeager, who retained air force pride by making the sole ground takeoff of the X-1 program on 5 January 1949.

May began to push the Skyrocket through numerous research runs covering a range of Mach numbers and altitudes and, on 12 July 1949, finally urged the D-558-II through Mach 1 before welcoming a new pilot to the program to share the ever-increasing workload.

One day after receiving the surprise offer in Brown's office at El Segundo, an apprehensive Bill Bridgeman approached the remote Muroc site, security clearances in hand. Like other major contractors, Douglas had its own test facilities at Muroc by this time, and Hangar 181 on South Base was home to the Skyrocket and its small test team. During the drive up into the High Desert, the ex-navy man had weighed up his options but still found himself unable to shake the feeling that he was simply not qualified for the job of flying such a high-performance aircraft. Now, as he passed through security checks and entered the shade of the hangar, he finally saw the cause of his uncertainty—the gleaming white Skyrocket tended to by a dozen or so Douglas technicians. Somewhat mesmerized by the sight, his reservations seemed to fade away. Bridgeman realized he had to seize this opportunity, but that would mean convincing the test team that, with his single year of test-piloting experience and not a single jet flight in his logbook, he was the right man for the job.

Al Carder was the Douglas coordinator for the D-558-II test project at Muroc. Under his watchful eye the small team of technicians, flight test engineers, and aerodynamicists worked around the clock to keep the Skyrockets flying. While the aircraft remained on the ground, Carder held the final say on most matters, including who would fly his multimillion-dollar charges. Bridgeman quickly sought Carder out, introducing himself and giving a brief summary of his piloting experience. Much to his relief, there was no immediate rejection, but Carder's manner made it clear that there was much work to do before Bridgeman would be allowed to take the Skyrocket aloft. Next, Bridgeman was introduced to Gene May, who seemed similarly unconvinced by the pilot's credentials. Nonetheless, May passed Bridgeman a manual for the Lockheed F-80 with a recommendation that he get checked out in the jet fighter at his earliest opportunity. As May disappeared, it seemed the interview was now over, and to Bridgeman's relief, he had passed the first stage of his initiation into the exclusive ranks of the Muroc rocketeers.

Taking his chance to sit in the Skyrocket's cramped cockpit, the rookie marveled at the range of unfamiliar instruments and controls that crowded every available surface. His eyes were immediately drawn to the all-important Mach meter, which he hoped would record his tentative steps through transonic and supersonic flight. To his left he found the turbojet's throttle and the ignition switches for the LR8 rocket motor's four chambers. Douglas had also fitted the D-558-II with a display of rocket seconds, showing the pilot how long his supply of propellants would last once the LR8 was running. As each of the motor's chambers was ignited, so the speed at which the remaining rocket seconds ticked away increased.

As he spent his first month at Muroc gaining jet and high-speed experience in a well-flown air force F-80 (earning him the right to add his name to a much-signed picture of the aircraft behind the bar at Pancho's), Bill Bridgeman began to feel more integrated into the Skyrocket operation. Al Carder had begun including him in the day-to-day program activities, such as rocket motor test runs and planning meetings, and on 2 August, Bridgeman made his first flight in the number one Skyrocket to gain a feel for the controls and handling. The rocket motor wouldn't be used on this occasion, just the turbojet and the extra boost of two JATO units to kick the research plane off the lake bed. As usual, the flight took place in the calm conditions just after dawn, with a nervous, caffeine-fueled Bridgeman squeezing himself into the cramped cockpit, cursing his tall, athletic frame. Already apprehensive about this first flight, the stakes were now raised as an air force F-86 roared low overhead—his chase aircraft, flown today by one of the *real* Muroc rocketeers, air force X-1 pilot Pete Everest. Releasing the brakes with the J34 whining behind him, Bridgeman began his long slow roll across the lake bed, easily outpaced by some of the ground crew as they rushed to the support vehicles. The Skyrocket seemed to take an age to pick up enough speed for him to ignite the JATOs and finally break contact with the ground, but as the spent boosters fell away, the sluggish performance of the Skyrocket disappointed Bridgeman. This was no F-80, he noted; it felt slow and heavy on the controls. In fact, the sleek research aircraft felt more like the bulky Liberators he'd flown back in the Pacific. After clawing his way to the planned altitude, he put the aircraft through a series of basic maneuvers to feel it out. After numerous dives, pull-ups, and control pulses, he decided to test out the aircraft's low-speed performance by lowering the flaps and gear and then allowing his airspeed to drop off until he approached stall conditions. After some slight buffeting, the ship began to rock gently from side to side—the dutch roll that May had warned him about. Increasing the power steadied the Skyrocket again, and with fuel beginning to run low, it was time to return home. Landing without incident, he watched Everest make a long lazy roll signifying the end of his chase duties, before disappearing in the direction of South Base. It was done. Bill Bridgeman had met the Skyrocket and, to the pilot's great relief, proven equal to the task.

With his aircraft returned in one piece, Carder suggested that Bridge-

7. Douglas Aircraft pilot Bill Bridgeman poses with a model of the D-558-II Skyrocket. Courtesy University of Southern California.

man should now visit the Santa Monica office to negotiate his price for flying the Skyrocket—a brief conversation as it transpired. With $12,000 remaining in the budget to cover the pilot's contract, Bridgeman would receive this in addition to his regular salary, if he was agreeable. Accepting the deal, he headed back to Muroc as an official member of the Skyrocket team; now the real work could begin.

His new routine consisted of meticulous preparation and planning for a single thirty-minute research flight in the D-558-II each week, while fitting in as many F-80 proficiency flights as possible. As he warmed to his new role inside the cockpit, Bridgeman found it difficult to adapt to the temperature extremes and constant winds of the Mojave Desert. The base's air-conditioned bachelor officer quarters and swimming pool were certainly an improvement over the tarpaper shacks of Kerosene Flats, but he still preferred to put distance between himself and the High Desert whenever possible. Following flights, he would often retreat down below to Malibu for the weekend, to surf and relax while the Douglas team examined the aircraft and reduced the experimental data.

Following four more experience-building turbojet-JATO research flights during August 1949, Al Carder decided that it was time to give the new man his first experience of rocket flight, using the Skyrocket's LR8. Fueling for rocket flights typically began in the early hours, and waking well before dawn, Bill Bridgeman dressed and headed to the Douglas hangar to observe preparations. The ever-present Carder was already hard at work, personally overseeing every step of the process to ensure that his ship was in optimum shape. To help prepare the aircraft and transport it out to its takeoff point, Douglas had constructed an elaborate trailer that tended to the Skyrocket's needs until the final moments before takeoff. With fueling complete, the trailer and its attendant convoy moved out from Hangar 181 and across the lake bed in a slow procession. Bridgeman rode in the radio car, intently scanning his flight plan, a document representing the culmination of a week's work by the eight Douglas flight engineers assigned to the Muroc operation. It detailed the data points required and the exact maneuvers the pilot would need to fly to collect them, with speeds, altitudes, and g-loadings followed precisely. On top of these piloting tasks, Bridgeman would need to remember to start and stop the Skyrocket's data recorders at the relevant points, or his efforts would be for naught. Every flight plan had to win approval from the eagle-eyed Carder to progress. But at this point, the experienced coordinator would always defer to the pilot; the man with his neck on the line always got the final say about what was possible without stretching the Skyrocket or its pilot too far during the brief flights.

As the convoy reached its remote destination near Muroc's North Base, Bridgeman left the radio car and, accepting the best wishes of his colleagues, squeezed into the cockpit aware that this time his aircraft's tanks were heavy with explosive propellant. As usual, the air force was taking care of chase duties, and for his first rocket flight, Bridgeman had warranted the attention of Mr. Supersonic himself, Chuck Yeager. As his F-86 circled overhead waiting for the Skyrocket to begin its ponderous takeoff roll, Yeager taunted, "You going to fly it or blow it up, son?" Bridgeman advanced the J34's throttle and willed the Skyrocket on as it slowly rumbled up to the speed at which he could light two of the LR8's chambers and take to the air. After rolling across the lake bed for seventy seconds, he flipped the first switch and was pushed back in his seat by 1,500 lbf. of rocket thrust. Seconds later he repeated the process, lighting a second chamber before pulling back on the Skyrocket's control yoke and heading skyward.

The exhilaration of takeoff was short lived, though, as Bridgeman shut off the rocket chambers to conserve valuable rocket seconds, relying on turbojet power alone for the long drag to altitude. As the Skyrocket slowly ascended, Bridgeman's eyes remained glued to the instruments, scanning temperatures and pressures to ensure that the J34 was coping and that the rocket motor was ready to go. His task was complicated by the early morning sunlight flooding the Skyrocket's cockpit, but suddenly shadow fell across the glass, allowing him to see his instruments more clearly. Noticing Bridgeman's difficulties, Yeager had moved his F-86 to provide shade for the rocket plane pilot—*Damn! He really is that good*, thought Bridgeman.

Upon reaching 31,000 ft., he could finally light the LR8 and see what it could do. Igniting the four chambers sequentially until the motor was generating its full 6,000 lbf. of thrust, the Skyrocket's personality changed markedly. Gone was the slow, heavy performance of the climb. Now the brilliant white rocket plane surged forward, leaving Yeager in its wake. The Mach meter's needle quickly reached 0.95—just shy of supersonic but the exact speed required by the flight plan to find the Skyrocket's buffet point while pulling a constant 3 g's, another data point filled on the aerodynamicists' performance curve. Soon the precious rocket seconds had all been spent, and the LR8 chambers cut out, leaving only the turbojet to power the aircraft to a safe lake bed landing. Again, Bridgeman had met the challenge, and with his first rocket flight successfully accomplished, he could now focus on the task ahead.

Bill Bridgeman was not the only pilot getting to grips with the D-558-II during the summer of 1949. While Douglas continued to extend the flight envelope in ships one and three, the second Skyrocket had moved to the neighboring hangar, home to the MFTU.

Rather than concentrating on the Skyrocket's speed or altitude envelope, NACA engineers were far more interested in examining how the swept-wing configuration performed across a range of flight conditions. Although subscale research had been conducted in Langley's wind tunnels, the transonic choking problem still rendered data between Mach 0.75 and Mach

1.25 unreliable. Subscale tests of rocket-boosted models had been carried out by Robert Gilruth's team at the Pilotless Aircraft Research Division (PARD) at Wallops Island, Virginia, but now the Skyrocket with its comprehensive array of instrumentation offered them a full-scale means to study stability and control across this range.

North American Aviation had already taken the bold move of modifying its original FJ-I straight-winged design into the swept-wing XP-86, dramatically improving the aircraft's performance. With early production versions of the F-86 (the P designation, for "pursuit" aircraft, having been replaced with the F, or "fighter," designation) entering squadron service with the air force in 1949, other manufacturers were eager to follow North American's lead, and the NACA was acutely aware that accurate swept-wing data would play a vital role in the design of this next generation of high-speed jets. While the number two Skyrocket still lacked its rocket motor and was therefore incapable of supersonic flight, the MFTU decided to push forward with its transonic research program until such time as Douglas was able to install the LR8. Gene May made two demonstration flights in the aircraft during 1948 to test the aircraft's airworthiness before handover, and now, after a long series of engine problems, the controls were finally ready to be taken up by NACA pilots.

Bob Champine studied aeronautical engineering in his native Minnesota before enlisting with the navy and seeing combat in aircraft including the F4U Corsair. After the war, he was stationed at Norfolk, Virginia, and as the end of his service time approached, he made the short flight to Langley to offer his services as a test pilot. Although just short of the required flying hours, Champine was accepted and taken under the wing of chief pilot Herb Hoover. Champine became aware of Hoover's involvement with the still-classified X-I program out at Muroc, but he was content to pursue his career at Langley until fate intervened. Following Howard Lilly's fatal crash in the Skystreak, the MFTU needed another pilot to help meet the growing demands of the X-I, D-558-I, and upcoming D-558-II programs. Champine accepted the opportunity, driving cross-country from Langley to the High Desert to take up his new posting. After making his first rocket flight in the second X-I on 23 November 1948, Champine then checked out in the second phase one Skystreak in early April 1949 before becoming the first NACA pilot to fly the D-558-II on 24 May, but this wasn't Champine's first experience in a swept-wing aircraft.

After deciding to move forward with the D-558-II in 1945, the navy recognized the need for a test vehicle to give pilots experience with the challenging low-speed handling characteristics predicted for swept-wing aircraft. Consequently, in 1946 it purchased two prop-driven Bell P-63 Kingcobras and had them fitted with 35° swept wings along with a variety of other measures to offset changes to the aircraft's center of gravity. These unusual aircraft were renamed L-39s, and before leaving Langley, Champine was given extensive opportunity to fly them. This experience would stand him in good stead as he now took on the NACA's stability and control investigations in the Skyrocket.

Following his initial familiarization hop in the D-558-II, Champine made a series of flights to calibrate the NACA instrumentation and examine the Skyrocket's stability at a range of speeds and g-loadings. Apart from a minor fire in the cockpit camera leading to a smoke-filled cockpit during his third flight, these flights passed without major incident, but that was all about to change. On 8 August, Champine embarked on his seventh research sortie, with the flight plan calling for him to place a constant g-load on the ship in a tightening turn. As he banked the aircraft to set up the test conditions, the Skyrocket's nose suddenly pitched up. Quickly regaining control but concerned that the aircraft may have sustained damage, Champine terminated the flight. Wind tunnel tests on swept wings had indicated the possibility of instabilities under certain conditions, but once Yancey's computers reduced the data from the flight, NACA engineers were shocked by the severity of the pitch-up that Champine had just experienced. Clearly the Skyrocket was exhibiting a behavior that demanded further investigation.

At this point, the unit's pilot roster was further bolstered by the arrival of John Griffith. A resident of Illinois, Griffith had flown for the USAAF in the South Pacific, earning two Distinguished Flying Crosses. He left the service in 1946 and returned to his studies in aeronautical engineering at Indiana's Purdue University. After graduating with honors, he joined the NACA Lewis Flight Propulsion Laboratory in Cleveland before transferring out to Muroc in August 1949. Upon reaching the High Desert, he quickly checked out in the XS-I, the D-558-I, and the D-558-II before beginning research flights in all three vehicles. During his fourth Skyrocket flight on 1 November, Griffith repeated the maneuvers that had led to Champine's pitch-up and got an even more spectacular result. Again, the aircraft's

8. A D-558-II being towed across the lake bed in 1949. Note the air intake on the aircraft's forward fuselage. Courtesy NASA.

nose snapped abruptly upward, but now as Griffith attempted to fly on at this high angle of attack, the Skyrocket began to roll uncontrollably. After regaining control of the tumbling aircraft and satisfying himself that no damage had been sustained, Griffith went on to initiate another pitch-up, with similar results.

With the Skyrocket's instrumentation returning a wealth of data from these initial pitch-up incidents, NACA engineers were able to determine what caused these abrupt losses of control. As the aircraft began to bank or approach a stall, the nose would rise with regard to the direction of flight, increasing the angle of attack and slowing airflow over the wings. Whereas at higher speeds the air tended to flow straight from the front to the back of the wing, it now moved outward, deflected toward the wing-tips by the sweep of the wing's leading edge. This phenomenon, known as spanwise flow, caused the wingtips to stall before the rest of the wing, cre-ating an increased upward pressure on the forward section of the aircraft. The result was a strong, sudden upward movement of the nose—a pitch-up.

This was then accentuated as, at high angles of attack, the Skyrocket's fin-mounted tail surfaces were effectively shielded from the airflow by the aircraft's wings, making them totally ineffective. The sudden, violent nature of such pitch-ups was extremely dangerous at transonic speeds, where an aircraft could easily exceed its structural limits and break up. But it was of equal concern to pilots flying at low speeds during approach and landing, when there was little opportunity to regain control.

Following these observations, a large proportion of the NACA research program on the Skyrocket concentrated on testing various structural methods of controlling or eliminating spanwise flow, including leading-edge slats and wing fences. The results of this research were then passed on to industry through a series of technical memorandums and used in the design of new swept-wing aircraft. For all its high-speed potential, research into pitch-up became one of the D-558-II's most notable contributions to aeronautical knowledge. The robust nature of the research aircraft and the skill of its pilots allowed it to repeatedly enter these dangerous conditions and bring back data where a production aircraft could not. Much of this work was carried out on the number three Skyrocket from 1950 onward. But for now, the number two aircraft returned to Douglas's El Segundo plant for conversion to the Skyrocket's ultimate all-rocket configuration.

As the decade drew to a close, things were changing out in the High Desert. December 1949 saw Muroc AFB officially renamed Edwards AFB in tribute to Captain Glen Edwards, who had lost his life testing Northrop's YB-49 flying-wing jet bomber the previous year. The ever-growing NACA facility had also received a new name, becoming the High-Speed Flight Research Station (HSFRS). The final month of the decade also brought changes within Douglas's Skyrocket operation, as Gene May decided it was time for him to retire from test-piloting, bowing out with a spectacular low-level demonstration of the Skyrocket staged for members of the press on 1 December.

Following his first rocket flight in the D-558-II, Bridgeman took on an increasing amount of research flying duties for both contractor and navy programs. Throughout 1950 he regularly flew the number one ship using either the turbojet-JATO or turbojet-rocket combinations depending on the flight plan. The Douglas flight engineers and aerodynamicists were gradually able to fill in their data plots as the Skyrocket returned reams of information on the sideslip, stability, and stall characteristics of the swept-wing configuration. On one occasion, Bridgeman inadvertently entered a spin, a situation the Douglas engineers believed was unrecoverable in the D-558-II. Thankfully, through a combination of piloting skill and the robustness of the aircraft, he was able to regain level flight. Both Bridgeman and Yeager (flying chase that day) chose not to report the incident to the ground. Yeager simply commented, "Oops," which was still enough to raise Carder's suspicions and lead to a summons for the two pilots to appear before the program coordinator the following day to explain themselves—it was hard to keep secrets in an aircraft as heavily instrumented as the Skyrocket.

But as successful as the ongoing research program was proving, the Skyrocket still suffered from the compromises made to accommodate its dual means of propulsion. Even when using JATO units or the LR8 rocket motor, the underpowered aircraft still struggled to reach test altitude with a reasonable load of propellants. Takeoffs remained inherently dangerous, with the constant possibility of a gear failure or low-level engine failure, spelling catastrophe for plane and pilot. The turbojet and its fuel tank took up precious volume within the Skyrocket's fuselage that could have been used for additional rocket propellants, and although the team knew that the aircraft was capable of far higher performance, Gene May summed it up when he said, "The ultimate performance of the aircraft will never be achieved with the limited supply of rocket fuel."

By 1949 the X-1 had long since proven the air-launch technique practical, and consequently, the NACA's earlier reservations had largely evaporated. If the Skyrocket could be air-launched, then its entire propellant load could be used to achieve higher performance at altitude rather than wasted on dragging the aircraft up through the lower layers of the atmosphere. Bell Aircraft was soon hoping to deliver its X-2, a swept-wing successor to the X-1 with predicted Mach 3 performance and capable of reaching altitudes in excess of 100,000 ft. The NACA felt that if the Skyrocket could move closer to its theoretical performance limits, it might contribute useful data to aid and complement the X-2. A decision was therefore made to convert the third Skyrocket for air-launch in its current turbojet-rocket configuration, while the NACA's number two ship would have its turbojet and attendant tankage removed to become an all-rocket, air-launched version. Douglas had initially hoped that Gene May might stay with the program long enough to carry out initial test flights of the air-launched versions, but with his departure, that task now fell to Bill Bridgeman.

This new phase of the program would require additional hardware, most notably a suitable launch aircraft. The navy had obtained four Boeing B-29s in 1947, designating them P2B-1s patrol bombers. One of these aircraft was now requisitioned by BuAer to become the Skyrocket's mothership, and Douglas was tasked with making the necessary modifications to the bomber so that it could carry the research aircraft beneath it in a similar fashion to the B-29–X-1 combination. The Skyrocket's longer fuselage and higher tail surfaces did complicate matters, however; large sections of the P2B-1s's structure had to be cut away, with the remainder strengthened to ensure that the aircraft would remain structurally sound with its heavy payload attached. Given the weight of the fully fueled Skyrocket, the slow climb to launch altitude would also place a tremendous strain on the bomber's four Wright R-3350 engines. These had proven temperamental during the B-29's service life, sometimes bursting into flames with little warning when pushed too hard. Clearly, the mothership would need very careful piloting, and fortunately for the program, the ideal candidate for the job was close at hand.

Some months earlier, the Douglas team out at Edwards had been joined by George Jansen, one of the engineering pilots whose technical knowledge had so inspired Bridgeman back at El Segundo. Jansen had traveled out to the desert to test the A2D Skyshark, an advanced turboprop attack fighter then under development for the navy. Unfortunately, on 19 December 1950, the prototype A2D crashed during a test flight at Edwards, killing its navy project pilot, Commander Hugh Wood. The Skyshark program went on hold while the incident was investigated, leaving Jansen at something of a loose end, but having taken a close interest in his friend's flight test program in the Skyrocket, he now volunteered to command the P2B-1s. Jansen, Bridgeman, and Carder thrashed out the finer points of how to take the Skyrocket to altitude and release it safely, and flight plans developed accordingly. However, for the research aircraft's pilot, there was another piece of new hardware to master.

Air-launching would mean that the Skyrocket could now attain much higher altitudes, especially the all-rocket number two aircraft. Like Everest before him, Bridgeman would soon be exceeding the Armstrong limit, and that meant a trip to the air force Aero Medical Laboratory at Wright-Patterson AFB for a two-week introduction to the partial pressure suit. By 1950 the newer David Clark T1 suit was available, and while marginally more comfortable than the earlier suits, the T1 still required some major adaptation on behalf of the pilot. Bridgeman underwent a series of pressure chamber runs to familiarize himself with the suit's inflation and learn the pressure-breathing technique. On his return to Edwards, Bridgeman would don the restrictive suit for ground tests whenever possible, practicing pressure breathing until it became second nature to him. If pressure was lost at altitude, he knew that the Skyrocket's already tight cockpit would become unbearably cramped as the suit's capstans and bladders inflated, but the system had earned the respect of all at Edwards since it had saved Everest's life the previous year.

The Skyrocket's air-launch campaign began with tentative steps using the number three aircraft still equipped with both turbojet and rocket propulsion. After a series of ground tests to check fueling and the all-important release mechanism, captive flights were flown to ensure that the combination was aerodynamically sound and that the Skyrocket's turbojet could be successfully started at altitude. With these technical hurdles cleared, the first drop took place on 8 September 1950, using turbojet power alone to remove the hazard of rocket propellants from an already risky equation. With Jansen at the controls, assisted by a crew of Douglas volunteers, the laden P2B-1S slowly circled its way to altitude above the dry lake. At 18,000 ft., Bridgeman—in regular flight garb, as he wouldn't be exceeding 25,000 ft.—and two of the mothership's crew made their way back to the Skyrocket. After helping him into the cockpit, locking the canopy, and making sure the research plane's internal oxygen supply was functioning, the crewmen retreated, leaving Bridgeman alone. As launch approached, he ran through the aircraft's systems, consulted the flight plan one final time, and started the turbojet. Soon Jansen began counting down steadily from ten to one, and then . . . release!

Falling rapidly from the gloom of the mothership's belly into the bright California sunshine, Bridgeman was now on his own. The drop had been clean, and now with a full tank of jet fuel, he hoped to make a previously impossible hour-long flight, with speed runs at various altitudes and Mach numbers on the way back down to the lake bed. Unfortunately, problems with the airspeed indicator thwarted these plans, forcing Bridgeman and the D-558-II into a premature return, but the first air-launch had been achieved safely, meaning they could now move forward with rocket flights.

The first air-launch using both the turbojet and the LR8 rocket motor occurred on 17 November 1950. Prior to this flight, fuel-jettison tests had been performed at altitude using colored water in place of the water-alcohol mixture and LO2, but today the tanks were full of the real thing. As with the X-1, emergency arrangements had been worked out between the mothership and research plane pilots before air-launch rocket flights began. If an emergency occurred in either the Skyrocket or the P2B-1S below 8,000 ft., the whole crew were to bail out immediately. At this stage, Bridgeman would still be in the bomber's forward section, so there wouldn't be enough time for him to enter the research aircraft and have a chance of gliding to safety. Above this altitude, Bridgeman was to quickly board the Skyrocket before Jansen cut him loose, and he would be on his own. Fortunately, though, on this first attempt, drop time approached without incident, and with its turbojet already running, the D-558-II fell away cleanly, allowing Bridgeman to concentrate on getting the LR8's chambers ignited. Under combined jet and rocket power, the ship quickly climbed to around 40,000 ft. before pushing over into a level speed run. Following the familiar buffet above Mach 0.9, the Skyrocket easily slipped through Mach 1.

By now the HSFRS was eager to obtain the number three aircraft in order to resume its research program, on hold since the second Skyrocket returned to El Segundo at the beginning of the year. All parties decided that Bridgeman should make one final demonstration flight before the aircraft was handed over, at which point Douglas would concentrate on the all-rocket number two aircraft. This final demonstration flight took place on 27 November, but far from being routine, it reminded all concerned that problems could strike at any time. Following the release from the mothership, Bridgeman ignited the LR8's chambers in quick succession and climbed away. Now, as he pushed over into level flight and approached Mach 1, he began to feel buffeting. Assuming that this was the regular disturbance that marked the Skyrocket's transition from transonic to supersonic flight, he continued as planned, but the buffeting persisted. Scanning his instruments, Bridgeman realized that the J34 turbojet was running way over nominal temperature and experiencing compressor stall. Quickly, he pulled back on the J34's throttle, but this caused the aircraft to pitch forward, momentarily interrupting the supply of propellants to the LR8 rocket motor. All four chambers now cut out, leaving the Skyrocket without power and losing altitude. This situation didn't unduly concern Bridgeman—the Skyrocket could easily glide back to base. But unfortunately, without the heating usually provided by the jet engine, fog began to appear between the panes of the cockpit, eventually obscuring his view completely. Flying blind, Bridgeman was relieved to hear Yeager's familiar voice. The chase pilot was now on his wing and able to guide him home. Following a number of unsuccessful attempts, Bridgeman managed to relight the J34 and restore visibility in time for a safe landing. The handover of Skyrocket number three still took place as scheduled, but the experience provided a cautionary note for the NACA pilots who would now fly it. Again, Bridgeman's wariness of the D-558-II had proven to be well founded. Now he moved on to the program's ultimate challenge—the newly returned all-rocket number two aircraft.

On its return to the El Segundo plant, the second Skyrocket underwent major modifications. Unlike the third D-558-II, which had simply required installation of the retractable hook system needed to secure it in the P2B-IS's bomb shackles, number two was stripped of its J34 turbojet and equipped with new tanks for the water-alcohol fuel and LO2 oxidizer for the LR8 engine as well as hydrogen peroxide to power the newly installed turbopump. Without the turbojet, there was no need for air intakes or the tailpipe, so these openings were covered to give the aircraft clean, unbroken lines. An expanded NACA instrumentation package rounded out the modifications, and once ferried back to Edwards AFB under the P2B-IS, Douglas engineers made final preparations for powered flights to begin.

Bridgeman would now map out the safe limits within which the higher-performance all-rocket aircraft could operate before it, too, was delivered to the HSFRS. He would be taking the Skyrocket to speeds and altitudes that it had never originally been designed for, and although Heinemann's team of engineers and aerodynamicists, as well as the flight planners out at Hangar 181, all believed that the D-558-II was up to the task, there was only one way to find out if reality matched the models.

As the first flight approached, Bridgeman's caution around the Skyrocket increased. Time and time again he'd proven equal to every surprise the aircraft had thrown at him, but now he had a nagging feeling that he was pushing his luck just a little too far. His working days became a blur of training sessions sealed in the Skyrocket as it hung under its static mothership on the Edwards ramp. Each day he donned his unwieldy T-1 pressure suit and forced himself to run through the flight plan while practicing the tiring pressure-breathing techniques he had learned back at Wright-Patterson. Reaching a peak of preparation as the end of the year approached, he felt ready to face whatever unknowns his aircraft or the atmosphere could deliver. Unfortunately, there was one eventuality for which Bridgeman couldn't train—a maddening series of aborts caused by various unforeseen technical failures on the Skyrocket. Three times during December 1950 he wound himself up ready for the first drop, and three times he returned as a frustrated passenger in Jansen's P2B-1s. As a new year rolled in, the Douglas team hoped for better luck, but again more aborts followed. The engineers worked furiously to fix one problem, only to have another appear, but it was hardly surprising that a new system, cold soaked at high altitude, would prove temperamental. Soon the reluctant Skyrocket became the subject of stinging flight-line humor—a hangar queen. Worse still, many on the base began to gossip that the problem lay with the man rather than the machine. Bridgeman had always projected a tone of respectful caution regarding the Skyrocket, but now some suggested he had become too cautious and developed abort fever, always finding a fault to save him from having to face his demons. Within Hangar 181, Carder and the team knew that such talk was ill founded. They saw the cockpit films showing the problems developing during each aborted drop attempt. Nevertheless, pressure began to mount from the navy and Douglas's own management to get the bird flying soon or risk losing the chance completely.

On 7 January it was time for launch attempt number eight. The eagerness of earlier attempts had largely evaporated. Now the demoralized crew trudged aboard the converted bomber expecting another fruitless round trip with their constant companion lashed in the bomb bay. Bridgeman went through the motions, donning his uncomfortable pressure suit and riding up to the point where he and his helpers trooped back to the waiting Skyrocket. Once secured in the cockpit, he ran through his predrop routine, watching and waiting for the inevitable glitch. Drop time approached, but still no problems. *Could this really be the one?* He allowed himself a moment of hope while continuing to pressurize the LR8 and run his last-minute checks, before he noticed, with his heart sinking, the rocket's fuel pressure slowly falling. Calling the abort, Bridgeman began shutting down the Skyrocket's systems in preparation for yet another trip home as a passenger. But any disappointments were short lived, as he now heard, to his horror, Jansen counting down to the drop. Bridgeman shouted frantically in an attempt to stop the countdown, but Jansen had his microphone key depressed and was therefore unable to hear his friend's pleas. With only seconds to go, the test pilot scrambled feverishly to prepare for flight before he and the Skyrocket dropped away from the mothership to meet their fate. As Jansen pressed the release button, Bridgeman fell away, and fortunately, he had done enough to light the LR8 cleanly, if a little late. As he climbed away from Jansen's aircraft, he uttered a phrase that would go down in Edwards folklore: "Damn it, George, I told you *not* to drop me!" With the eventful launch now behind him, Bridgeman was finally able to see what the modified Skyrocket could do by taking it up to 40,000 ft. and pushing over into level flight. The rocket plane quickly accelerated through Mach 1 but experienced some dutch roll that would need further investigation. Nosing over into a shallow dive, the D-558-II now achieved a program record speed of Mach 1.28 before the propellants were gone and Bridgeman began the long glide back to Edwards.

The launch mix-up had caused much consternation among the entire team, tempering postflight celebrations. In the confusion, Bridgeman had neglected to start the NACA data recorders and consequently returned empty-handed. But the incident could have been far more serious, and Al Carder knew it. He insisted that before another launch attempt took place, a "go" light should be installed in the P2B-1S, which would be illuminated by the Skyrocket's pilot *only* when everything was ready for a drop.

Bridgeman's luck with aborts failed to improve much, as the next four attempts all saw the Skyrocket returning beneath its launch aircraft, but on 5 April a second successful drop was finally made. The flight plan was similar to the previous launch, but this time before Bridgeman could push over, the Skyrocket began to roll, a motion that increased in severity as the aircraft leveled out and continued to accelerate. At Mach 1.36, the shaken pilot decided enough was enough and shut the rocket motor down rather than risk a complete loss of control. Returning to Edwards, Bridgeman demanded answers before he would risk pushing the Skyrocket further. The following day, a team of engineers from El Segundo arrived at Edwards in an attempt to understand and rectify the situation. As they examined the aircraft and flight data, they soon identified the cause of the stability problems. With the Skyrocket now operating at altitudes where the atmospheric pressure was far lower, the LR8's rocket plume was expanding enough to interfere with the aircraft's rudder, leading to the unpredictable rolling. With the problem identified, the engineers now proposed a fix; a retractable pin could be fitted within the Skyrocket's fin and deployed at high speeds to lock the rudder. After a short delay for this fix to be installed, it was time to try it out.

Time was now running short for the Douglas team. Their original contract had called for three powered demonstration flights of the modified number two Skyrocket before it was handed over to the NACA, but Douglas and BuAer remained committed to testing the aircraft's performance limits and maybe achieving one last record before the air force began flying the improved x-1s. On the third contract flight, Douglas decided to send the Skyrocket higher and faster as a demonstration of the potential they felt the aircraft still offered. On 18 May, Bridgeman dropped away from the P2B-1S and rapidly lit all four rocket chambers. Pulling the yoke back, he began a steep climb to around 55,000 ft. before pushing over into a level speed run. With the rudder locked, there was no repeat of the wild instability that had dogged the previous flights, and Bridgeman was able to let the LR8 run until its propellants were exhausted. When the data was reduced, Douglas revealed that the Skyrocket had hit Mach 1.72 at an altitude of 62,000 ft. Bill Bridgeman was now the fastest man alive, and the program had truly hit its stride. Any further flights would require a contract extension and the agreement of the waiting HSFRS, but under pressure from a jubilant navy, the NACA yielded. On 11 June, Bridgeman repeated his previous flight plan, pushing the unofficial speed record out to Mach 1.79, but worryingly, even with the rudder locked, he still experienced some stability problems. Eager to push things further still, the flight engineers now suggested that he try a quicker push-over at .25 g's rather than the previous .8 g's, to improve the Skyrocket's acceleration and hopefully find more speed.

The next flight took place on 23 June, with Bridgeman following the revised plan, but as he made the gentler push-over, the rocket plane began to roll uncontrollably from side to side by as much as 80° in each direction. Fearing that neither he nor the Skyrocket could take much more of the severe punishment, Bridgeman shut off the LR8 at a peak speed of Mach 1.85. He now found himself in a high-speed dive, heading away from the lake bed, and only through desperate efforts was he able to get the D-558-II back into level flight and turned around. Bridgeman would later confess that he had briefly considered using the escape capsule during this wild ride, but knowing that he would have no control over the detached nose section, he elected to stay and take his chances with the ship.

Clearly the reduced push-over at .25 g's hadn't produced the results the flight planners had intended. While the maximum Mach number had been significantly increased, they had almost lost Bridgeman and the aircraft in the process. With one final chance to achieve maximum speed, the next flight plan reverted to the original push-over at .8 g's, hopefully avoiding the severe oscillations of the previous flight. Douglas had also finished installing an LO2 top-off system in the mothership, meaning the Skyrocket would have full supply of oxidizer when dropped, hopefully earning a few more valuable seconds of powered flight. On 7 August 1951 Bridgeman dropped from the P2B-1S and followed the flight plan through the push-over at .8 g's, before reducing the load still further to .6 g's, when he noticed the Skyrocket's left wing beginning to drop. Quickly returning to a load of .8 g's, the aircraft continued to thunder onward until the propellants were exhausted at a new unofficial speed record of Mach 1.88 at 67,000 ft. In his postflight report, Bridgeman stated that he felt the Skyrocket could potentially reach Mach 2 under perfect conditions, but now that challenge would await another pilot.

One final flight remained for Bill Bridgeman. While most of the program's recent focus had been on achieving maximum speed, BuAer had also given Douglas permission for an attempt at a new unofficial altitude record. The Skyrocket's performance had now been stretched so far beyond the original design goals that each new trip into the unknown carried significant risk, but following lengthy consultations between Heinemann's team at El Segundo, Douglas's chief aerodynamicist Charles Pettingall, and Al Carder's team out at Edwards, an acceptable flight plan was finally devised. On 15 August, Bridgeman dropped away from the Superfortress, lit all four of the LR8's chambers and pointed the nose toward the darker skies above. The Skyrocket raced upward, but rather than push over around 50,000 ft. as he had on his speed runs, Bridgeman held his climb angle and continued to streak toward the heavens. The Skyrocket's left wing again showed a tendency to dip, but this was controllable, and as the last of the propellants were exhausted, the LR8 fell silent with the aircraft still coasting upward on its vast arc. Finally pushing over at the peak of his climb, Bridgeman was rewarded with a breathtaking vista of California spread out beneath him. Back on the lake bed, NACA radar tracked him to a peak altitude of 79,494 ft., finally surpassing the long-standing mark set by the *Explorer II* balloon in 1935. Bill Bridgeman had certainly come a long way since the interminable PBY patrols of 1941. Now as he returned the Skyrocket to Rogers Dry Lake for the final time, he finally allowed himself a moment to reflect on his achievements.

In these prespaceflight days, stories of record-breaking flights out in the High Desert generated huge public interest. Bridgeman became something of a celebrity for a short time, with his Skyrocket exploits being celebrated on the cover of *Time* magazine, but he remained modest about his achievements. When asked about how it felt to break records, he always maintained that he had gained far more satisfaction from his first solo flight during naval training than he did from his time in the D-558-II. On the subject of risk, he calmly stated, "I truly believe it is safer 'upstairs' than it is on Wilshire Boulevard during the morning and evening rush hours."

Bridgeman was honored by his peers, receiving the prestigious Octave Chanute Award in 1953, but he was ready to move to a new challenge. After making more than eighty flights during his two years with the Skyrocket, it was now time for Bridgeman to take on another experimental Douglas aircraft, the turbojet-powered X-3. As the all-rocket ship two moved across to the NACA hangar, Bridgeman remained available to pass his experiences on to the Skyrocket's eager new pilot, Scott Crossfield, in whose hands the D-558-II would soon write another chapter of aviation history.

4. The Race to Mach 2

Even before Bill Bridgeman set his unofficial speed and altitude records in the Skyrocket, events on the other side of the world had begun to exert an influence over activities at Edwards AFB. On 25 June 1950 the Korean People's Army flooded south over the thirty-eighth parallel, the informal demarcation line that had divided the Korean peninsula since the end of the Second World War. With the Soviet Union and China actively supporting the Korean People's Republic, the United Nations moved to defend the southern Republic of Korea, deploying a predominantly American force to repel the Communist invaders.

The Korean War represented the first real combat test for the recently formed USAF, whose primary goal was the establishment of air superiority over the peninsula, allowing allied bombers to operate uncontested. Initially, the air force's first-generation jet fighters, such as the Lockheed F-80 and Republic F-84, proved effective, but allied pilots soon began to encounter a dangerous new adversary in the skies above Korea—the MIG 15. The small, swept-wing fighter could easily outrun and outclimb the straight-winged Shooting Stars and Thunderjets, and in the hands of experienced Soviet pilots, MIG 15s began to inflict significant losses on the UN coalition, until the USAF was able to deploy its new F-86 Sabre, which proved an effective counter to the Soviet fighter.

The appearance of the MIG 15 challenged Western preconceptions regarding the state of Soviet technology, leading to increased defense spending that ended a period of postwar austerity for the U.S. aviation industry. Air force planners now used lessons learned during jet-versus-jet combat over Korea to develop specifications for a new generation of high-speed fighters and interceptors capable of matching anything the Soviets might deploy in the coming years. Even as research into the supersonic characteristics of swept wings continued out at Edwards, North American Aviation was busy developing its successor to the F-86. Designated the F-100, the new design incorporated features directly derived from the high-speed research programs, offering pilots supersonic speed in level flight and marking the start of the advanced "Century Series" fighters that would remain air force mainstays for decades to come.

This renaissance in military aviation helped hasten major changes in the Antelope Valley. During early 1951 an organizational shake-up within the still-nascent air force brought Edwards under the purview of the newly formed Air Research and Development Command (ARDC) rather than the AMC, with the base being designated as the Air Force Flight Test Center (AFFTC) during June 1951. Experimental aircraft programs, including the ever-expanding X plane series, became the responsibility of the Wright Air Development Center (WADC), a division of the ARDC. Having taken command of the base in August 1949, Brigadier General Boyd had already overseen the relocation of flight-testing activities from Wright Field to Edwards, and he now initiated a major redevelopment program with the aim of creating a large, modern facility equal to the air force's ever-increasing flight test needs. During his time in command, Boyd's predecessor, Colonel Signa A. Gilkey, had authored a master plan outlining what he felt to be necessary changes, including the rerouting of the Santa Fe railroad, which still ran across the lake bed, and the construction of an entirely new base two miles northwest of the existing site, allowing for improvements to infrastructure while also providing service personnel with better housing and schools for their families. While planning for its future on the ground, the AFFTC was also awaiting a new family of high-speed research aircraft that promised to eclipse the navy-sponsored Skyrocket's recent achievements.

Within a month of Yeager's historic Mach 1 flight in October 1947, the air force had instructed Bell Aircraft to begin design studies for an advanced version of the X-1. While it would retain the same general layout and control systems as the original aircraft, the advanced X-1 would remedy many of its predecessor's shortcomings and offer significantly improved performance. With RMI's long-awaited turbopump now available and the aircraft's fuselage stretched by four and a half feet, the advanced design would feature large, thin-walled cylindrical tanks made from lightweight aluminum, increasing the amount of propellants that could be carried.

Another new addition was the raised cockpit featuring a bubble canopy to give pilots far better visibility than the original x-1's flush design. As the cockpit would now be entered from above, pilots would no longer be exposed to the freezing slipstream while using a ladder or elevator to reach a side-mounted hatch; however, the new design did not include an ejection seat, making the pilot's prospects for escape little better than those in the original aircraft. Finally, Bell's engineers revised the location and access for internal systems, including research instrumentation, greatly simplifying operations for ground crews, with the aim of reducing maintenance times.

The air force initially ordered four of the advanced vehicles (designated x-1a through x-1d), with each intended to investigate a different area of research, but the x-1c weapons test bed was subsequently canceled when it became clear that the f-86 Sabre could perform this role. A Boeing b-50 bomber—an updated version of the Superfortress, featuring structural improvements plus more powerful Pratt and Whitney Wasp Major radial engines—was acquired to carry the heavier rocket planes to their launch altitudes in place of the b-29.

As work progressed on the first of the advanced x-1s, Bell finally completed the much-delayed x-1-3, following years of funding problems and technical delays. *Queenie* (as the aircraft became known, thanks to three years as a partially complete hangar queen) belatedly reached Edwards in April 1951, and although the x-1-3 fulfilled Bell's original design vision for the x-1, it now represented something of an interim step between generations, featuring the original cockpit layout alongside the low-pressure turbopump-driven fuel system. Lighter with an increased propellant capacity when compared to her siblings, Bell hoped that *Queenie* might be capable of speeds in excess of Mach 2.

With the first of the advanced x-1s, the x-1d, arriving in the High Desert as the x-1-3 was being prepared for its demonstration flights during July 1951, it appeared that the air force and the NACA were on the cusp of a new era of high-speed research. Fate, however, held different plans.

The x-1-3 finally took to the skies for its maiden flight on 20 July 1951, with Bell test pilot Joe Cannon at the controls. *Queenie* was dropped at an altitude of 30,000 ft. for a short glide flight to investigate its basic stability and handling, and following a clean release, Cannon put the aircraft through

9. Bell Aircraft test pilot Jean "Skip" Ziegler with the Bell x-1a.
Courtesy Niagara Aerospace Museum.

a series of maneuvers, closely observed by air force chase pilots General
Boyd and Pete Everest (now chief of flight test operations at Edwards). As
expected, *Queenie* handled every bit as well as its x-1 siblings, but as Cannon approached the lake bed, he was caught out by the aircraft's tendency
toward poor pitch control at low speeds and, consequently, stalled into a
heavy landing. When the aircraft slammed down, its nosewheel support
structure broke, and *Queenie* slid to an unceremonious halt. Cannon had
joined the x-1 nosewheel club. With only a single glide flight behind it, the
unlucky aircraft was once again confined to the hangar, awaiting repairs.

Attention now turned to the powerful x-1D, as the first advanced x-1
was carried aloft beneath the b-50 mothership to make its inaugural glide
flight a mere four days after *Queenie*'s inauspicious debut. Bell had selected
its chief test pilot, Jean "Skip" Ziegler, to perform the x-1D's demonstration flights. A usaaf veteran and one of the first pilots to fly the demanding Himalayan transport route known commonly as the Hump, Ziegler
had joined the Curtis-Wright company following the war, earning a repu-

tation as a tenacious air racer, before moving to Bell in 1948. Following a brief spell as a test pilot for North American Aviation, he returned to Bell Aircraft, becoming project pilot for the x-5 variable geometry test bed as well as the advanced x-1s and the forthcoming x-2.

Tall and quietly spoken, Ziegler was highly regarded by his Bell colleagues, but the 24 July flight marked his first experience in one of the company's air-launched research aircraft. After dropping away from the B-50, he put the gleaming aluminum x-1D through a nine-minute glide flight, investigating the general handling and low-speed characteristics of the longer, heavier design. Although the aircraft flew well, it had inherited its predecessors' tricky landing characteristics, and Ziegler managed to break the nosewheel, sending the x-1D to join *Queenie* in the workshop. Fortunately, Bell's maintenance crew were becoming increasingly adept at repairing nose gear, so they were able to get the aircraft fixed and airworthy within a month of Ziegler's landing mishap.

During this delay, the air force had been forced to look on as Bill Bridgeman soared beyond 79,000 ft. in the D-558-II, and although the sight of a navy-sponsored aircraft setting new records in the skies above Edwards must have been a bitter pill to swallow, the x-1D now offered a means to rectify the situation. In order to get things moving, the AFFTC accepted the aircraft from Bell without any further demonstration flights and wasted no time in making its intent clear. On 22 August, just one week after Bridgeman set his altitude record, the x-1D was airborne again, slowly circling to altitude beneath the B-50. As the first air force pilot to fly the aircraft, Pete Everest had been given the unambiguous instruction to "see what [the x-1D] could do wide open." Although he would later claim that Bell had conducted half a dozen test flights in the aircraft prior to the air force's acceptance, the reality was that with only a single tentative glide flight behind it, the x-1D was now fully fueled and gunning for a record that, in Everest's words, would be hard to beat.

After entering the rocket plane's cockpit while the B-50 climbed toward its launch point, Everest wasted no time in powering up the aircraft's systems and scanning its many gauges. As he checked the tank pressures, the test pilot was dismayed to see the needle denoting his nitrogen pressure falling rapidly. Although the x-1D no longer needed nitrogen to force its propellants to the XLR11 engine, it still used a smaller supply of the gas to provide pressure for other vital systems. Knowing that the problem could signal the end of the day's flight attempt, Everest climbed out of the X-1D to confer with air force flight engineer Jack Ridley and his Bell counterpart Wendell Moore, both of whom had accompanied the pilot into the B-50's bomb bay. With the nitrogen pressure still falling, the trio accepted defeat and aborted the launch, but as Everest leaned back into the cockpit to pressurize the oxygen tank in order to jettison the oxidizer, he was instantly flattened by a sudden, violent explosion. In the confusion that followed, Everest and his shocked colleagues managed to scramble away from the stricken rocket plane, but when General Boyd, flying chase in close proximity, reported flames streaming from the X-1D's fuselage, Jack Ridley knew that he needed to act fast in case a second explosion occurred, taking the B-50 and its crew with it. As fire began to rise into the bomb bay, Ridley was left with no choice but to pull the release handle, sending the X-1D to a fiery fate on the desert floor.

When accident investigators examined the charred and twisted remains of the aircraft, they found little to indicate the cause of the catastrophic explosion. Attention focused on possible leaks within the fuel system, with the accident board theorizing that alcohol vapor may have formed around propellant lines before being ignited by a spark from the aircraft's radio. The loss of the X-1D caused ripples of unease around the base. The NACA was about to begin flying the second Skyrocket, and Bell was almost ready to resume its contractor demonstration phase in the X-1-3. As both aircraft used similar fuel systems and near-identical rocket motors to the X-1D, concerns were raised that they too might be at risk. However, with no definitive cause for the explosion identified, it seemed that little else could be done to improve safety beyond increased vigilance by the maintenance crews.

On 9 November 1951 the now-repaired *Queenie* returned to the skies beneath the same B-50 that had recently survived the X-1D explosion. Today the X-1-3 was carrying a full load of water-alcohol mixture and LO2, but as this was merely a captive fuel jettison test, distilled water had been substituted for the turbopump's volatile hydrogen peroxide propellant. As the B-50 reached altitude, Joe Cannon began the test by jettisoning the distilled water, but as this streamed away in the research plane's wake, he noticed that the aircraft's nitrogen pressure was falling away rapidly, leaving the frustrated pilot with no option but to abort the remaining tests, as he could no longer pressurize the water-alcohol mixture or LO2 tanks. Descending back toward Edwards with the still-fueled *Queenie* lashed below it, the B-50 crew were faced with a nerve-racking landing, and much to everyone's relief, they managed to bring the combination down without incident. However, there still remained the small matter of jettisoning the x-1-3's propellants.

After briefly calling at the propellant loading area to replenish *Queenie*'s nitrogen supply, the B-50 headed for a remote part of the ramp where the x-1-3's tanks could be safely emptied. With fire crews in attendance, Cannon reentered the research plane's cockpit, but just as he began to pressurize the LO2 tank, a muffled explosion shook *Queenie*. The aircraft's LO2 tank had been torn apart, and now its contents, along with a full load of fuel, were pouring onto the ramp, rapidly enveloping both the x-1-3 and the B-50 in a thick cloud of freezing vapor. Knowing that he needed to get away from the rocket plane as quickly as possible, Cannon emerged from the mist, bravely shouting warnings to the nearby ground crew. Unfortunately, as he fled, the pilot was floored by a second powerful explosion, sustaining serious burns as he tumbled into the flood of propellants. Bell's alert ground crew managed to pull their man clear before rushing him to the base's hospital, but as fire crews moved in to bring the blaze under control, it quickly became obvious that the x-1-3 and the B-50 had been damaged beyond repair. Although it took Joe Cannon almost a year to recover from the injuries he had sustained that day, he was eventually able to return to duty as Bell's chief of flight test.

All hopes of reaching new speeds and altitudes in the x-1-3 and x-1D had been reduced to charred scrap in a few short months, with neither aircraft having logged a single powered flight. As in the case of the x-1D, the accident board were unable determine the exact causes behind *Queenie*'s explosion, but in the aftermath of these incidents, the AFFTC faced a lengthy delay before the next advanced x-1 would be delivered.

Following the frustrating events of 1951, the new year began with a change of command at Edwards AFB. When Boyd moved on, he was replaced by Colonel J. Stanley Holtoner, who arrived just as funding to begin the redevelopments outlined in Gilkey's master plan was released. The AFFTC would soon be moving away from South Base, and the winds of change now sweeping across the dry lake also reached one of its more notorious locations. A bitter dispute erupted between Colonel Holtoner and Pancho Barnes as the air force sought to obtain her land as part of the base-expansion plans. To make matters worse, allegations of salacious activities at the Happy Bottom Riding Club led the air force to ban its servicemen from the property, removing much of Barnes's clientele.

Against this backdrop, only one rocket plane program remained operational at Rogers Dry Lake, and it wasn't under air force control. At the HSFRS, the D-558-II was about to enter a new phase of its long career, and an ambitious young NACA pilot believed that the aircraft still had some potential left to tap.

In the winter of 1950 twenty-nine-year-old Albert Scott Crossfield arrived in the High Desert, armed with a degree in aeronautical engineering, four years' experience as a naval aviator, and a seemingly unshakable belief in his own abilities. Born in Berkley, California, during 1921 to a Mexican mother and an American father, Crossfield had been raised in the then small Southern Californian coastal town of Wilmington. Obsessed with flight from an early age, he battled through years of poor health as a child following a severe bout of pneumonia, with doctors telling the would-be aviator that he would likely never pass a pilot medical.

Isolated during his convalescence but encouraged by family friend Charles Leinesch, the young Scott Crossfield (he had dropped Albert, his father's name, by this time) became increasingly absorbed by aeronautics, developing into a skilled model aircraft builder by his early teens. It was not unusual for him to contact the NACA, seeking relevant research papers in an effort to improve his groundbreaking designs, but Crossfield was still determined to fly and began taking on chores at the local airport in exchange for flight tuition. Unfortunately, his plans suffered a setback as the economic misfortunes of the Depression hit the Crossfields, and the family were forced to sell their comfortable Wilmington property, relocating to a rundown farm in rural Washington State.

Father and son set about transforming their land with characteristic gusto, and the combination of hard work and fresh air gradually strengthened Scott's constitution. In addition to his farm chores, he soon sought out the local airport, secretly resuming flight tuition whenever his meager funds allowed, until another intervention by Leinesch helped shape the teenager's career trajectory. Having moved to Seattle, the family friend now worked for the Civil Aeronautics Board, and he invited Crossfield to a test flight of Boeing's new Clipper flying boat on Puget Sound. The young man was immediately impressed both by the aircraft and its test pilot, Eddie Allen. This admiration only deepened when Leinesch explained that Allen wasn't merely a pilot but also an experienced engineer; rather than simply flying the planes, he also helped to shape and improve them. Immediately recognizing that this would be the ideal role for his talents, Scott Crossfield set about planning out his own aeronautical education accordingly.

Following high school graduation, he moved to Seattle to study aeronautical engineering at the University of Washington, but as the United States ramped up preparations for war during 1941, Crossfield put his studies on hold to become a parts coordinator on Boeing's busy bomber production lines. Although this role helped him to gain a greater understanding of how aircraft were constructed, the young pilot still felt he could be of more value flying, rather than building, warplanes. In the wake of the Japanese attack on Pearl Harbor on 7 December, Scott decided to enlist in the USAAF, but he initially failed his medical due to a rapid heartbeat (a problem that persisted throughout his career, possibly as a result of his childhood illness). After taking medication prescribed by the sympathetic doctor, he passed the medical on his second attempt and deployed to Arizona for basic training. Unfortunately, the sudden influx of eager new recruits led to delays as the army struggled to place the cadets, and never happy to waste valuable time, Crossfield headed back to Seattle to resume his work at Boeing while awaiting further orders.

Having waited for a month or two, he decided that the USAAF had missed its chance and enlisted in the navy, quickly progressing through flight training at Corpus Christi, Texas. Suitably qualified and still intent on seeing combat, Crossfield was dismayed to receive orders to remain in Corpus Christi as an instructor. Although he felt he possessed a talent for nurturing prospective aviators through their training, Crossfield still yearned to play his part in the Pacific campaign, and following persistent requests, a transfer eventually arrived. He now found himself assigned to a dive-bomber squadron rather than his longed-for fighter assignment, but by the time his new outfit was ready to deploy, hostilities were drawing to a close, meaning Scott Crossfield ended his naval service without ever tasting combat.

Frustrated, he headed back to Seattle to resume his studies, joining the naval reserve to fly F4U Corsairs and perform in the unit's display team to maintain his proficiency. Crossfield also took on a role overseeing and programming activities at the University of Washington's wind tunnel facility, sometimes undertaking research on behalf of his former employer, Boeing. The many hours spent supervising these tests gave Crossfield a valuable insight into the inner workings of a research program, experience that would serve him well as he left student life behind and started looking for permanent employment.

A freshly graduated Scott Crossfield weighed up his likely career options and decided that the NACA might provide a suitable home for his talents. After speculative applications failed to yield any offers, he adopted a characteristically direct approach, confidently journeying to Ames during November 1950, to introduce himself and to offer the facility his services. But the brash young engineer was soon taken aback to learn that a message from his wife, Alice, had beaten him to California. It seemed that one of his earlier applications had generated some interest after all, and he now needed to report to the HSFRS for an interview with chief pilot John Griffith.

After catching the first available train into the High Desert, Crossfield was met by Griffith at Mojave station and driven to the HSFRS, where the two pilots discussed Crossfield's qualifications and experience. Once Griffith had given him an overview of the station's current research projects, Crossfield was introduced to the chief of flight operations (and his prospective boss), Joe Vensel. Famously a man of few words, Vensel had lost much of his hearing as a result of prolonged exposure to the constant drone of engines while flying in open cockpits. In an effort to preserve what remained, he was rumored to turn his hearing aid down during lengthy meetings, but somehow he retained an uncanny knack for hearing all the important points.

Griffith and Vensel gave Crossfield a guided tour of the NACA research fleet, which included the second Bell x-1, the Douglas Skystreak, and the tailless Northrop x-4, leaving the would-be research pilot in no doubt that he had found the ideal place to pursue his career. Finally, Crossfield was introduced to the station's director, Walt Williams, who explained that Bob Champine's recent return to Langley had left the HSFRS in desperate need of new pilots to cope with the growing number of research programs—a situation that was about to become even more acute as, unbeknownst to

Crossfield, Griffith was soon to leave for the Chance-Vought Company. Having convinced all present that he was the right man for the job, Scott Crossfield headed back to Seattle to tie up his affairs in the Pacific Northwest before returning to California with his family to begin a new life in the Antelope Valley.

Within weeks of becoming the HSFRS's newest research pilot, Crossfield set about getting acquainted with the station's stable of unusual aircraft, although sometimes his enthusiasm and hard-charging attitude got the better of him. During his first flight in the X-4, he attempted, against advice, to loop the aircraft, but he only succeeded in flaming out both engines, leading to an emergency landing. While still a student in 1947, Crossfield had heard rumors of Slick Goodlin's alleged bonus demands to push the X-1 through Mach 1 and had immediately written to Bell offering to do the job for free (ironically, John Griffith had done exactly the same thing). Some three years later, on 20 April 1951, he finally got his chance to fly the rocket plane.

While the B-29 launch aircraft circled to altitude, Crossfield gingerly entered the X-1's cockpit as, in an uncanny echo of Yeager's historic supersonic flight, he was nursing three cracked ribs sustained while trying to push a mechanic out of a hangar window some days earlier. The flight got off to an uncertain start as the bullet-shaped rocket plane dropped away from the B-29 when the incorrectly trimmed horizontal stabilizer caused the rocket plane to pitch upward, stall, and fall onto its back, leading chase pilot Pete Everest to dryly note, "Well, that's certainly a new way to launch." After regaining control, Crossfield lit the XLR11's chambers in quick succession, pulling away to Mach 0.9, but the drama was far from over for the rocket plane rookie. With his propellants exhausted, Crossfield began to descend toward the lake bed, but the already restricted view from the X-1's cockpit now became worse still as the canopy froze over. Chase pilots Everest and Ridley closed around the X-1 to guide the NACA pilot home, but the proud Crossfield was determined to make a smooth landing on his debut in order to avoid a broken nosewheel. With no other means at hand to clear the glass, he loosened his straps, reached down to remove his right shoe, and then used his sock to clear a hole in the ice. After landing with all gear intact, Crossfield emerged from the X-1, sporting one bare foot. Clearly, his first flights were destined to be eventful.

His mixed luck continued the following month as, following four turbojet-powered familiarization flights, Crossfield made the first NACA rocket flight in the number three Skyrocket (now bearing the new tail number NACA 145). Dropping away from the P2B-1S at 35,000 ft. with 145's turbojet already running, he lit the LR8 rocket motor and pulled the nose up. As he reached the desired test altitude, Crossfield pushed over, but in a repeat of Bridgeman's final flight in the same aircraft, the J34 turbojet overheated and flamed out. The negative g's caused by the sudden loss of thrust interrupted the flow of propellants to the rocket engine, causing all four chambers to cut out. Without power or cockpit heating, the Skyrocket's canopy froze. Again, the alert chase pilots rushed to Crossfield's assistance, but as he descended into warmer air, the ice thawed enough that he was able to brush it away with his hand. Although no socks would be required for a safe landing today, Crossfield was beginning to acquire a colorful reputation around the Edwards flight line.

Following Bridgeman's final flight, the second Skyrocket had been transferred into NACA hands on 31 August 1951, receiving the tail number NACA 144. Having now assumed responsibility for the entire Skyrocket operation, the NACA elected to ground the number one aircraft pending its modification for air-launch, leaving NACA 144 and NACA 145 available for research. During 1950 Jack Russell had cemented his previously temporary move to the HSFRS, allowing the station to secure his unparalleled practical knowledge in maintaining and testing temperamental rocket motors on a permanent basis. With Russell's guidance, the HSFRS ground crews were able to refine maintenance operations to a point where rocket planes could be turned around between flights within days, rather than weeks. The P2B-1S launch aircraft also moved over to the HSFRS, where it gained the nickname *Fertile Myrtle* due to the number of rocket-powered "children" it dropped.

Crossfield would now take NACA 144, then the world's fastest and highest-flying aircraft, through a series of high-speed stability tests. Wherever Bridgeman had noted problems, Crossfield would now follow, gathering data to help NACA engineers determine the causes of—and hopefully solutions to—stability issues. On 28 September, less than a month after the NACA received the aircraft and only weeks after the air force had lost the X-1D, Crossfield dropped away from the P2B-1S for his first familiarization

10. The second D-558-II Skyrocket being loaded beneath its P2B-1S launch aircraft, Fertile Myrtle, in 1953. Courtesy NASA.

flight in 144, taking the rocket plane only as far as Mach 1.2 before problems with the LR8 rocket motor halted further progress.

Crossfield made three more flights in 144 before winter rains flooded the lake bed, but ever-resourceful, he soon came up with a plan to keep the Skyrockets flying. If a temporary drag chute system could be fitted to shorten the research planes' rollout after landing, then it should be possible to use the main paved runway at Edwards AFB rather than the flooded lake bed. The enterprising test pilot helped design and construct a prototype system before conducting its first test flight using NACA 145. As the moment of truth approached, Crossfield chopped the throttle, touched down smoothly on the runway, and flicked the switch for chute release, but rather than slowing down, the Skyrocket just kept rolling. The chute had failed to deploy. After running out of runway, Crossfield finally brought 145 to a halt, at which point the drag chute abruptly fell from its canister. No further runway landings were attempted, and although the failure was later traced to an electrical problem rather than any deficiency in Crossfield's engineering skills, some within the HSFRS may have taken a quiet satisfaction in the pilot's discomfort. With his self-belief and blunt opinions, Crossfield did not always make himself popular around the NACA hangar, on one occasion chastising D-558-II crew chief Don Borchers for refusing to let the Skyrocket fly without further modification, adding pointedly that there were "too many damn mechanics playing engineers" at the HSFRS. After center director Walt Williams backed his crew chief's decision, an angry Crossfield returned to the workshops, almost injuring himself with a high-pressure line. Borchers didn't miss his chance. "Too many damn engineers playing mechanics," he retorted.

Once the winter rains had abated, the Skyrocket faced a further delay, as the installation and calibration of additional instrumentation kept aircraft 144 and 145 grounded until June 1952. The resumption of flights marked the start of a two-pronged assault on the stability question, with each aircraft examining the flight regime that best suited its capabilities.

Extensive research in the NACA's wind tunnels had produced a number of promising wing modifications that, it was hoped, might eliminate the pitch-up problem, including permanently deployed leading-edge slats to increase lift and prevent wingtip stalling; wing fences, vertical plates running from fore to aft across the wings to prevent spanwise flow; and chord extensions, which added extra area to the outer wing panels. The turbojet- and rocket-powered NACA 145 was used for full-scale testing of these modifications, but ultimately, none were able to fully calm the Skyrocket's tendency to pitch-up, as they couldn't address the aircraft's underlying problem of wing downwash on the high-mounted tail surfaces at high angles of attack. Consequently, the NACA was able to instruct the aviation industry that the horizontal tail surfaces of supersonic swept-wing aircraft were far more effective when placed low on the fuselage to avoid similar problems. North American's F-100 was among the first designs to benefit from this advice, and low-mounted, all-moving tails became a common feature on many fighters of the era.

While he continued to fly pitch-up investigations in NACA 145, Crossfield also began pushing the all-rocket NACA 144 toward its maximum performance in search of high-speed instabilities. By now, Crossfield was juggling test commitments on numerous programs, but the situation eased when the HSFRS expanded its pilot roster to manage the station's increasing workload. Walt Jones and Stan Butchart had joined the pilots' office early in 1951, although Jones's stay in the desert proved relatively brief (due in part to a personality clash with Crossfield). Later in 1951, Joe Walker transferred across from the NACA Lewis Flight Propulsion Laboratory. Having obtained a physics degree before joining the USAAF, Walker had gone on to fly P-38s during the war before joining the NACA following his return to the United States. Former naval aviator John B. McKay (forever known as Jack) also joined the HSFRS soon after, having completed his degree in aeronautical engineering and served a brief internship at Langley.

In the spring of 1952, Bell's X-2 finally appeared at Edwards AFB for a series of glide tests flown by Skip Ziegler and the AFFTC's Pete Everest. Crossfield was slated to fly the much-delayed triple-sonic rocket plane once it transferred to the NACA. Anticipating the upcoming challenge, he saw the Skyrocket as a means to obtain experience on swept-wing stability at higher speeds while the new aircraft awaited its troublesome engine, and his thoughts inevitably turned toward an attempt on Mach 2. Bill Bridgeman had shared his opinions with the NACA pilot, regarding a perceived link between the Skyrocket's longitudinal stability problems at high speed and low angles of attack. Armed with this insight, Crossfield found himself able to balance 144 on a stability knife-edge as he pushed toward higher Mach numbers, and by August 1953 he had edged the aircraft to Mach 1.878, slightly below Bridgeman's fastest mark but still tantalizingly close to Mach 2. With NACA Headquarters uninterested in pushing faster for the sake of the record books, the navy saw an opportunity to mark the fiftieth anniversary of the Wright brothers' first powered flight by improving on the Skyrocket's existing speed and altitude marks.

The second of Bell's advanced X-1s, the X-1A, had made a handful of initial demonstration flights with Ziegler at the controls during the spring of 1953 before returning to Wheatfield for modifications. Now, with the X-1A due back at Edwards within weeks and the AFFTC hoping to begin its high-speed test flights before the year was out, the navy had only a narrow window of opportunity in which to act. The D-558-II had already surpassed its initial 1946 performance estimates on numerous occasions, but Marion Carl now returned to the High Desert, determined to coax the aircraft higher and faster still. With the HSFRS's well-drilled team on hand to offer the marine pilot their full assistance, former naval aviator Scott

Crossfield eagerly shared his knowledge of the Skyrocket with Carl, just as Bridgeman had with him.

Ostensibly, the navy flights were being made to test a new full pressure suit being developed by the David Clark Company. Whereas the partial pressure suits then in use relied on the vicelike grip of capstans to prevent the body from expanding, the new suit could apply a more even pressure via a single pressure garment, a far less uncomfortable solution for the pilot. Carl was to test the prototype suit at high speeds and altitudes, while hopefully exceeding Bridgeman's unofficial records in the process. Ever the engineer, Crossfield became involved in final modifications to the suit, offering Clark's team the use of his house (and his wife's sewing machine). The close working relationship that developed between Crossfield and Clark during this period would pay dividends in coming years, but for now the pilot remained focused on supporting Carl's record attempts.

After two familiarization flights in the dual-powered NACA 145 during July 1953, Marion Carl felt ready to shoot for the altitude record in the all-rocket NACA 144. Attempts on 14 and 18 August were aborted prior to launch following technical problems, but on 21 August the marine pilot finally dropped away from the P2B-1S, lit the LR8's four chambers, and accelerated away in a steep climb. Holding the Skyrocket on profile as the last available propellants left the tanks, he soared silently onward to an altitude of 83,235 ft., besting Bridgeman's previous mark by nearly 4,000 ft. With an unofficial altitude record now in the bag, Carl turned his attentions to the speed record, making his first attempt on the final day of August. Like Bridgeman and Crossfield before him, Carl found that the Skyrocket's stability became marginal at Mach 1.5, with the aircraft rolling wildly in the thin air. Undaunted, he was back in the Skyrocket two days later for a second attempt, but after only managing a maximum speed of Mach 1.728 at 46,000 ft., the navy chose to cancel any further attempts on the speed record. But although Marion Carl had failed to reach Mach 2 in the D-558-II, Scott Crossfield remained convinced that he could still make the breakthrough if given the chance.

On 16 October 1953 the Bell X-1A arrived back at Edwards following a spell at the company's Wheatfield plant, where the frequency control vibrations Skip Ziegler had noted during its initial demonstration flights the previous April had been investigated. During the intervening months, Ziegler had been killed when the second x-2 exploded during fueling tests. This explosion, the third in a Bell rocket plane, led the company to replace the complex tube bundles used for nitrogen storage in the advanced x-1 and x-2 series with simpler spherical tanks in its remaining rocket planes. With these modifications complete, the x-1A was ready to fly, but a new pilot was needed to complete Bell's demonstration phase. With Ziegler gone and Cannon still recovering following the x-1-3 explosion, Larry Bell asked the AFFTC whether an air force pilot might be available to make these flights. As he had already assigned himself as project pilot for the x-2, Pete Everest was unable to fill the vacancy himself. Fortunately, Chuck Yeager had recently returned to Edwards AFB, having attended Air Command and Staff School, and given Yeager's considerable experience in the original x-1, Everest had no hesitation in assigning his fellow West Virginian to fly the x-1A.

As the air force looked to recover from recent setbacks, the NACA continued to push on with its high-speed stability program in NACA 144. Following Marion Carl's record attempts, nozzle extensions were fitted to the aircraft's LR8 rocket motor to prevent its rocket plume from interfering with the rudder at higher altitudes (one cause of the wild instabilities pilots had experienced). These extensions also had the added benefit of increasing thrust by as much as 6 percent in the thin stratospheric air. HSFRS mechanics also moved 144's tank regulators into the cockpit, allowing the pilot to increase pump inlet pressures during flight, thus further boosting engine performance for a limited period. Together, these changes dramatically increased NACA 144's thrust, and on 14 October, just two days before the x-1A's return, Crossfield took the aircraft to a new unofficial speed record of Mach 1.96.

With Mach 2 now so close and Yeager preparing to take the x-1A aloft at the earliest opportunity, Crossfield (with Walt Williams's tacit approval) devised a plan to ensure that he and the Skyrocket got there first. Rather than go through official channels and risk a refusal by NACA administrator Dr. Hugh Dryden, Crossfield called on the navy's liaison officer at Edwards, casually mentioning that the freshly modified Skyrocket represented a golden opportunity for the navy to beat its air force rivals to Mach 2 before the upcoming Wright brothers memorial dinner. Crossfield's carefully planted suggestion was duly passed up through BuAer channels, and soon Dryden received an official request from the Pentagon. Crossfield would get his shot at Mach 2—now he would need to make good on his promises.

By mid-November 1953, preparations were underway to get both the X-IA and the Skyrocket in the air for attempts on Mach 2. Yeager hoped to take the X-IA aloft before the end of the month for initial familiarization flights before he pushed on toward the aircraft's maximum performance. At the HSFRS, preparations for Crossfield's attempt were kept low key. The NACA had no wish to get into a publicized contest with its landlord, but a successful flight would offer a welcome boost to the committee's profile, strengthening its hand when it came to funding requests for future projects. During January 1953, ground was broken on a new site for the NACA station as part of the air force master plan. The expanding civilian operation, now approaching two hundred employees, had outgrown the makeshift hangar and lean-to arrangement it had occupied since the late forties at South Base. The new site, located just north of the planned air force redevelopments, would offer a permanent, purpose-built home for the station's flight research activities. As the day of the Mach 2 attempt drew close, Dryden's assistant Walter Bonney arrived at the HSFRS to oversee events. Should Crossfield's attempt prove successful, Bonney would be in place to manage press coverage and ensure that the right people in Washington heard all about it.

Within the hangar, NACA 144 underwent last minute modifications to gain every possible advantage during the maximum speed attempt. Gaps between the aircraft's panels were taped over, and the entire airframe was waxed and polished to remove even the smallest drag-generating imperfections. The steel propellant jettison pipes, located to either side of the LR8's nozzles, were replaced with lighter aluminum versions, and these were canted inward so that they would burn away when no longer needed, removing a few more vital pounds of weight. NACA project engineer Herman Ankenbruck had devised an optimum parabolic flight profile for the attempt, with Crossfield needing to maintain an exact climb angle up to 72,000 ft. before pushing over into a shallow dive and accelerating until his propellants were exhausted.

During the early hours of Friday, 20 November, Hangar 182 was abuzz with final preparations. Maintenance crews had devised a method of cold soaking the D-558-II's tanks for many hours before final fueling, meaning they could maximize the density of propellant that could be carried, and now, in the blustery cold of the predawn, NACA 144 hung below *Fertile Myrtle* shrouded in mist from its supercooled oxidizer. At one point, progress had been halted by a potentially serious accident, as the turbopump's concentrated hydrogen peroxide propellant was being loaded. A frozen valve had led to launch panel operator Jack Moise being showered with the highly reactive liquid, but thankfully, a quick-thinking colleague turned a firehose on him, preventing serious injury. As the morning progressed, Crossfield pulled on his restrictive partial pressure suit and made final preparations for his big moment. Although he was suffering from a mild bout of flu, nothing was going to keep Crossfield out of the Skyrocket's cockpit today. It was midmorning before fellow NACA research pilot Stan Butchart lifted the heavily laden P2B-1S away from the Edwards runway to begin the long climb. As they rose, the bomber's LO2 top-off system ensured that the Skyrocket's oxidizer tank remained full right up to the moment of launch, meaning maximum burn time for the LR8.

After more than an hour spent slowly circling to altitude, *Fertile Myrtle* finally reached 32,000 ft., and Crossfield, now safely enclosed within 144's cramped cockpit, conducted final checks before priming the LR8 rocket motor. As Butchart counted down to release, the time for talking was over. Falling from the P2B-1S, Crossfield lit the rocket motor and headed for the record books.

In spite of his flu, Crossfield executed the profile perfectly, pushing over at exactly 72,000 ft. and then accelerating onward as he dived back toward the desert floor. As the turbopump fed the last of the propellant into the four combustion chambers, Crossfield saw the Mach meter's needle nudge just beyond Mach 2. Final confirmation would not come until he had the aircraft safely back on the lake bed, where Roxanah Yancey's computers could reduce the flight data, but Crossfield was in little doubt that he had just made history. As the Skyrocket decelerated, he made a series of maneuvers to gain valuable high-Mach stability data before turning for home and gliding down to a smooth dead-stick landing. As NACA 144 rolled to a halt, a crowd converged to check that pilot and aircraft were both unscathed after their record-breaking exploits. Spotting Walter Bonney, the jubilant Crossfield commented, "I don't think you've wasted your time coming out here." More than two years after Bill Bridgeman had

11. Station director Walt Williams (left) in conversation with Scott Crossfield (center) and director of flight operations Joc Vensel (right) in 1953. Courtesy NASA.

stated his belief that the Skyrocket could reach "the magic Mach 2," calculations confirmed that Scott Crossfield had indeed become the first person to travel at twice the speed of sound, pushing the Skyrocket to a peak velocity of Mach 2.005 at 65,000 ft.

As word of the new unofficial speed record was released to the press, Crossfield got a taste of the transitory fame previously afforded to other test pilots. But even as the press hailed a new "Fastest Man Alive," the AFFTC team were working hard to ensure that the NACA pilot's moment of glory would be short-lived.

Crossfield's flight was a bitter blow for the X-1A team. With the numerous delays that had dogged the advanced X-1 program, the air force could only look on, no doubt regretting the premature loss of the X-1D. When assigned to the program, Yeager had insisted on having the same team who helped him push the original X-1 through Mach 1. Duly assembled, Jack

Ridley and Dick Frost were hard at work making final preparations for his first flight in the x-1a. Yeager was no fan of the NACA research pilots, suggesting that they were engineers first and pilots second. He reserved special criticism for Crossfield, whom he would later describe as arrogant and unwilling to take advice. But as they pushed on with their own program, there was little to gain by dwelling on their rival's success; a philosophical Jack Ridley simply told Yeager, "We'll take 'em on Mach 3."

With his extensive experience in the x-1, Chuck Yeager may have been the obvious choice to pilot the x-1a, but in private he voiced doubts about taking the program on. Neither Bell nor the NACA could state with any certainty how the rocket plane might behave much beyond Mach 2, but wind tunnel data suggested that the x-1a was likely to lose directional stability at Mach 2.3, potentially leading to a total loss of control. The lack of an ejector seat worried Yeager, but given the speeds and altitudes where problems looked likely to occur, he figured that successful escape would be a remote possibility anyway. Unlike Bell's own test pilots, Yeager would receive no bonuses for taking on the risky high-speed flights. With a wife and young family to consider, he took the unusual step of demanding that Larry Bell draw up a private life insurance policy to provide financial support in the event of a tragic accident (although arrangements between service personnel and companies were not considered legal). He also made his wife, Glennis, promise that she wouldn't sell his Model A Ford to another pilot should he fail to return. Yeager had bought the car for $100 from the widow of Major Neil Lathrop, former head of flight test operations at Edwards AFB, who died in a crash of the prototype Martin XB-51 jet bomber the previous year. Lathrop had himself obtained the car from the widow of Captain Joe Wolfe, killed while testing Boeing's revolutionary B-47 bomber in 1951. With that history, Yeager didn't want the Model A being seen as a jinx on any pilot who owned it.

Whatever concerns he held in private regarding the x-1a, Yeager remained committed to the project, and on 21 November 1953, one day after Crossfield broke Mach 2, he made his first familiarization flight in the aircraft. To his relief, the x-1a offered no surprises when compared to its predecessors, and on his second flight, some two weeks later, he took the aircraft supersonic for the first time, reaching Mach 1.5. This exceeded Yeager's fastest flight in the original x-1 over five years earlier; both pilot and aircraft would now be entering new territory as the high-speed research program began in earnest. The air force was still hoping the X-1A might exceed Crossfield's mark before the fast-approaching Wright brothers memorial dinner, and when Yeager made the aircraft's first high-Mach flight on December 8, reaching Mach 1.9 at 60,000 ft., all looked set for a new record on the next flight.

Given the NACA and Bell's preflight warnings that the X-1A could depart from controlled flight beyond Mach 2.3, a small speed increment seemed the sensible course of action for the fourth flight, but again the AFFTC team chose to go all out, with Yeager and Ridley privately referring to the record attempt as Operation NACA Weep. On 12 December, Yeager dropped away from the B-50 mother ship at just over 30,000 ft., igniting three of the XLR11's rocket chambers before beginning his climb out. Reaching 45,000 ft., he lit the final chamber and pushed through slight buffeting approaching Mach 1 before the rocket plane resumed its steep climb. Dazzled by the morning sunlight, Yeager was finding it difficult to read his instruments and consequently overshot his planned push-over altitude, reaching 70,000 ft. before he brought the silver rocket plane into level flight. As the X-1A raced onward with its propellant supply far from exhausted, Yeager broke Mach 2 at an altitude of 76,000 ft., but in spite of the preflight warnings, he continued to press on. As the X-1A thundered on beyond Mach 2, Yeager was about to learn in the most dramatic of fashions that the engineers' concerns had been well founded.

Some ten seconds after entering level flight, the X-1A's nose began to yaw before its left wing slowly dipped. Yeager attempted to correct, but this caused the aircraft to roll rapidly in the opposite direction before again snapping back into a left roll. Recognizing the danger too late, Yeager cut off the XLR11, but he was now traveling in excess of Mach 2.4. In the tenuous upper atmosphere, the X-1A departed violently, tumbling in all axes. There was little the test pilot could do as he was thrown around the cockpit with such force that his helmet cracked the canopy. Yeager's T-1 pressure suit inflated instantly, keeping his body encased in its lifesaving grip, while the semiconscious pilot struggled to regain control of the tumbling rocket plane. On the ground, Bell and air force crews nervously awaited news from Yeager, while some miles above, chase pilots Jack Ridley and Arthur "Kit" Murray scanned the skies for the silver rocket plane. Tumbling rapidly, the X-1A now entered denser air below 40,000 ft. and decelerated back below Mach 1. As the tumbling stabilized into an inverted spin, the aircraft continued down to 30,000 ft., ever closer to the mountains below. Regaining his faculties, Yeager was able to recover the x-1A, first to a normal spin, then into level flight. Audibly shaken, he called, "I . . . uh . . . got in bad trouble. I'm down to 25,000 over Tehachapi . . . I don't know whether I can get back to base or not. I can't say much more . . . I gotta save myself." As the chase pilots rushed to his assistance, Yeager jettisoned his remaining propellants and headed for the pale expanse of Rogers Dry Lake, still some sixty miles distant. With Murray now on his wing, Yeager coaxed the rocket plane back to a lake bed landing, now joking that he wouldn't still be in the x-1A if he'd had a means of escape.

Yeager's survival had been extremely close. After departing from controlled flight, the x-1A had fallen over 50,000 ft. in less than a minute, with Yeager experiencing forces of up to 8 g's. The aircraft carried small movie cameras aimed out from its nose and across its right wing, and the rapidly revolving blur of sky, desert, and mountain they captured that day, along with still images from the chase aircraft, give some idea of the chaotic ride Yeager endured as he tumbled earthward. Had he failed to recover control when he did, it is unlikely he could have avoided disaster. Although badly shaken and suffering the petechiae (ruptures to blood capillaries) associated with pressure suit inflation and high g-forces, Yeager sustained no serious injuries during his dramatic flight. The x-1A also survived the ordeal without major structural damage, a testament to the strength of the aircraft. As Yeager commented during his glide back to Edwards that day, "I don't think you'll have to run a structural demonstration on this damned thing!"

Within days, Yeager flew to Washington DC for a round of air force press engagements, and on 17 December it was his achievement, rather than Crossfield's, that was celebrated at the memorial dinner to mark the fiftieth anniversary of the Wright brothers historic flight. As Yeager and the air force reveled in the success of Operation NACA Weep, Yancey and her fellow computers at the HSFRS toiled on, reducing the data from the x-1A's instrumentation. The jagged graphs they produced detailed the rocket plane's wild gyrations as it had tumbled down through the stratosphere. Using this information, NACA engineers sought the exact causes behind the x-1A's instability and what this might mean for the next generation of high-speed military aircraft.

In fact, the x-1A data confirmed theoretical work published by William H. Phillips of the NACA's Langley laboratory in 1948. Whereas an aircraft's control surfaces would dampen any tendency to roll or yaw in the denser air at lower altitudes, the x-1A's control surfaces had been unable to damp these movements in the thinner air as it continued to accelerate above 70,000 ft. The lateral and longitudinal instabilities of the aircraft then became coupled, meaning the x-1A began to simultaneously pitch, roll, and yaw until the aircraft entered denser air, where its wings and tail surfaces could again begin damping the motions into a recoverable spin.

In short, by overshooting his planned altitude and continuing to accelerate beyond the recommended Mach 2.3, Yeager had pushed the rocket plane to a point where the laws of inertia outweighed aerodynamics, and he almost paid with his life. This phenomenon became known as inertial coupling, and it would present many challenges to high-speed flight in coming years.

Neither the x-1A or D-558-11 would ever match the speeds they had attained during the final months of 1953; indeed, it would be some time before another pilot would push beyond Yeager's Mach 2.44 mark. The air force quietly chose to heed earlier advice, limiting flights in the x-1A to a maximum of Mach 2. The NACA Skyrockets continued to fly, but attention turned back to research with NACA 144 gathering additional data on high-speed stability while NACA 145 investigated the transonic characteristics of new shapes for external stores. The x-1A was due to be transferred to the HSFRS in 1954, but before letting the aircraft go, the AFFTC had one more record to shoot for. Given the atmosphere of intense interservice rivalry, Marion Carl's unofficial altitude record of 83,235 ft. became the air force team's focus during their final months with the x-1A.

Following his near-fatal brush with inertial coupling, Yeager left the Mojave to take on new challenges. Pete Everest became the new owner of Yeager's Model A Ford, buying it for the now customary $100, but as chief of flight test operations at the AFFTC, other commitments (including the much-delayed x-2 program) ruled him out of the high-altitude x-1A flights. Instead, Everest selected one of the AFFTC's most experienced pilots, Captain Arthur "Kit" Murray. Having graduated from General Boyd's flight test school at Wright Field, Murray had become the first air force test

12. Chuck Yeager and Arthur "Kit" Murray shake hands in front
of the Bell X-1A. Courtesy USAF.

pilot permanently assigned to Muroc, flying most of the first-generation
jet fighter and bomber types. Having served as chase during many earlier
rocket plane flights, Murray now took over the X-1A, and during the spring
of 1954 he completed a single familiarization flight before embarking on a
series of attempts to better Carl's record.

Murray's time in the X-1A was dogged by numerous aborts due to technical issues, but on 28 May he was finally able to launch, easily exceeding Carl's mark by reaching 87,094 ft. Seven days later, he soared to an improved peak altitude of 89,750 ft., although the X-1A again fell victim to inertial coupling as it climbed through 80,000 ft. Fortunately for Murray, as his speed was far lower than Yeager's had been, he endured a less violent ride as the X-1A tumbled over the top of its steep arc and back down toward the High Desert. Regaining control at 65,000 ft., Murray was able to bring the aircraft back for a safe landing. On 26 August, Kit Murray made the final air force flight in the X-1A, using experience gained during his previous flight to retain control as he coaxed the aircraft to a new unofficial altitude record of 90,440 ft., the highest that any of the X-1 series aircraft would ever reach. Murray later described the unusual sensations associated with high-altitude flight: "I begin to feel weightless. . . . I'm climbing so steeply I can't see the ground, and I feel confused. I have a sense of falling, and I want to grab something for support." At the apogee of his flight, Murray was clearly able to see the curvature of the distant horizon and the inky blackness that lay above.

Following this flight, the record-breaking X-1A was transferred to the NACA, moving northward to the recently completed and renamed High Speed Flight Station (HSFS). The AFFTC's attention now turned to the last of the advanced X-1s, the X-1B. Having arrived at Edwards in July 1954, the X-1B offered the air force little beyond what they had already achieved in the X-1A. As it would be prone to the same high-speed instabilities as its sibling, the AFFTC decided to use the aircraft for pilot familiarization flights during the later months of 1954. Air force pilots Jack Ridley, Kit Murray, Robert Stephens, Horace Hanes, and Richard Harer all flew the rocket plane before AFFTC commander Brigadier General J. Stanley Holtoner took his turn in the X-1B on 26 November (Holtoner, promoted to brigadier general in late 1952, followed Boyd's example of flying as many of the aircraft that came through the test center as possible).

Having been granted permission to take the aircraft beyond Mach 2 as part of his preparations for piloting the X-2, Pete Everest would make the final air force X-1B flights. Despite the X-1A's tangle with inertial coupling and the severe reservations of some colleagues, Everest believed he had learned enough from Yeager and Murray's experiences to keep the rocket plane in check at high speed. Following one familiarization flight, Everest made his high-speed attempt on 2 December 1954. After dropping away from the B-50 and firing three of the XLR11's chambers, he climbed to around 50,000 ft. before leveling off to ignite the final chamber and accelerate through Mach 1. Now supersonic, the X-1B continued to streak upward, but unlike Yeager, Everest hit his push-over point exactly and quickly moved past Mach 2 in level flight at 65,000 ft. As the X-1B continued to surge forward with all chambers burning, the aircraft began the rolling, fishtailing motions that had preceded Yeager's loss of control. Gently applying what little aerodynamic control he could, the coolheaded Everest steadied the rocket plane before carefully shutting off the rocket chambers one by one to avoid any sudden changes in momentum. As he decelerated back through Mach 2 and turned toward the lake bed, Everest felt vindicated. He had reached a maximum speed of Mach 2.3 and demonstrated that, with sufficient care, it was still possible to retain control of the aircraft. He later stated that he felt the NACA's pilots would be able to safely reach Mach 2.5 in both the X-1A and the X-1B, but the HSFS had little interest in pushing the advanced X-1S so close to the edge. The day after Everest's high-speed flight, the X-1B was handed over to the NACA, marking an end to the air force's involvement with the historic Bell X-1 family of aircraft.

Before it began research flights, the HSFS sent both the X-1A and X-1B to Langley for the installation of additional instrumentation, and before they returned to the desert, the aircraft also traveled north to Buffalo, where Bell fitted them with basic catapult ejection seats similar to those used in the Northrop X-4. Joe Walker was assigned to perform high-speed and high-altitude stability tests in the X-1A following the aircraft's return during the early months of 1955, but the first of these flights did not take place until 20 July, when he reached a modest Mach 1.45 at an altitude of 45,000 ft. A second flight attempt later that month was canceled, leaving Walker waiting until 8 August for another chance to fly the X-1A.

That morning, as Stan Butchart and Jack McKay piloted the B-29 mothership on its slow drag to altitude, Walker entered the rocket plane to begin his predrop preparations. Finally reaching their 31,000 ft. launch altitude north of the town of Victorville, over an hour after leaving Edwards, Walker began to pressurize the X-1A's LO2 tank, but a huge blast suddenly shook both aircraft. The jolt was so severe that Butchart thought another aircraft had collided with them, but Kit Murray, flying chase in an F-86, quickly reported that there had been an explosion in the X-1A. Although his canopy had been shattered by debris from the blast, Murray doggedly remained on station, describing how the initial fire had been extinguished but the rocket plane was now venting vapor. Fortunately for Walker, two of the B-29 crewmen, Charles Littleton and Jack Moise (veteran of the hydrogen peroxide spill on the day of Crossfield's Mach 2 flight), rushed to release the X-1A's canopy and pull him clear. In Walker's words, "The crew was looking from above—waving for me to get out of there. I was trying to dump cockpit pressure, the crew waving for me to hurry up. I remember being impressed with the fact I wasn't getting out as fast as I should." After exiting the rocket plane, Walker briefly passed out from lack of oxygen before crawling forward to the bomber's cockpit, where Butchart was able to help.

Although the launch aircraft had sustained no major damage, Murray reported that the X-1A's landing gear had deployed, rendering a landing with the rocket plane still attached extremely hazardous. To add to the problems, the X-1A's tanks still held some of their water-alcohol mixture and hydrogen peroxide. Walker and NACA crew chief Dick Payne unsuccessfully attempted to jettison the remaining propellants as the B-29 circled down to lower altitudes, and after discussions with Joe Vensel on the ground, the chief of flight operations reluctantly informed the B-29 pilot, "Butch, you might as well drop it." Butchart headed out over the Edwards bombing range to release the X-1A. "It fell nicely. Then it came up into a loop and flipped back and augered in," Walker later recalled.

Another Bell rocket plane had been lost following a mystery explosion, adding to the earlier losses of the X-1D, the X-1-3, and the second X-2. Knowing that the nitrogen tube bundle (suspected to be at fault in the earlier incidents) had been replaced in the X-1A, investigators now looked elsewhere for a cause. In the weeks following the accident, the twisted remains of the X-1A were examined alongside the still intact X-1B in the HSFS hangar. When the X-1B's LO2 tank was examined, an oily substance was discovered that was subsequently identified as tricresyl phosphate (TCP), an organic compound used in a material known as Ulmer leather. Bell had used Ulmer leather gaskets to seal the inspection hatches and other joints in the tanks of all their rocket planes, but as investigators spoke with the Linde Air Products Company, they learned that Ulmer leather was regarded as extremely hazardous when used with LO2, since the material, once cooled, became explosive under even modest impacts. Bell's Wendell Moore, who had witnessed the X-1D explosion firsthand, decided to conduct a basic experiment by immersing samples of Ulmer leather in LO2 and striking them with a hammer. Moore's explosive results led the air force and the NACA to conduct further investigations, with a test rig constructed by HSFS engineer Don Bellman yielding thirty explosions from thirty attempts.

When additional tests revealed that the shock of pressurizing the oxygen tank—the point at which all the explosions had occurred—was enough to initiate a reaction, investigators reached the inescapable conclusion that Bell's decision to use Ulmer leather had been behind the losses. The company's unfortunate oversight had cost the air force and the NACA four valuable research aircraft, one launch aircraft, and more importantly, Ziegler and Wolko's lives. In the aftermath of this discovery, the Ulmer leather gaskets were immediately removed from the X-1B and the remaining X-2 in order to prevent further losses.

The final hectic months of 1953 represented the end of an era at Edwards AFB. Since 1946 the skies above the Antelope Valley had played host to the first round of research aircraft, with the air force–funded X-1 and the navy's D-558-1 Skystreak both born out of a necessity to understand the dynamics of transonic and supersonic flight. The conservative, turbojet-powered Skystreak was well suited to transonic research, allowing the rocket-powered X-1 to push through Mach 1, giving a broader picture of the challenges faced at high speeds. The navy-sponsored D-558-11 Skyrocket had proven an imperfect yet still effective overlap between the transonic and supersonic realms, helping engineers to understand the aerodynamic behavior of swept wings. Once converted into its ultimate all-rocket configuration, the Skyrocket had probed speeds up to Mach 2 and entered the upper reaches of the stratosphere, where it was soon joined by the air force–sponsored X-1A, with both aircraft returning vital data on new areas of instability.

Although the NACA played a key technical role in all these programs, the fierce rivalry between the air force and navy led the press to portray events in the High Desert as an ongoing duel, with competing designs vying to snatch new records for their respective services. Beyond the X-1A, this duel gave way to increased cooperation, but that didn't mean the rivalries or sense of competition disappeared from the lake bed.

November 1953 had also marked the sad demise of a local legend. Even as the lawsuit with the air force dragged on, Pancho Barnes's Happy Bottom Riding Club burned to the ground under what some would see as suspicious circumstances. Much had changed since flight research had come to the High Desert in 1946, but fresh challenges lay ahead for the pilots of the AFFTC and the HSFS. And on 18 November 1955 a new research aircraft finally scored its first rocket contrail across the blue desert sky.

5. Facing the Heat

Even as Stan Smith's team at Bell Aircraft worked on their design for the xs-1 during 1945, new data was already suggesting that swept wings might offer significant advantages over straight wings when it came to high-speed flight. Bell had used straight wings on the xs-1, knowing that the first U.S. jet fighters would share this configuration, but as allied engineers gained access to a wealth of aerodynamic data from captured German research facilities, the prevailing attitudes began to change. While Douglas Aircraft's Smith and Root inspected these sites on the navy's behalf, the usaaf had also deployed its own experts, including Theodore von Kármán and Bell Aircraft cofounder Robert Woods. With Robert Jones's 1945 naca report on swept wings having already raised interest at Wright Field, the newly available German data convinced Ezra Kotcher that the configuration should be tested as part of the atsc research aircraft program.

At Kotcher's request, Bell carried out wind tunnel tests using both forward- and rearward-swept wings on models of the xs-1, but although the results appeared promising, the xs-1 would require a major structural redesign to accommodate the new wings. Consequently, Bell concluded that a purpose-built swept-wing design would be a more practical solution, just as Douglas's Ed Heinemann had when faced with a similar request from the navy. But whereas Douglas and BuAer had designed their swept-wing D-558-11 to research the transonic–low-supersonic speed range, Bell and the atsc chose a more ambitious goal.

On 11 September 1945 Bob Stanley gathered a small group of colleagues to sketch out basic aims and requirements for the newly christened Bell Model 52. With construction of the xs-1 now well underway, Stanley hoped to approach Wright Field with this new proposal early the following month, and encouraged by the newly available swept-wing research, the group (which included xs-I project engineer Stan Smith, aerodynamicist Paul Emmons, and pilot Jack Woolams, among others) now envisioned an aircraft that would easily surpass the xs-I's performance, representing a massive leap in aviation technology.

With a projected maximum speed in excess of Mach 3 and the ability to soar beyond 100,000 ft., the Model 52 would encounter oven-like temperatures due to atmospheric friction, meaning it could perform research into structural heating as well as high-speed stability. To achieve these lofty ambitions, Bell would need to push the very limits of aeronautical engineering; from propulsion to materials to aerodynamic control, little about the Model 52 would be conventional. When NACA and ATSC representatives visited Buffalo on 14 December to review the xs-I mock-up, a preliminary development contract for the new swept-wing research plane was duly signed. The Model 52 now took the USAAF designation MX-743 but became known more commonly as the xs-2 (this was later shortened to x-2 as the air force expanded the experimental series beyond high-speed studies).

Following his work on the xs-I, Paul Emmons now accepted the formidable challenge of designing the aerodynamic configuration for Bell's new triple-sonic rocket plane. After much trial and error with various wing layouts, he eventually settled on a low-mounted, 40° swept wing with a span of thirty-two feet and a t/c of 10 percent. The wing featured a biconvex airfoil section, meaning that both the upper and lower surfaces used the same arced profile. Initial work into biconvex airfoils had been carried out in Europe during the early 1930s, with the profile promising very low drag at supersonic speeds. The British Miles M.52 research plane had been designed with biconvex section wings, and although that aircraft was canceled during construction in 1946, the British government had shared much of the revolutionary research plane's design data with Bell Aircraft during 1944. Though never officially acknowledged, it seems plausible that this may have played a part in Bell's decision to adopt a biconvex wing section for the x-2.

The x-2's fuselage was approximately thirty-eight feet long and essentially circular in section, with dorsal and ventral tunnels and the cramped cockpit taking up much of its sharply pointed nose. As with the Douglas D-558 designs, the entire nose section could be explosively jettisoned in an emergency, before being stabilized by a drogue parachute. As the capsule reached a safe altitude, an alarm would sound, at which point the pilot would jettison the canopy and bail out to descend under his own parachute.

A dual-chamber Curtis-Wright XLR25 rocket engine capable of producing 15,000 lbf. of thrust (two and a half times that of the XLR11) would provide the necessary propulsion, with a basic throttle system offering a range of power settings between 50 and 100 percent during flight. Like RMI's motors, it would use water-alcohol propellant and LO2, with a maximum fuel load of 13,800 pounds, allowing for just under three minutes of powered flight. The majority of the X-2's fuselage would be taken up by three large tanks, two of which (front and rear) would carry LO2, while the center tank would hold the water-alcohol fuel. The tanks themselves were to be integral to the fuselage, using the aircraft's structure as their outer skin, while being separated by bulkheads and balsa wood discs, saving a considerable amount of weight when compared to various arrangements used on the X-1 family. Propellants would be fed to the XLR25 by a powerful turbopump.

The aircraft's all-moving horizontal stabilizer would sit near the base of a large vertical fin and feature a t/c of 8 percent with the same 40° sweep as the wing. Unlike its predecessors, the X-2 was designed for air-launch from the outset, with the rocket plane's dimensions and maximum weight being closely constrained by the capabilities of its carrier aircraft. As all landings would be on the dry lake bed at Muroc, Bell eschewed regular landing gear in favor of a simpler center skid and nosewheel arrangement, reducing weight and complexity. Emmons's design placed NACA instrumentation in a readily accessible, pressurized bay located behind the cockpit capsule, with slide-out shelves and removable baskets, making the task of installing or removing instrumentation a relatively quick process—a huge improvement over the ad hoc arrangement of the original XS-1.

As the X-2's skin was expected to reach temperatures as high as 630°F, Bell could not use standard aluminum structures, as these would soften in the intense heat, losing structural integrity. Emmons and his team settled on heavier but more heat-resistant stainless steel for the aircraft's wings and tail, with a copper-nickel alloy called K-Monel being preferred for the fuselage. The cockpit would be cocooned in a thick glass-wool blanket, providing the pilot with additional thermal protection, while the heavy, removable canopy featured thick quartz glass specially treated to block the expected harmful UV rays at high altitude. Whereas the X-1s had all used unboosted controls, Bell felt that power assistance would be necessary at the X-2's higher speeds, and to provide this, they commissioned a revolutionary electric-powered control system from the Bendix Corporation. Essentially an early analogue fly-by-wire system, the pilot's control inputs would be converted into electrical signals and transmitted to servos, which would then move the relevant control surfaces. Mechanical devices would offer physical feedback, or "feel," through the controls for the pilot.

Subscale models of the X-2 were extensively tested in NACA wind tunnels throughout the design process, and Bell also benefitted from free-flight testing using rocket-propelled models. The NACA Pilotless Aircraft Research Division (PARD) at Wallops Island, Virginia, fired subscale models out over the Atlantic at supersonic speeds, with data on preprogrammed control pulses being returned via telemetry. Tests were also undertaken to determine likely g-loads during separation of the nose section, with a full-scale model of the capsule being tested atop an A-4 rocket at White Sands Missile Range during August 1948.

By the time Yeager broke Mach 1 in October 1947, the air force had already signed a contract with Bell for two X-2 aircraft, allowing construction on the new rocket plane to begin in earnest. Unfortunately, this process would prove to be far more challenging and time consuming than Bell could ever have imagined.

Somewhat optimistically, Bell Aircraft had originally forecast that the first of the two X-2s would be delivered in late 1949, but almost from the outset, delays began to push this date further and further into the future. Very little about the X-2's construction was proving simple or conventional, and consequently, the company and its subcontractors were confronted by daunting technical challenges.

With Bell Aircraft in the process of developing its own Rocket Propulsion Division, it had originally intended to develop the X-2's engine in-house, but once the enormity of the triple-sonic rocket plane's development challenges became clear, it decided to look elsewhere. The contract was eventually awarded to the Curtis-Wright Corporation in April 1949, with delivery of the first units scheduled for March 1950. The XLR25 engine featured two combustion chambers, with a smaller 5,000 lbf. unit located atop the larger 10,000 lbf. unit, and was relatively compact considering its predicted power output. Both chambers were regeneratively cooled, with the system's turbopump using the same water-alcohol fuel as the engine, rather than the concentrated hydrogen peroxide favored by RMI. The team at Curtis-Wright included some former assistants of the liquid-rocket pioneer Robert Goddard, and the XLR25 itself featured many of Goddard's innovations. But given the scale and ambition of the new throttleable motor, developmental problems seemed inevitable. Progress was initially hampered by industrial action rather than technical issues, when Bell workers walked out during the summer of 1949, forcing Curtis-Wright to shelve work on the XLR25 in favor of other projects. Once work did resume, combustion problems in the larger 10,000 lbf. chamber led to numerous explosions, a problem that continued to dog the program for years. Development of a reliable and reusable high-speed fuel pump also proved time consuming, and these delays were further compounded by exhaustive testing of the new electrical control system designed to throttle the engine via a complex series of valves.

Back at Bell's Wheatfield plant, fabrication of the x-2's K-Monel skin demanded new welding techniques, which weren't fully perfected until 1949. The x-2's tiny cockpit brought problems of its own, with space being so limited that new small-scale instruments had to be developed to fit its diminutive instrument panels. The cockpit was so cramped that pilots were forced to use their left hand to activate controls on the right-hand panel and vice versa, and as it was unheated, they would need to don a bulky winter flying suit and gloves over their T-1 partial pressure suit to endure the cold temperatures at altitude, further restricting movement. As they sought to reduce the aircraft's weight, Bell fatefully decided to remove a second, larger parachute from the crew capsule, retaining only the smaller stabilizing chute, which was not capable of lowering the capsule to the ground at a survivable speed.

Elsewhere, the x-2's groundbreaking electrical-power control system also ran into problems. By 1950 Bendix had produced a functioning test system, featuring cockpit controls linked up to the various control surfaces, but during tests, the system struggled to keep up with pilot inputs and became overloaded. The artificial feel device on the control stick was also causing trouble, occasionally producing violent movements. When Pete

Everest and Scott Crossfield (project pilots for the AFFTC and the NACA, respectively) attended a demonstration of the system on one occasion, the unhappy Crossfield suggested that Everest should test the system's stick responses for himself. In Crossfield's words, "When Everest pulled on the stick, the electrical units took hold, the stick whipped violently, and Everest, a small man, was thrown clear of the cockpit." Although Everest never corroborated this account, it became increasingly obvious that the power control system was still some way from maturity, leading Bell to abandon the Bendix system in favor of more conventional boosted controls.

By late 1950, Bell had completed both airframes, but as Curtis-Wright was still some way from delivering a functioning XLR25, the decision was made to proceed with glide tests while engine development continued. While the first x-2 remained in Wheatfield awaiting its rocket motor, the number two aircraft was rolled out on 11 November 1950 to begin eight months of ground testing ahead of its debut flight.

As the x-2 slowly crept toward flight, Bell also set to work modifying the rocket plane's Boeing B-50 carrier aircraft. Alongside the structural changes needed to fit the x-2 below the bomber, a LO2 top-off system was installed to ensure that its oxidizer tanks remained full, and a nitrogen supply was also fitted, providing a source of high-pressure gas to spin up the x-2's turbopump prior to launch. Sway braces were fixed below the bomber's wings to prevent the x-2 from rocking while attached to the parent aircraft, and new glazed fairings beneath the bomber shielded the x-2's cockpit area. During the climb to altitude, the pilot would enter the x-2 and secure the cockpit canopy with the assistance of support crew. The fairings offered protection from the freezing airstream while also providing light, meaning x-2 pilots wouldn't be subjected to a sudden burst of sunlight as they dropped away—something that x-1 and D-558-11 pilots had found disconcerting. During loading operations, the B-50's gear would be placed on large hydraulic jacks, raising the bomber. The x-2 would then be towed into position on a specially constructed ground-handling dolly, before the bomber was slowly lowered down over the rocket plane. The x-2 would then be winched into place and attached to the B-50's bomb shackles. A series of captive flights of the mated aircraft were made between July 1951 and March 1952, allowing Skip Ziegler to test predrop procedures before the x-2 moved to Edwards AFB on 22 April.

13. The Bell x-2 on the ramp at Edwards AFB. Courtesy USAF.

Finally, on 27 June 1952, almost five years after the production contract had been signed, the x-2 was ready for its first glide flight, with Ziegler due to make a series of pull-ups and low-speed stall tests to test the aircraft's stability. By 1952 many early questions regarding the performance of swept wings had been answered by the D-558-II, but although Bell had used the navy's two L-39 swept-wing test aircraft during the x-2's development, the new rocket plane still presented some unique challenges. Landing using the x-2's unorthodox gear would require great care, with some engineers expressing concerns that the aircraft's tail-high attitude during slide-out might lead to instability. The simple castering nosewheel was also relatively small and could not be steered, leaving the pilot with few options for controlling the aircraft's direction on the lake bed.

As the B-50 strained for altitude, the tall Ziegler squeezed himself into the rocket plane's tiny cockpit, having to bend his legs uncomfortably simply to place his feet on the rudder pedals. With his trademark ace-of-spades-adorned helmet forced tight against the canopy, Ziegler was fortunate that he did not require an uncomfortable pressure suit for today's low-altitude test. Following release, the flight itself was brief and for the most part uneventful, with the x-2 handling well at low speeds as it glided back toward the lake bed. During his final approach, Ziegler lowered the aircraft's flaps, released the main skid, and jettisoned the nosewheel cover, allowing the gear to extend. As he eased the x-2 down, the narrow skid contacted the hard-packed surface, causing the aircraft to pitch wildly. In an unhappy echo of the x-1's landing problems, the nosewheel collapsed almost immediately, leaving Ziegler as a helpless passenger while the x-2 thundered unsteadily across the lake bed. Desperately using the rocket plane's ailerons in an attempt to steady his slide-out, Ziegler only succeeded in driving his right wingtip into the dirt before the aircraft rolled to the left and slowed to a stop after a wild one-thousand-foot ride.

Following the dramatic climax to its debut flight, repairs and modifications kept the x-2 grounded until 10 October, when Ziegler managed to keep the aircraft steady as it slid out, thanks to a new wider main skid and small retractable whisker skids fitted beneath the wings. Buoyed by the apparent progress, Bell invited Pete Everest to make his glide flight later that same day. After launching from the b-50 at 30,000 ft., Everest found the second x-2's temporary unpowered control system heavy as he performed a basic handling checkout during his descent to the lake bed, but the AFFTC pilot was generally satisfied with the rocket plane's airworthiness. Lining up for landing, he deployed the main skid and nosewheel, but when he attempted to extend the new whisker skids, only the left unit deployed. Fortunately, the force of the x-2's initial contact with the lake bed was enough to jolt the right skid free, allowing Everest to complete his slide-out without further incident. With its initial glide flights successfully completed, the white dart-like research aircraft was secured beneath the b-50 on 18 October for the long trip back east, where its long-awaited XLR25 motor would be installed.

Although the XLR25's development had proven so problematic that Curtis-Wright had announced its intention to cease rocket motor development once its x-2 commitment was fulfilled, the company had finally wrung enough bugs from the engine to allow delivery of the first units to Bell. Curtis-Wright had come under increasing pressure since missing its original 1950 delivery date, with the NACA suggesting that twin XLR11 motors could be installed in the x-2 instead (an idea that would later resurface for the x-15), but as 1953 dawned, it seemed that the troublesome XLR25 might finally deliver. After installation in the second x-2, the XLR25 underwent thorough static testing before Bell began a series of captive flights below the B-50 in order to test in-flight fueling and jettison procedures.

On 12 May the B-50 and x-2 combination headed out from Buffalo to test the LO2 top-off system over nearby Lake Ontario. As the morning's testing began, the x-2's LO2 tanks were topped off to replace oxidizer that had boiled away during the climb, and once this had been completed, the x-2's jettison system was used to vent some of the LO2. In the B-50's bomb bay, Skip Ziegler crouched over the x-2's open cockpit, checking tank pressures as its LO2 tanks were topped off for a second time, but as he pressurized the rocket plane's tanks for a second jettison test, disaster struck. A Bell chase pilot saw a huge red fireball erupt from beneath the B-50, an explosion powerful enough to roll his aircraft and thrust the bomber upward by as much as 200 ft. After seeing a wing panel fly past his cockpit, the chase pilot confirmed that the x-2 had been completely destroyed, leading the B-50 crew to comment, "We've lost the beast." As the flames dissipated in the B-50's bomb bay, Ziegler was nowhere to be seen. A second Bell crewmember, Frank Wolko, had been observed leaving the B-50 following the explosion, but no parachute was spotted. The blast had wreaked serious damage on the bomber, but pilots Bill Leyson and David Howe battled heroically to nurse the crippled B-50 back for an emergency landing (the aircraft was later scrapped as a result of the damage it had sustained). Extensive searches on Lake Ontario failed to find any trace of Ziegler or Wolko. Fragments of the balsa wood fuel tank separators bobbing on the lake's surface were all that remained of the x-2.

Within hours of the explosion, Bell and the air force began separate investigations into the incident, questioning whether there may be a link to the earlier losses of the x-1D and the x-1-3. The absence of substantial wreckage made it difficult to pinpoint an exact cause, but both parties suspected that leaking oxygen vapor may have been ignited by an electrical fault. During a conference held at Wright-Patterson AFB on 9 June, the WADC and Bell Aircraft expanded on this hypothesis, suggesting a number of improvements to ensure that the remaining x-2 and advanced x-1s (the x-1A and x-1B) would be spared a similar fate. Jack Ridley, speaking on behalf of the AFFTC, stated that these aircraft would remain grounded until Bell installed an alternative to the fragile tube-bundle assemblies used for nitrogen storage. It was also decided that the x-2 program should proceed using the remaining aircraft once it had received all necessary modifications.

Back at the HSFRS, Scott Crossfield was becoming increasingly dissatisfied with the ongoing saga. As NACA project pilot for the x-2, Crossfield had watched impatiently as delay after delay had pushed the prospect of a Mach 3 flight ever further into the distance. In a characteristic display of self-assurance, he approached station director Walt Williams with a proposal; given his skills as both an aeronautical engineer and research pilot, Crossfield suggested that Williams should assign him to Bell as the NACA's point person for the x-2. Once installed at the Wheatfield plant, he would personally supervise progress on the rocket plane, driving through required changes while making sure that the aircraft made it into the air before becoming obsolete. While Williams was sympathetic to Crossfield's case (he had temporarily assigned technician Don Borchers to perform a similar role during the xs-1's development), he was not prepared to lose his senior pilot for an indefinite period of time while the station remained heavily committed to other programs. Williams was perhaps also mindful of the impact that Crossfield's blunt opinions might have on the HSFRS's sometimes uneasy relationship with Bell. Although his suggestion went no further on this occasion, Crossfield felt convinced that future research aircraft programs would need an experienced engineer-pilot at the center of the design team from the outset, a conviction he would soon put into practice.

Others within the NACA used the enforced hiatus to suggest modifications to the aircraft's design. Engineers at Ames felt that the now six-year-old x-2 might benefit from new high-temperature alloys in areas that might be particularly affected by kinetic heating at high speeds. In light of recent encounters with high-speed instability in both the x-1A and the D-558-II, it was also suggested that the x-2's nose should be retrofitted with small canard surfaces to improve stability, but modifications would cost money and cause further delays. By early 1954, preliminary studies were already underway for a new hypersonic research aircraft that could exceed Mach 5, and while it was still hoped that the x-2 might contribute useful data on structural heating and high-speed, high-altitude flight to help shape the proposed aircraft, patience and funding were beginning to run out for the Bell rocket plane.

On 15 July 1954 the remaining x-2 (still minus its engine) arrived at Edwards AFB beneath a new B-50 launch aircraft. Following Ziegler's death, Bell could no longer field a test pilot with rocket plane experience, so Pete Everest volunteered to conduct the demonstration flights rather than risk an additional delay. Having first spied the swept-wing rocket plane during a visit to Bell in December 1949, Everest had become fascinated by the x-2. He had marveled at the powerful potential of its sleek design, instantly recognizing that this aircraft, rather than the x-1, would offer the ultimate test of his piloting skills. Remembering this encounter, he would later recount, "Silently I promised myself to plan my future, if possible, someday to fly the x-2." Now, almost five years later, he was ready to take the aircraft aloft in an attempt to realize its long-promised potential.

Following an extensive series of ground tests and the installation of NACA instrumentation, the first x-2 was finally carried aloft on 5 August 1954 for Everest to make an initial glide flight, checking out the aircraft's general handling through the hydraulic control system. As the AFFTC chief of flight test operations, Everest handpicked his test crew, selecting one of his top multiengine pilots, Captain Fitzhugh Fulton, to take charge of the B-50. Although Chuck Yeager was about to leave Edwards, Everest asked him to fly chase for the day; the original Mach buster still had the sharpest eyes in the business, plus a wealth of experience when it came to experimental aircraft—invaluable commodities during a first flight.

After being released by Fulton at 30,000 ft., Everest put the first x-2 through its paces as he glided earthward. All went well until he deployed the landing gear as he approached his lake bed runway. As the nosewheel cover jettisoned, Yeager spotted that the wheel itself had not extended fully and appeared to be turned 45° to the right. To complicate matters, the right whisker skid had also failed to deploy. Bell engineers reassured Everest that his nosewheel should straighten out on contact with the lake bed, with the right whisker skid likely to be jarred loose as on his previous glide flight back in 1952. On touchdown, however, the nosewheel jammed in its skewed position, breaking the supporting structure, while the whisker skid remained resolutely stowed beneath the x-2's wing, causing the aircraft to roll right before flipping back in the opposite direction. After a few cycles of this wild ride, the left wingtip gouged into the lake bed, bring-

14. The Air Force Flight Test Center's Frank "Pete" Everest.
Courtesy USAF.

ing the rocket plane to an abrupt halt. As support vehicles rushed to the scene, a shaken Everest emerged from the x-2's cockpit to survey the damage. To compound matters, a displaced circuit breaker in the instrument bay meant the NACA instrumentation had failed to record any data that could have shed light on the landing instabilities. Bell engineers reluctantly concluded that the aircraft would need to return to Buffalo for structural repairs and a closer examination of the troublesome gear.

With the x-2 grounded, NACA and WADC representatives met at Wright-Patterson AFB to discuss the limited progress and to outline the next steps once the x-2 returned to flight. Bell expected that the subsonic demonstration phase would require three glide flights (Everest's 5 August flight constituting the first of these) followed by four powered flights, reaching a maximum velocity of Mach 0.9. Following air force acceptance of the aircraft, the AFFTC would carry out a short series of envelope-expansion flights, allowing them to establish the x-2's performance limits, before the rocket plane would be loaned to the NACA to begin a more thorough research program. Of course, these plans were entirely dependent on Bell's ability to prove that the x-2 and XLR25 could operate reliably.

The x-2 returned to Edwards on 16 January 1955, still minus its engine. Bell had hoped to fit the long-overdue XLR25 while the aircraft was back at Buffalo, but further delays at Curtis-Wright meant that the x-2 would remain engineless for its two remaining glide flights. Following captive flights to check fueling and jettison procedures, the next glide was scheduled for 8 March, for which the x-2 would be fully fueled, allowing Everest to test the emergency jettison procedures while still attached to the launch aircraft, before dropping away once the aircraft's tanks were empty.

As the heavily laden B-50 reached 30,000 ft., the x-2's LO2 tanks were topped off, ready for the jettison test to begin, but problems soon emerged. Given the positioning of the aircraft's oxidizer tanks, it was vital that the LO2 was jettisoned evenly to avoid upsetting the x-2's center of gravity. Unfortunately, as Everest watched his cockpit gauges, he realized that the two tanks were emptying at different rates. Worse still, the entire jettison process took ten minutes, far more time than the x-2 would have in an actual emergency. With the rocket plane's landing gear unable to bear the additional weight of unjettisoned fuel on landing, it was clear that the system would need to be improved before powered flights could begin. With the tanks now empty, Everest gave Fulton the go-ahead to drop, and the x-2 fell away cleanly from the B-50. The brief glide was taken up with stall tests both in clean configuration and with the gear and flaps down, before Everest rolled into his approach, easing the temperamental rocket plane onto the lake bed at a speed of 160 mph. Today, both the nosewheel and whisker skids deployed correctly, and the x-2 initially slid straight and true. But as the aircraft crossed one of the lake bed's many oil-based runway markings, it skidded out of control, flipping left and then right, before finally coming to a halt. Mercifully, Everest escaped uninjured, but the test pilot's patience with the x-2's unpredictable landing characteristics was wearing decidedly thin. In response, Bell engineers stiffened the whisker skids, hoping this might stabilize the x-2 during slide-out, but taking no chances, Everest insisted that a metal crash bar be fitted across the instrument panel, figuring that if he was going to be a helpless passenger, he may as well have something secure to hang on to. On 6 April the x-2 launched on its final planned glide flight, and while the descent was routine, the landing situation remained perilous. Rather than solving the problem, the stiffened whisker skids actually increased the rolling motions, throwing the hapless Everest around the cramped cockpit as he hung grimly to his crash bar. When the x-2 finally came to rest, skewed some 90° to its original direction of travel, the furious test pilot finally snapped. Pete Everest would not fly the x-2 again until Bell solved the landing problems once and for all.

After the aircraft made yet another return to Wheatfield on 8 April, Bell engineers and their air force counterparts began a painstaking analysis of the landing issues. The root of the problem appeared to lie in an earlier air force requirement for the x-2's main gear strut to comply with fighter specifications, forcing the aircraft into a 7° tail-high attitude on the ground—an unstable position during slide-out, as Everest and Ziegler had both found. The strut was now shortened, lowering the x-2's stance to 3°, while the skid's width was increased from twelve to twenty-one inches to provide a larger footprint. As these modifications were being made, technicians finally installed the long-awaited XLR25, and hopes were high that after almost a decade in development, the x-2 might now reveal its true capabilities. However, following its return to Edwards on 21 July, an incident involving another Bell rocket plane grounded the seemingly jinxed x-2 yet again. After the Bell x-1A had exploded during launch preparations on 8 August 1955, the cause behind the mysterious losses of the second x-2, the x-1D, and the x-1-3 was finally traced back to Bell's use of Ulmer leather gaskets in the LO2 tanks of their rocket planes. Given that the remaining x-2 and the x-1B still carried similar gaskets, both aircraft headed back to Buffalo for safer replacements to be fitted.

At Wright-Patterson AFB, the WADC watched events with growing exasperation. In the ten years since the program's inception, the x-2s had managed a sum total of six unpowered glides. The second aircraft had been lost in an explosion, while the first had experienced a series of landing accidents. Air force investment in the x-2 program had now exceeded $16 million, and following the recent contract award for the hypersonic x-15 to North American Aviation, cancellation once again seemed a real possibility. As the rocket plane returned to the High Desert, Bell received a stark ultimatum from the WADC—if the x-2 failed to make a powered flight before 31 December 1955, the program would be axed.

With time running out, Bell abandoned its earlier plans for an additional glide flight to test the landing gear modifications, electing instead to move straight to powered flights. On 25 October 1955 the x-2 was once again heading for altitude beneath the B-50, but as Everest began his predrop checks, he spotted a nitrogen leak, halting plans for a powered flight on this occasion. After rapidly jettisoning his propellants through the x-2's new, wider jettison pipes, Everest dropped away to make a glide flight, salvaging something from the day's efforts. Having performed a series of pull-ups and rolls during descent, he brought the x-2 in for a perfect landing on its revised gear, providing some consolation for the disappointed Bell team. A second powered flight attempt was scheduled for 18 November, with the flight plan calling for acceleration up to Mach 1.5, giving Everest his first opportunity to test the x-2's transonic handling.

As flight day dawned, the nervous ground crew swarmed around the x-2 as it hung beneath the B-50, clouds of vapor swirling across the Edwards ramp as its propellant tanks were slowly filled. With the air force's end-of-year deadline fast approaching, Everest took up his place in the B-50's glazed nose, silently hoping that today, finally, the x-2 might deliver on a decade of unfulfilled promise. Following takeoff, Fulton slowly circled the bomber up into the clear winter skies above Rogers Dry Lake as Everest headed back to take his place in the waiting rocket plane. By the time the B-50 reached 30,000 ft., the test pilot was finishing the final items of his hour-long, 127-item preflight checklist. With everything looking good, Everest made a final call to Fulton—go for drop. As the gleaming white research plane was released, Everest noted how rapidly he was falling, which was hardly surprising given that the fully fueled x-2 weighed in at twenty-five thousand pounds, twice its more familiar dry weight. Fearing that the aircraft might stall, he quickly ignited the XLR25's smaller 5,000 lbf. thrust chamber—the only one that would be used for today's flight. After diving to build up speed, he pulled back on the stick and accelerated upward through 45,000 ft. at Mach 0.95, experiencing mild buffeting as he edged through the transonic zone. Keen to investigate the disturbance, Everest cut the engine, allowing the x-2 to decelerate, before relighting the XLR25 to race forward again. Unfortunately, as he did this, the engine coughed loudly and quit, prematurely curtailing the x-2's first powered flight. Once safely back on the lake bed, technicians discovered that a small fire had occurred in the engine bay, grounding the aircraft temporarily for repairs to be carried out. To Bell's relief, the air force overlooked the incident, accepting that the x-2 had indeed made a powered flight ahead of the 31 December deadline, earning the program its reprieve.

Three attempts at a second powered flight took place during December 1955, but a combination of bad weather and technical issues meant the x-2 would not fly again until 1956. In light of the ongoing delays, the air force informed Walt Williams that the aircraft would now not be available to the HSFS until April 1956. With the all-rocket D-558-II having answered many key questions regarding the high-speed stability of swept wings and development of the hypersonic x-15 now occupying key engineers, initial NACA enthusiasm for the x-2 had waned considerably by this time. The HSFS still hoped that the aircraft might return useful data on heating and stability at Mach 3, but the continuing delays made it difficult to predict when, or even if, this might actually occur. As 1955 passed, the NACA were not the only ones looking toward the future. Pete Everest had now been at Edwards AFB for six years, and his tour in the High Desert was drawing to a close. New orders meant he would leave for Armed Forces Staff College in July 1956, but his commitment to the x-2 remained strong. As the new year arrived, he still hoped to reach Mach 3 before departing.

The x-2's second powered flight eventually took place on 24 March 1956, but on this occasion, Everest was unable to light the 5,000 lbf. chamber, limiting performance to Mach 0.91 at 45,000 ft. using the 10,000 lbf. chamber alone. Once again, the air force informed the NACA that they would require additional time to test the troublesome rocket plane, but fortunes improved the following month when on 25 April, Everest finally took the x-2 supersonic. Using both of the XLR25's chambers, he raced to Mach 1.4 at an altitude of 50,000 ft., performing stick pulses to gather the first data on the aircraft's high-speed stability. Further supersonic flights followed on 1 and 19 May, with the latter flight reaching Mach 1.8 during a shallow dive from 60,000 ft. At long last Curtis-Wright appeared to have wrung the bugs from the XLR25. And with the aircraft demonstrating reasonable stability beyond Mach 1, confidence in the x-2 began to grow; it was time to push on into uncharted territory.

While the x-2 had introduced many new technologies (some more successfully than others), the flight program also benefitted from a new development on the ground that would go on to play a huge role in the future of aviation—computer-based flight simulation. During 1952 the AFFTC had purchased a Goodyear L3 Electronic Differential Analyzer (GEDA) on advice from the NACA. The GEDA was a large analog computer capable of rapidly performing previously time-consuming calculations, and as the x-2 slowly inched toward flight, a young HSFS engineer named Richard E. Day (along with his supervisor Joe Weil) began using the GEDA to model the aircraft's flight characteristics. Working alongside AFFTC engineers, Day programmed the huge computer with the necessary flight equations, developing a basic simulation of the rocket plane using available wind tunnel and free-flight model data. By adding a cathode ray tube display and a rudimentary control stick, the simulated x-2 could be flown by engineers and pilots. As actual flight data became available, Day was able to refine this model to a point where he could test how the x-2 might behave under various conditions, with a reasonable degree of accuracy. The GEDA could now be used to offer x-2 engineers and pilots a glimpse into the aircraft's potential stability at higher speeds and altitudes, allowing them to plan upcoming flights and devise piloting techniques accordingly. As the x-2's initial forays beyond Mach 2 loomed, Everest regularly trained using the GEDA, testing each control input under a variety of parameters in an attempt to ensure that the x-2 would remain stable during flight.

During the final weeks of May, it was time to see if the simulations matched up to reality, as Everest embarked on the first of the x-2's planned speed-expansion flights. The GEDA had indicated that stability was likely to decay alarmingly as the rocket plane approached Mach 3, prompting

NACA engineers to urge caution. However, with time now running short, the AFFTC favored a bolder approach. After an aborted launch attempt on 21 May, the B-50 and X-2 combination climbed into clear desert skies the following morning, with hopes high that a sixth powered flight would finally take the X-2 beyond Yeager's Mach 2.4 mark of December 1953. After separating from Fulton's B-50 at 30,500 ft., Everest lit both chambers and slowly eased back on the stick as 15,000 lbf. of thrust pinned him into his seat. Soaring upward, the motor performed flawlessly, easing the X-2 beyond Mach 2 until, after 136 seconds of powered flight, the rocket plane reached Mach 2.53 at an altitude of 58,370 ft. While the aircraft remained controllable, Everest was acutely aware that he needed to fly very gently at these speeds and altitudes lest he trigger the instabilities that Dick Day and his colleagues had predicted via the GEDA. Precision flying was no easy task in the X-2, as the cockpit instrumentation often lagged behind the aircraft's actual performance. Even reading the cockpit's miniature instruments while the vehicle vibrated from the XLR25's incredible power proved challenging, but Everest trained himself to think ahead of the data, often adding 5,000 ft. to his indicated altitude to stay within the flight plan. Through his careful efforts, Pete Everest had just become the fastest man alive, but the test pilot felt sure that the X-2 could push on toward Mach 3.

Determined to leave the program in safe hands following his departure, Everest selected two highly regarded test pilots from the AFFTC's Fighter Test Division to take on the envelope-expansion flights following his departure. Captains Iven Kincheloe and Milburn Apt joined the program, and their preparations soon involved GEDA sessions, planning meetings, and chase duties for Everest's flights. Both men typified the new breed of postwar test pilot; whereas their predecessors often relied on stick and rudder skills honed during combat, Kincheloe and Apt came to the job with solid engineering credentials alongside their proven piloting ability. The Michigan-born Kincheloe had gained his bachelor of science degree in aeronautical engineering from Purdue University, becoming an air force cadet while pursuing his studies. In 1948 the young Kincheloe had met Chuck Yeager while visiting Wright-Patterson AFB, taking the opportunity to sit in the cockpit of the historic X-1. Following his air force graduation, he participated in a test program for the new F-86E at Edwards AFB, and this first taste of test-piloting convinced Kincheloe that this was the career path he should follow. First, though, a combat tour in Korea sharpened the young aviator's skills, with Kincheloe gaining double-ace status during his one hundred combat missions. On returning to the United States, the would-be test pilot was assigned as a gunnery instructor at Nellis AFB. After repeated requests to transfer to the AFFTC, Kincheloe's dream looked as though it was beyond his reach, until he learned of an air force exchange program with the Empire Test Pilots School in England. Seizing this opportunity, Kincheloe spent the next ten months in Farnborough before transferring back to the United States and the AFFTC. The road to Edwards had been a long one, but Kincheloe quickly established a reputation as both an excellent flyer and a sharp technical mind, leading to his selection for the x-2.

Hailing from Kansas, Milburn "Mel" Apt had joined the USAAF in 1941 before earning a science degree from the University of Kansas in 1951. Following this, he went on to attend the Air Force Institute of Technology at Wright-Patterson AFB, earning a second degree in aeronautical engineering before transferring to the USAF Test Pilot School (USAF TPS) at Edwards, from which he graduated in 1954. Having shown himself to be an excellent pilot and engineer, Apt was assigned to perform research into the inertial coupling problem that had been plaguing North American's new F-100 fighter. During the course of this high-priority program, he routinely pushed the aircraft into known areas of instability, helping to gather vital data that eventually allowed the NACA to recommend structural alterations to the F-100, dramatically improving the aircraft's operational safety. Apt's reputation as a cool and capable test pilot with experience in high-speed instability made him an excellent choice for the x-2 program.

Kincheloe, or "Kinch" as he was commonly known, became the third person to pilot the x-2, on 25 May 1956, when he took the rocket plane out to Mach 1.14 before engine problems curtailed his checkout flight. Far taller than Everest, the new pilot found the x-2's cockpit to be an uncomfortably tight fit, but the x-2 proved easier to fly than he had expected. By now the earlier landing problems had been well and truly banished, so much so that the whisker skids had been removed and a narrower center skid had been installed. With Kincheloe checked out, attention now turned to Everest's final flights, as the senior pilot prepared to push the x-2 further into the unknown.

Having won permission to delay his departure from the AFFTC until 15 July, Pete Everest had little time left to fulfill the ambition he had nurtured since first seeing the x-2 in 1949. In an attempt to wring maximum performance from the rocket plane, both of the XLR25's chambers were fitted with nozzle extensions to optimize thrust at high altitude. Bell also shortened the tank sensor probes, allowing as much propellant as possible to be used. Everest now hoped to push the x-2 to a point where friction from the thin high-altitude air would generate temperatures capable of destroying lesser aircraft. Having received a fresh resin-based white coating, technicians applied a rainbow of Tempilaq temperature-sensitive paint stripes at various points on the x-2. Each stripe would melt at a known temperature, offering engineers a visual reference to the thermal punishment the rocket plane endured as it approached Mach 3.

Everest's all-out speed attempt on 12 July did not go as planned. Straining against the g-forces as he reached Mach 1.5, he slowly pushed the aircraft over into level flight to make his speed run but, in doing so, inadvertently overcontrolled, entering negative-g conditions, under which the remaining propellants rose in the tanks, exposing the now-shortened fuel probes. Concluding that the propellants had been exhausted, the XLR25's control system prematurely shut the rocket motor down. Everest returned to base, cursing himself for letting his colleagues down and wasting his last chance at Mach 3. After the x-2 was examined and fixes were made to prevent a reoccurrence of the problem, the Bell team rallied behind the test pilot in an attempt to win him one more flight. The contractor team, having worked with Everest for so long, regarded him as one of their own; if they had to work double shifts to get him back in the air within days, then that was what they would do. Encouraged, Pete Everest made a final appeal to his superiors—one more week, one more chance, and then he would hang up his spurs and leave the High Desert behind him.

On 23 July, Everest made his familiar journey to the AFFTC in the battered old Ford Model A he would soon pass to Kincheloe for the customary $100 fee. Passing through the doors of the test center below the stenciled legend "Through these portals pass the oldest and boldest pilots in the world," Colonel Frank Everest knew that it was now or never. His week-long extension had been stretched by technical glitches, but today everything was ready. His test team colleagues had been hard at work through the small hours of the morning, precooling the tanks to ensure the maximum possible propellant load. As the x-2 hung below its launch aircraft, gleaming white amid the swirling clouds of frozen oxygen, Everest pulled on his restrictive pressure suit and thick coveralls before making his way out to the flight line where Fitz Fulton and his crew were waiting. Just before 7:00 a.m. the b-50 left the runway, with Everest taking up his usual position in the bomber's nose before making his way back into the bomb bay during the slow climb to altitude. Having reached 30,000 ft. some thirty miles east of Edwards, Fulton released the x-2, and firing both of the xLR25's chambers, Everest pulled back on the stick to begin his climb.

Preparations for today's flight had included numerous sessions with Dick Day and the GEDA, and as he headed for the thinner air and darker skies above, Pete Everest knew that the x-2's stability would become increasingly marginal as he pushed beyond Mach 2.5. Any overcontrol or inadvertent stick movement might easily send the rocket plane tumbling, and with this in mind, Everest gently inched the stick forward at 55,000 ft., aiming to level out around 65,000 ft. As he thundered onward through the stratosphere, he noticed flecks of paint flying past his canopy as the intense 500°F heat blistered the aircraft's nose. After 139 seconds the xLR25 quit, its propellants exhausted. "Bingo!" Everest exclaimed to his team below; he had hit Mach 2.87 at 68,000 ft., faster than any human had traveled before him. Cautiously, he made a series of planned stick pulses, attempting to validate the GEDA's predictions, remaining ever vigilant for signs that the x-2 might depart from control. Decelerating now, Everest began his descent, performing maneuvers at each prescribed data point before, somewhere over Bakersfield, he entered a long sweeping turn until the pale expanse of Rogers Dry Lake lay before him. Everest brought the x-2 in for a perfect landing, and as the convoy of support vehicles converged on the scorched rocket plane, he was able to reflect on his achievements. Although he had fallen just short of his Mach 3 target, Everest had extended his own unofficial speed record by a significant margin. He remained convinced that the x-2 still had more to give, but that was for another pilot to prove on another day. As Everest would tell his parents, "I've accomplished my mission at Edwards, and you can't stand still." It was time for Kincheloe and Apt to pick up the reins and push on during the air force's final months with the x-2.

15. The U.S. Air Force x-2 flight test team in July 1956.
Left to right: Stu Childs, Charlie Bock, Fitz Fulton, Pete Everest,
Iven Kincheloe, and Milburn Apt. Courtesy USAF.

The aircraft's handover to the NACA had by now been delayed numerous times. Originally, the HSFS had expected to receive the x-2 during December 1955, but delays and modifications during the demonstration phase had conspired to push the date further and further into 1956. As the x-2 finally began to prove itself in Everest's hands, a new and final handover date of 1 November had been agreed to. If the air force wished to push the x-2 further, it had only a few months remaining in which to do so. Although he had been closely involved in the program, attending preflight briefings and acting as a chase pilot during Everest's final flights, Milburn Apt had still not made a checkout flight in the x-2 due to other testing commitments. Consequently, Kincheloe would be the first of the two pilots to push at the edges of the x-2's performance envelope, with a series of high-altitude expansion flights that the air force hoped would significantly exceed 100,000 ft.

On 30 July, a mere seven days after Everest's record-breaking Mach 2.87 flight, the x-2 was tested, polished, primed, and ready to go again. Kincheloe had completed a GEDA simulator session the previous day, and fully briefed on the x-2's instabilities and the appropriate piloting techniques to counter these, he felt ready for the challenge ahead. Unfortunately, as Kinch prepared for his second x-2 flight, a leak in his emergency oxygen bottle meant a return to Edwards as a passenger in the b-50. An effort the following day was also aborted due to throttle problems on the xlr25, and a third planned attempt on 2 August was scrubbed due to maintenance on the b-50. Kincheloe finally dropped away from the b-50 at the fourth attempt, on 3 August, for a final checkout before his maximum-altitude attempt. Igniting both xlr25 chambers, Kinch pulled back on the stick, bringing the x-2 into a 30° climb. As the aircraft became supersonic around 40,000 ft., he steepened his climb still further. The x-2's controls would become increasingly ineffective as its altitude increased, so it was vital to establish the correct profile early. Today's flight peaked at 87,750 ft., with a maximum velocity of Mach 2.5. Having gathered stability data during both the climb and the descent, Kincheloe was happy with the aircraft's performance, and once safely back on the lake bed, he felt confident to go for maximum altitude on the next flight. Following further sessions with the GEDA, an attempt was made on 8 August, but this was thwarted by an electrical fault in the turbopump, leading to an early engine shutdown and a maximum altitude of 70,000 ft.

Following the frantic pace of recent months, the program now paused as the x-2's fuel system and engine received a full overhaul. While grounded, the x-2 and its b-50 carrier aircraft were officially signed over to the air force. Bell Aircraft's extended demonstration phase was over, but the AFFTC now had only two months in which to complete the envelope-expansion phase before the rocket plane moved to the HSFS. With the aircraft's overhaul complete, Kincheloe received a final simulator briefing in preparation for a second maximum-altitude attempt, scheduled for 7 September. The flight got off to an inauspicious start, as the b-50, now sporting both ARDC and AFFTC crests, ran into engine trouble as it strained for altitude. With one engine smoking due to a broken oil line, Fulton and his crew decided to drop the rocket plane at a lower-than-planned 29,000 ft. As the gleaming x-2 fell away from the bomber, Kincheloe lit both chambers and rotated into his initial 30° climb. Within twenty seconds, he had pushed through Mach 1 with barely a perceptible bump and pulled the stick back farther, steepening his climb to 45° as he blazed through 56,000 ft. At this angle, it was impossible for the pilot to see the horizon ahead due to the X-2's nose. But a low-tech solution now came to the aid of this most technically advanced of aircraft—before the flight, lines had been drawn on the inside of the canopy using a red grease pencil. During the climb, Kincheloe could match these lines to the horizon to ensure that he was on profile. Soaring ever higher, the pilot could now do little to alter the aircraft's trajectory, and when the XLR25 cut out at around 90,000 ft., the laws of physics were firmly in control as the X-2 carved a huge ballistic arc over the High Desert.

With the sky around him darkening to a deep-purple hue, Kincheloe had left 99 percent of the earth's atmosphere behind him as his altitude peaked at 126,200 ft. Four minutes after falling away from the B-50's bomb shackles, Kinch had become the first human to see our home planet from over 100,000 ft., but there was little time for reflection as the X-2 began a slow bank to the left. From his work with Dick Day, Kincheloe knew that any attempt to correct this movement might trigger the sort of wild tumbling that Kit Murray had endured during his high-altitude flights in the X-1A. Fighting his piloting instincts, Kincheloe left the controls, trusting that he would be able to recover the aircraft safely once it descended into the denser air below. Speed built up quickly as the X-2 entered the downward leg of its arc, with the aircraft falling sixteen miles in a minute as Kincheloe strained to maintain a constant angle of attack. Finally, at 40,000 ft., he was able to pull the still-supersonic X-2 into level flight before turning for home and a routine lake bed landing. The Bell X-2 now held both the unofficial speed and altitude records, returning valuable data on stability and control as well as structural heating at each of these extremes. While the exact details of Kincheloe's flight weren't immediately released, the air force did concede that Kit Murray's previous record of 90,440 ft. had fallen. As *Aviation Week* revealed more details in the weeks following the flight, Iven Kincheloe was hailed as "the first of the spacemen." While his record altitude fell far short of accepted boundaries for true spaceflight, in the eyes of the press the handsome, blond Kincheloe personified bravery and technical excellence—an impression that the air force was more than happy to promote.

With the X-2's handover fast approaching, work at the HSFS intensified in preparation for the forthcoming research program. The NACA recruited

Bell technicians in order to help operate the aircraft, and HSFS employees, including rocket-shop stalwart Jack Russell, received extensive training on the XLR25 engine. Thanks to Dick Day's diligent work with the GEDA, it was now possible to model the X-2's stability across a wide variety of flight conditions, allowing Joe Vensel's Flight Operations Division to plan flight profiles that could verify these predicted behaviors. Following Crossfield's departure, the station's chief pilot Joe Walker took over the X-2, and as the November 1956 deadline drew closer, Walker undertook simulator sessions and ground runs of the XLR25 to thoroughly familiarize himself with the aircraft's systems. However, before the NACA pilot would get his chance to fly the X-2, the air force had one more goal to chase.

By 24 September, Milburn Apt had completed his GEDA sessions and received his custom-fitted David Clark pressure suit. Although he had yet to fly the X-2, the AFFTC decided that Apt should proceed directly to an envelope-expansion flight, without the customary checkout; on his first sortie, Apt would follow an "optimum maximum energy flight path," seeking a maximum Mach number for the aircraft. NACA instrumentation had revealed that Everest's Mach 2.87 mark had been attained during a slight climb rather than level flight. Given the aircraft's lagging cockpit instrumentation and a slight thrust misalignment on the XLR25, minor deviations from the optimum flight path were not unusual, but now Apt would try to fly a profile that, if all went well, should carry the X-2 to Mach 3. An initial attempt scheduled for 25 September was delayed, as Kincheloe, who would fly chase for Apt, was out of town attending the premiere of the movie *Toward the Unknown*. This somewhat melodramatic depiction of life at Edwards featured a dramatized version of an incident that had occurred during December 1954, when Apt had saved the life of fellow pilot Captain Richard Harer after an in-flight drag chute test went badly wrong. In the movie, an intervention by chase pilot Major Bond allowed General Banner to land safely after he was unable to jettison his drag chute. In real life, Harer had been less fortunate, becoming trapped in his burning F-94 following a crash landing on the lake bed. As chase pilot, Apt had landed alongside Harer's aircraft before braving the flames to smash the jammed cockpit canopy and rescue his semiconscious colleague. Apt received the Soldier's Medal, the highest decoration for noncombat valor available to the air force, for his actions during this incident. *Toward the Unknown* climaxed with William Holden's Lincoln Bond narrowly surviving a high-altitude bailout from a tumbling Bell x-2—clearly, nobody was too worried about bad omens ahead of the program's thirteenth powered flight. While awaiting Kincheloe's return, preparations continued at Edwards, and following final checks, the x-2 was signed off as ready for flight on 26 September, meaning an attempt could be made the following morning.

As Milburn Apt made his predawn arrival at the AFFTC, the x-2 was undergoing final fueling and instrumentation checks as it hung half hidden below the B-50. The pilot seemed in good spirits, joking with Kincheloe while donning his pressure suit, and at 7:15 a.m. Apt walked out to the flight line, receiving good-luck handshakes and backslaps from the ground crew, before climbing aboard the bomber. Some fifteen minutes later, Fulton eased the heavily laden B-50 into the air as Apt took a final look at his flight plan before heading back to the bomb bay at 7,000 ft. Although he had sat in the x-2's cockpit many times, things were different today as technicians helped tighten his straps before lowering the heavy tinted canopy into place. Smaller and stockier than Kincheloe, Apt still found the x-2's cockpit cramped, with his shoulders pressed tight against the aircraft's sides. As the x-2's LO2 tanks were topped off, the last of the bomber's crew retreated, leaving Apt to make his predrop checks. The x-2 had shown an almost infinite capacity to produce glitches, but today everything seemed to be going to plan as he moved smoothly through the checklist. Reaching 30,000 ft., Apt was joined by Kincheloe, who was flying low chase in an F-86 in order to assist with final checks and monitor ignition following the drop. At 8:45 a.m. Fulton wheeled the B-50 to the east as Apt successfully primed the XLR25. Everything was now ready, and in the rocket plane, Apt counted down before exclaiming, "Drop it!"

Falling from the B-50, Apt immediately spurred both chambers of the XLR25 into action, with Kincheloe confirming, "Ten's going good," and then, "Five's going," before easing the stick back until the x-2 entered a 33° climb as per the flight's complex profile. Pinned back by the acceleration, Apt maintained his climb, passing Mach 1 at 43,000 ft. and Mach 2 some 7,000 ft. later, before he began to push the aircraft's nose over at 67,000 ft. As he brought the x-2 into level flight at 72,000 ft., Apt was traveling at Mach 2.2 with plenty of propellant still available. Now he eased the still-accelerating x-2 into a shallow dive, using gravity to assist the thundering

XLR25 in a final push for maximum speed. Hurtling onward, Milburn Apt watched as the cockpit's Mach meter edged up to and beyond 3.0. Engine cutout had been expected at the 130-second mark, but the XLR25 continued to roar, pushing the X-2 ever faster until the propellants were finally exhausted after 145 seconds. Diving through 67,000 ft., Apt was now traveling at an incredible Mach 3.196, and with the powered phase of the flight now complete, the pilot reported, "The engine has cut out, and I'm beginning to turn." Beyond some garbled shrieks that some interpreted as "There she goes," that was the last communication Milburn Apt would make.

Turning for home, Apt and the X-2 tumbled from the precarious tightrope of stability they had been treading. With the aircraft's angle of attack increasing during the left turn, the control surfaces became unable to counter the increasing yaw and roll, sending the X-2 into the wild gyrations of inertial coupling. Apt was thrown violently about the tiny cockpit, experiencing terrific g-loads as he desperately fought to regain control using the techniques he had practiced. As the X-2 careened toward the desert, it entered an inverted spin, and following two unsuccessful recovery attempts, Apt pulled the T-shaped handle between his legs to initiate cockpit separation at around 40,000 ft. As four explosive bolts released the aircraft's nose, the capsule pitched downward, subjecting Apt to severe negative g-forces until the parachute deployed to stabilize the descent. The exact sequence of events that followed is still unclear, but it seems that Apt, at least partially incapacitated as a result of his brutal ordeal, was unable to leave the capsule as the escape alarm sounded. Although he had released his restraints and jettisoned the canopy, Apt was still in the nose section as it hit the desert floor, and he was killed instantly. Less than three minutes after he had dropped from the B-50, Milburn Apt was dead—the thirteenth test pilot to perish at Edwards since 1950. The crumpled remains of the X-2's cockpit was found amid low desert scrub northeast of Edwards AFB, some five miles away from the rest of the aircraft.

In the aftermath of the X-2's destruction, a detailed investigation attempted to piece together what had happened during Apt's final moments. As with so many aviation accidents, a number of factors seemed to have contributed to the aircraft's destruction. There appeared to have been some confusion within the AFFTC as to the flight's objective, with Apt being advised to "stay within the envelope of knowledge," while being asked to fly a profile with the clear potential to exceed previous speeds. Base commander General Holtoner later commented, "I think that every supervisory guy from me on down has criticized himself, because if we had told this boy to stop at a specific speed, this wouldn't have happened."

While some cited Apt's lack of experience in the x-2 as a contributing factor, he had in fact flown the complex profile almost perfectly, but this, combined with the additional seconds of thrust from the XLR25, had carried the x-2 well beyond the envelope of knowledge and into the uncertain stability predicted by the GEDA. Dick Day had worked closely with Apt prior to the flight, and their training sessions had covered not only the planned profile but also appropriate recovery techniques should the x-2 depart from controlled flight. Given his earlier work on the F-100, few pilots at Edwards during this period could match Apt's knowledge and experience regarding inertial coupling, but investigators questioned why the pilot had turned for home while still traveling at such high velocity when the GEDA had predicted the x-2's likely departure from controlled flight under these conditions.

The scenario that eventually emerged from the investigation suggested that Apt may have been unaware of his exact speed given the tendency of cockpit instrumentation to lag behind the aircraft's actual performance, although the cockpit camera had clearly recorded a Mach meter reading in excess of Mach 3 when control was lost. The additional thrust and perfect profile meant the x-2 was traveling farther from Edwards than Apt had expected. As he continued along an easterly flight path, Apt was faced with a dilemma requiring a split-second decision—continue until the aircraft had decelerated enough to turn safely and risk not being able to reach Edwards or begin turning above the planned speed and risk losing control of the aircraft. Apt chose the latter, perhaps feeling he could cope with any instability. In reality, the loss of control was so sudden and so severe that the situation must have appeared unrecoverable to the pilot, prompting him to trigger the escape system.

Days before Apt's fatal flight, Everest had stated that he regarded the x-2's escape system as unsatisfactory and would have avoided using it due to the g-forces associated with separation (Everest later estimated these forces to be in the region of 14 g's). Both Bill Bridgeman and Scott Crossfield had earlier reached similar conclusions regarding the Douglas Skyrocket's escape capsule. All three pilots felt that the safest option was to stay with the aircraft until it decelerated and descended to lower levels, where control might be regained, as Yeager had demonstrated in the x-1a (although Yeager had little option as the x-1a offered no means of escape). Crossfield was especially critical of the escape capsule concept, later describing it as a "way to commit suicide to keep from getting killed." In Apt's case it seems likely that he was temporarily incapacitated by the separation forces, and this may have prevented him from leaving the capsule before it hit the ground. Bell's original plan to fit the capsule with a parachute large enough to be capable of lowering the nose section at a safe landing speed had been dropped as a weight-saving measure early in the x-2's development process.

Following the official inquiry, the air force received strong criticism for its decisions surrounding the x-2's final flight. In a damning 1957 document, AFFTC historian Ronald-Bel Stifler questioned not only why Apt had been pushed straight into a maximum-performance flight without the usual build up but also why the air force had felt the need to push so hard for Mach 3 when the NACA were soon due to undertake similar flights. Some at the HSFS felt that the air force had ignored warnings to proceed more cautiously as the x-2 approached Mach 3. The question of whether the air force prioritized a speed record over pilot safety remains contentious to this day, but in any event, the x-2 program was now over.

Given its long, troubled gestation and the eventual loss of both aircraft, along with the lives of Ziegler, Wolko, and Apt, the Bell x-2 gained an unhappy reputation as something of a jinxed aircraft. While its decade-long development rendered the aircraft, by the time it finally flew, partially obsolete in terms of its original mission, perhaps it is better to remember what the x-2 represented at the program's outset. Designed at a time when Mach 1 still remained to be conquered, Bell's promethean ambitions for their swept-wing rocket plane were frustrated by immature technologies. Stability augmentation systems that might have tamed the x-2's instabilities began to mature during the late 1950s. But ultimately, Bell had tried to reach too far too soon—put simply, the x-2 was ahead of its time. As the remains of the wrecked rocket plane were given an ignominious desert burial, another research aircraft would soon pick up the torch of high-speed, high-altitude research, building on the hard-earned lessons of the Bell x-2.

6. The End of the Beginning

Milburn Apt's fatal flight in the X-2 brought the first decade of rocket plane research over Rogers Dry Lake to an unhappy conclusion. During the ten years that had passed since the X-1's arrival at the Muroc AAF, the base had changed almost beyond recognition. Gone were the tarpaper shacks and rattlesnake-infested temporary hangars that had awaited the early flight test contingents from Wright Field and Langley. As Colonel Gilkey's master plan had gradually come to fruition along the lakeshore, the often-inadequate postwar facilities had given way to a modern, fully equipped test center. Although much of the infrastructure at South Base was simply demolished, the two largest hangars were painstakingly transported across the two miles of lake bed that separated the old and new facilities. With the final elements of the AFFTC redevelopment nearing completion, 1956 proved to be a high-water mark for the so-called golden age of flight-testing at Edwards AFB, with one hundred major projects reaching their conclusion.

The NACA operation had also grown steadily since moving to its new home just north of the main air force base. The hangars that flanked Building 4800 on Lilly Avenue now played host to an extensive fleet of experimental and service aircraft. With the HSFS's workforce now numbering in excess of two hundred, activities were organized under four main divisions: Research, Flight Operations, Instrumentation, and Administration. While the recently completed facilities offered significantly improved working conditions, the new multifloor building and ever-expanding workforce inevitably meant that some of the communal feel of the old facility was lost.

The modern hangar spaces were a welcome upgrade for the station's technicians and mechanics, but the long hours demanded by rocket plane operations did not change. Following every flight, the aircraft would be towed back to the hangar, where filmstrips from research instrumentation were removed. These were passed to Yancey's computer department, who then began the painstaking task of reducing the data. The operations engineer and crew chief would then take over as the aircraft's guardians until it left for the next flight. Crew chiefs held the final say over all maintenance matters relating to their assigned aircraft, and until they were satisfied that the vehicle was fit to fly, it would remain firmly on the ground.

The turnaround process began with detailed postflight inspections of all systems. Any anomalies or faults that had cropped up during the flight would be thoroughly investigated, and fixes would be made as required. The often-temperamental rocket motors received special attention, with their components being dismantled, purged, cleaned, and replaced where necessary before the motor was reassembled in the station's rocket shop. The refurbished unit would then undergo static firings on the test stand, ensuring that everything was working correctly before reinstallation in the aircraft. The motor and propellant system were then tested together, with the aircraft chained down to the ramp as technicians put the rocket engine through its paces. This process often saw the seemingly bizarre sight of Jack Russell or one of his colleagues leaning inside the aircraft's engine compartment to tweak valves or listen for faults, even as long tongues of flame sporting their characteristic Mach diamonds screamed from the nearby nozzles.

Alongside the crew chief, the operations engineer would assume responsibility for all aspects relating to the forthcoming flight, including research requests and flight planning. While technicians toiled to make the aircraft ready, flight planners would also be hard at work devising a detailed plan to satisfy the engineers' data requests for the upcoming flight. Flight planners worked closely with the station's research pilots, briefing them on the exact flight conditions and maneuvers that would be required, while also devising contingency plans should problems occur in the air. Minor structural modifications to the aircraft might be requested in the form of change orders depending on an upcoming flight's objectives, and at this point, the crew chief could call on the HSFS's masters of improvisation, the fabrication shop. Finally, when all was ready, checked, and rechecked, the crew chief and operations engineer would sign the necessary documentation, declaring that their charge was ready to fly once more.

Undoubtedly, the losses of both the X-1A and X-2 had come as a blow to NACA research plans, but while the air force would make no more rocket plane flights until the arrival of the x-15 later in the decade, the HSFS still had two serviceable aircraft at its disposal. In late 1954 the station had taken delivery of the Bell x-1B, and following an extensive period of modifications at Langley and Bell's Buffalo factory (including the replacement of the potentially deadly Ulmer leather gaskets), the sole surviving advanced x-1 made its first NACA flight on 14 August 1956. Although Pete Everest had proven that the x-1B could safely reach speeds in excess of Mach 2, the NACA planned to use it to investigate how dynamic heating might affect upcoming service aircraft operating between Mach 1.5 and 2. While back at Langley, the x-1B had been fitted with three hundred thermocouples, allowing researchers to obtain a detailed picture of temperatures across the airframe as the friction from air molecules caused kinetic heating at high speeds. The job of performing these heating flights fell to HSFS pilot Jack McKay. By 1956 McKay was an experienced rocket plane pilot, having flown a large proportion of the D-558-II research program during its later years (McKay would make the final flights in all three Skyrockets). The former naval aviator was also no stranger to midair emergencies and had recently survived one particularly close call while preparing for a flight in the second Skyrocket.

On 22 March 1956 McKay had been preparing to drop away from the P2B-1S launch aircraft, when the bomber's number four engine failed. Pilot Stan Butchart attempted to feather the engine's propeller (turning its blades edge on to the airstream to reduce drag), but unfortunately, the blades refused to stay in place, rotating back to their original positions. In this state, the propeller was likely to overspeed and eventually disintegrate, potentially damaging both the bomber and the rocket plane still hanging beneath it. Butchart made a second attempt to feather the prop with the same result and, knowing that the engine only carried enough oil for three attempts, was left with a problem.

Deciding that he would need to drop the Skyrocket as soon as possible, he was about to let McKay know the situation, when the research pilot's voice crackled over the radio. As he had worked through his preflight checklist, McKay had broken a valve in the D-558-II's cockpit and was now requesting an abort. Butchart was fast running out of options but elected to make a final attempt to feather the prop. When this failed, he had no choice; "Jack, I've got to drop you," he told the startled McKay, who now scrambled to get the Skyrocket ready for an unplanned glide flight as Butchart reached for the emergency jettison handle.

As McKay fell away safely, the overspeeding propeller finally gave way, disintegrating with such force that one of the blades sliced clean through the bomber's fuselage right where the Skyrocket's cockpit had been moments earlier. McKay was able to glide home safely, but Butchart and his copilot that day, a young NACA pilot named Neil Armstrong, were left fighting the bomber's damaged controls all the way back to Edwards. McKay himself had been at Butchart's side the previous year when the X-1A had exploded beneath its B-29 launch aircraft, so it may have come as some relief to the pilot when he dropped away safely in the X-1B to make his familiarization flight. Unfortunately for McKay, his first experience in a Bell rocket plane didn't pass entirely without incident, as, having taken the aircraft out to Mach 1.12 at 45,000 ft., he joined the ever-growing nosewheel club following a heavy landing.

After undergoing what were now routine repairs to the damaged gear, the X-1B made five more heating research flights with McKay in the pilot's seat, concluding with a flight to Mach 1.94 on 3 January 1957. Although flights like these may have seemed less glamorous than the headline-grabbing record attempts that took place over Edwards during the fifties, McKay's work in the X-1B helped validate earlier laboratory results and also returned interesting information on variable heating rates across the aircraft's structure. This was exactly the sort of research that the NACA had hoped to perform in the X-2, but the X-1B had proven itself an able, if limited, replacement. With the heating research program now complete, the aircraft now underwent further modifications in order to test a new control system intended to combat the high-altitude stability problems that had affected both the X-1A and X-2.

During their high-altitude flights, Kit Murray and Iven Kincheloe had entered a zone where dynamic pressure was so low that their standard aerodynamic controls became largely ineffective. With the upcoming X-15 aiming for altitudes in excess of fifty miles, engineers knew that a new means of nonaerodynamic control would be needed to adjust the aircraft's attitude. The solution appeared to lie in a reaction control system (RCS) using sets of small thruster nozzles positioned at key points around the aircraft. These would use steam, produced by concentrated hydrogen peroxide passing through a catalyst screen, to generate small directional jets. At low

16. NACA technicians testing the early reaction control system fitted to the
Bell X-1B. Courtesy NASA.

dynamic pressures, these would be enough to nudge the aircraft in the
opposite direction, allowing the pilot to control the vehicle's attitude accurately. Designers at North American planned to fit the X-15 with three sets of
thrusters, allowing control of pitch, roll, and yaw, and the NACA were keen
to test an RCS in flight, gathering data to aid the new rocket plane's development while also investigating the piloting challenges of the new system.

The X-1B was fitted with small wingtip extensions carrying both upward
and downward thrusters to assess how well the system could control the
aircraft's roll at high altitude. McKay made way for Neil Armstrong, who,
following in Joe Walker's footsteps, had transferred to the HSFS from the
NACA's Lewis Flight Propulsion Laboratory in his native Ohio. Armstrong
had joined the NACA in 1955, having seen combat over Korea as a naval aviator. Following the end of hostilities, he had returned to Purdue University,
completing a bachelor's degree in aeronautical engineering.

During his brief spell at Lewis, Armstrong had conducted icing research
and studied heat transfer using air-launched rockets, but when Scott Crossfield's departure opened up a spot in the HSFS pilots' office, he transferred

across to the High Desert and began checking out in the station's extensive roster of aircraft. In preparation for the RCS research flights, HSFS technicians had constructed a rudimentary simulator known as the "iron cross," consisting of two intersecting I beams mounted on a pivot, giving the approximate shape of the X-1B. By using small gas thrusters on the front and rear of the central bar, the pilot—in the rudimentary cockpit atop the "fuselage"—could control the pitch of the apparatus, while thrusters at the "wingtips" offered control over roll attitude.

After many hours on the iron cross, Armstrong made a low-altitude checkout flight in the X-1B on 15 August 1957. Like McKay, Armstrong experienced difficulties when it came to landing the X-1B, porpoising after his initial touchdown before finally coming down heavily on the nosewheel, with predictable results. In typically honest fashion, he later admitted, "It didn't really fail; I broke it." Following this unfortunate rite of passage, Armstrong made a second flight on 27 November, reaching 60,000 ft. Although he briefly tested the RCS, the dynamic pressure was still too high for the thrusters to overpower the aircraft's remaining aerodynamic stability, rendering the test largely ineffective. Two more flights followed during January 1958, but routine checks on the X-1B after Armstrong's fourth flight, on 23 January, revealed fatigue cracks in the aircraft's LO2 tank. HSFS technicians welded the affected areas, but concerns over safety persisted. And in June 1958 the NACA decided to retire the aircraft.

Although Armstrong's flights in the X-1B had been curtailed before he was able to test the RCS at high altitude, the program still contributed to the X-15's development. The control system was subsequently removed from the X-1B and installed in an early Lockheed F-104, capable of reaching similar speeds and altitudes as the rocket plane without the time-consuming preparations and support.

Armstrong's 23 January flight marked the end of the advanced X-1 program, with the X-1B as the sole survivor of the three aircraft produced. Although the X-1A had pushed the boundaries of flight beyond Mach 2 and 90,000 ft., the legacy of Bell's second-generation X-1s was marred by the explosions that had destroyed both the X-1D and the X-1A. There was, however, still one operational X-1 in the HSFS hangars—the very first of Bell's famous rocket planes to have flown over Rogers Dry Lake now returned to the skies, albeit in a highly modified guise.

The second Bell X-1, veteran of both Slick Goodlin's Muroc demonstration flights and the NACA's transonic research program, had been grounded during November 1951 as a precaution after metal fatigue was detected in some of the high-pressure propellant system's spherical nitrogen tanks. At the time it of its withdrawal, it seemed that the NACA X-1 would soon become surplus to requirements as the finally finished X-1-3 and the first of the advanced X-1s, the X-1D, arrived at Edwards. Unfortunately, when both of these new rocket planes were lost before making powered flights, the second X-1's retirement now looked likely to hinder the NACA's planned high-speed research program. With this in mind, HSFS engineers examined the possibility of modifying their aging X-1 to give it a new lease on life. Both the X-1-3 and the X-1D had featured the turbopump-driven low-pressure propellant system, and although plans for the X-1's modification did not stretch as far as replacing the aircraft's heavy spherical propellant tanks, the HSFS decided to install a turbopump, allowing the aircraft's system to run at low pressure, negating the need for the suspect nitrogen spheres.

Beyond improving the propellant system, the NACA were also interested in testing how well extremely thin, low-aspect-ratio wings would cope with loads during high-speed flight. With the agreement of the air force, the Stanley Aviation Corporation (formed by Bob Stanley, following his recent departure from Bell Aircraft) was awarded a contract for a new 4 percent t/c wing, but the manufacture of this new unit posed a new set of challenges. With a maximum thickness of less than three and a half inches at the root, the new wing would need to be extremely stiff, while also hosting two hundred pressure orifices and 343 gauges ranged across its span to record every detail of structural performance. While Stanley Aviation worked on the razor-thin wing, the HSFS rectified one of the X-1's major shortcomings—the lack of an effective means of escape for the pilot. The station's engineers fitted the aircraft with a surplus ejection seat from the Northrop X-4, but this in turn required a new jettisonable canopy in place of the X-1's original flush design. Rather than fitting a clear bubble canopy as Bell had with its advanced X-1's, the HSFS chose a more substantial design based on the revised canopies of the D-558 series.

To complement its improved propellant system, the aircraft also received a new engine when Jack Russell's rocket team modified a surplus LR8 from one of their D-558-IIs to replace the original XLR11, making this the only member of the x-1 family not to carry the "army engine." (In the station's rocket shop, the xLRII was casually known as the "dash II" or "army engine" as opposed to the LR8—the "dash 8" or "navy engine." As a rule air force engines used odd numbers in their designations, while navy engines used even numbers.) In the spring of 1954 the much-modified rocket plane gained a new designation; it would henceforth be known as the x-IE.

With all major modifications completed by mid-1955, the x-IE underwent months of static testing at the HSFS before taking to the air in December that year. Following the dramatic end to his involvement with the x-IA, Joe Walker now became program pilot for the newly rejuvenated rocket plane. Crossfield's recent departure from the HSFS had finally settled a somewhat tense situation regarding the title of chief pilot at the station. With Walker having joined the NACA in 1945, the former USAAF pilot understandably considered himself the senior man following his transfer to the HSFS in 1951, but in Crossfield's opinion his earlier arrival made him the number one pilot. Both men were known for their headstrong natures, leading to an often strained professional relationship. The x-IE now offered Walker the opportunity to gain valuable high-speed rocket plane experience while he awaited the x-2's arrival from the AFFTC. Following an unsuccessful launch attempt on 3 December, the x-IE made its first glide flight on 12 December, with a first powered flight, albeit curtailed by turbopump problems, following just three days later.

The x-IE did not fly again until April 1956, when Walker resumed testing of the aircraft's systems and slender wings. By now, the white rocket plane had been adorned with the nickname *Little Joe*, along with an illustration of two dice showing deuces. In all, the x-IE would fly eleven times during 1956, with Walker taking the aircraft to Mach 2 for the first time on 31 August, making it the only one of the original x-1's to reach this milestone. Unfortunately, although progress was being made, the x-IE was blighted by technical problems, with both its turbopump and reconditioned engine proving frustratingly unreliable, leading to frequent aborts or in-flight mechanical failures. There were also the almost-obligatory landing problems, with one incident in May 1957 seeing the aircraft slide across the lake bed on its belly before a sheepish Walker commented, "Well, I'm down." As the dust settled, ground crews were shocked to find that the aircraft's nosewheel had been sheared off completely during the heavy landing, and the main gear had also sustained considerable damage, keeping the x-1e grounded until August that year.

The aircraft received additional modifications throughout its operational life, including two small ventral fins fitted during late 1957 to improve the aircraft's directional stability at high speeds. After twenty-one flights in the x-1e, Joe Walker bowed out of the program in September 1958 in order to concentrate on training for his upcoming role as the NACA's senior x-15 project pilot. Jack McKay was an able replacement, stepping in to complete the research program having made his first familiarization flight on 19 September.

Following the loss of the x-2 in September 1956, the HSFS hoped that the makeshift x-1e might pick up some of the ill-fated rocket plane's research program, and this led to a new proposal to extend the x-1e's performance. HSFS engineers Hubert "Jake" Drake and Donald Bellman had calculated that by increasing pressure in the LR8's chambers and changing from a water-alcohol fuel to a more volatile but powerful fuel known as U-Deta (a combination of unsymmetrical dimethylhydrazine and diethylene triamine), the aircraft would have the potential to reach Mach 3. When another landing accident grounded the x-1e in June 1958, the station's technicians carried out the necessary modifications, and McKay began tests of the improved system during October, with two high-chamber-pressure flights followed by a low-speed, low-altitude test of the new fuel on 6 November.

In anticipation of forthcoming high-speed flights, the aircraft returned to the HSFS workshops to have an improved ejection seat fitted, but routine inspections now revealed cracks in the x-1e's oxidizer tank. Faced with the prospect of expensive repairs and the inescapable reality that, as with the x-1b, the aircraft's mission could now be carried out just as effectively and far more conveniently and economically using an F-104, the HSFS chose to retire the x-1e. McKay's 6 November flight proved to be not only the final flight of the x-1e program but also the final flight by any member of the x-1 family; Bell's groundbreaking supersonic design had by now been caught and surpassed by new service aircraft, direct beneficiaries of its pioneering supersonic research. Although its flight career had been relatively brief, one of the x-1e's more significant contributions had been to maintain rocket plane proficiency among the station's technicians as they awaited the arrival of the x-15.

In the twelve years that had passed since Jack Woolams first dropped away from the B-29 over Pinecastle Field, Bell's bullet-shaped rocket planes had vanquished the so-called sound barrier, flown higher than any vehicle before them, and helped elevate pilots such as Chuck Yeager and Pete Everest to legendary status. Although the program had been badly affected by the losses of three of the six aircraft in what were later found to be to be avoidable circumstances, the X-1 had vindicated Ezra Kotcher's belief in the practicality of rocket research aircraft and had helped bring about a speed revolution in aviation. Following its retirement, the X-1E was proudly placed on display in front of Building 4800 at the HSFS, where it remains to this day as a reminder of the pioneering early years of high-speed flight research.

During the mid-1950s, the NACA's Hartley Soulé had loosely defined the current and planned research aircraft programs in terms of three discrete rounds, charting the anticipated evolution from atmospheric to exo-atmospheric (or space) flight. Round one covered the early research aircraft, developed to return data on transonic and supersonic flight, as well as to perform research into exotic, tailless, delta-wing, and variable-sweep-wing configurations. Round two would bring hypersonic flight, with speeds in excess of Mach 5, and altitudes reaching the lower fringes of space. Finally, round three would see the development of a winged, rocket-boosted space plane capable of performing a controlled lifting reentry from orbital speeds of Mach 25 before making a conventional runway landing.

The retirements of the X-1B and X-1E during 1958 spelled the end of round one. In the twelve years that had passed since the X-1's first powered flight over Muroc AAF, the ever-expanding complex played host to a variety rocket-powered research aircraft and witnessed Machs 1, 2, and 3 fall in reasonably quick succession. These early rocket planes flew far beyond previous altitude records, reaching the upper limits of the stratosphere where their pressure-suited pilots became among the first humans to see the curvature of the earth. While piloted spaceflight was still regarded as a somewhat distant curiosity by mainstream science and the general public, a new rocket plane capable of making brief leaps beyond the atmosphere was already in the final stages of production.

Out at Edwards AFB the AFFTC and the HSFS were readying themselves for the upcoming assault on hypersonic flight, creating the complex infrastructure that would be required to support the x-15 program—round two in Soulé's orderly progression toward space. Already, companies, including Bell Aircraft, were refining plans for the next step, producing preliminary designs for winged vehicles that would be capable of leaving the atmosphere, circling the globe at terrific velocities before flying home under the control of an entirely new type of pilot—the astronaut. But even as the rocket plane pioneers of Rogers Dry Lake stood ready to take their first steps into this new frontier, events way beyond the windswept expanses of the Mojave Desert were generating political shockwaves more powerful than any sonic boom.

During the later years of the Second World War, an incredible new weapon had burst forth from the research institutes of Nazi Germany. On 8 September 1944, as Allied forces continued their push across western Europe, the cities of Antwerp and London were both rocked by sudden explosions. In the words of American war correspondent Edward R. Murrow, a new wonder weapon was inflicting "death from the stratosphere." Wernher von Braun and his expert team of rocket engineers at the Peenemünde research center had ushered in the age of the ballistic missile through the first operational launches of the A4 rocket, or the V-2 as it became more commonly known. Although von Braun's new wonder weapon arrived too late to affect the outcome of the war, it became the first human-designed object to leave the atmosphere, briefly arcing into the vast realm of space that lay beyond before falling back to strike its earthly targets without warning. As Hitler's Reich fell, many of the Peenemünde team surrendered to U.S. forces and were spirited away to a new life in New Mexico, along with large quantities of materiel relating to the A4 program. With the war in Europe finally over, two new dominant global powers began to emerge. Soon political boundaries were drawn between the United States and the Soviet Union, plunging the world into a new type of conflict—the Cold War.

Initially, the United States sought to project its military might through the use of conventional air power. As the only nation capable of wielding the awesomely destructive atom bomb, the United States pressed forward with the development of larger, intercontinental-range bombers under the control of the newly formed Strategic Air Command (sac). By 1949, however, America's brief nuclear monopoly ended as the Soviet Union tested their first atomic weapon, but although they now had the atom bomb, the

Soviets lacked an effective means to deliver it. While SAC's huge Convair B-36 bombers represented the very visible "big stick" of American foreign policy, the Soviets were initially left fielding their far less capable Tupolev Tu-4, a copy of the wartime Boeing B-29. Although the United States had retained the services of key members of von Braun's Peenemünde team, the U.S. military were slow to grasp the potential of rocketry as a means of delivering long-range weapons. By contrast, the Soviets, having missed out on much of the top German rocketry talent, chose to integrate what German knowledge they had secured with the skills of their own rocketry pioneers, such as Sergei Korolev and Valentin Glushko. As America developed a new generation of jet bombers and cruise missiles, the Soviet Union ploughed an increasing amount of its available resources into long-range missile programs, with the visionary Korolev offering increasingly capable short- and intermediate-range weapons, utilizing new concepts and technologies far beyond those of von Braun's original A4.

Having had a profound effect on supersonic flight research out at Edwards, the outbreak of the Korean War in 1950 also changed the fortunes for rocketry in the United States. The army now moved von Braun's team from their virtual exile in New Mexico to Huntsville, Alabama, where they began work on the new Redstone intermediate-range missile. Elsewhere, the air force and Convair began development of the Atlas intercontinental missile, which it was hoped would be capable of delivering an atomic warhead directly to Soviet territory from the mainland United States.

Against the background of this Cold War military buildup, a new peaceful scientific initiative was proposed in 1952. The International Geophysical Year (IGY) was to run from July 1957 until December 1958, offering scientists from forty-six countries the opportunity to conduct coordinated research into a wide range of physical phenomena, including studies of the upper atmosphere. During 1955 both the United States and the Soviet Union announced their intentions to orbit small scientific satellites during the IGY, but these efforts were initially low-key scientific endeavors with funding to match. However, as this interest in artificial satellites began to coincide with the emergence of powerful new missiles, thoughts turned to the potential propaganda value such a launch might bring.

Like von Braun, Korolev had long been fascinated by the prospect of space travel, and the Soviet chief designer now calculated that his proposed R-7 missile, far more powerful than its American counterparts, as it had been built to carry bulkier Soviet nuclear weapons, might soon offer the means to place objects in orbit. Consequently, his OKB-1 bureau began work on an ambitious orbital laboratory named Object D to be sent aloft during the IGY. The U.S. Army, U.S. Navy, and U.S. Air Force had all submitted proposals to launch a satellite, but President Eisenhower's administration—eager that the project be perceived as a peaceful, scientific venture in keeping with the aims of the IGY—chose the navy's Project Vanguard, which would use hardware developed from scientific sounding rockets rather than repurposed missiles.

As the IGY dawned in the summer of 1957, Korolev's efforts were mired in problems. A series of failed test launches had left his R-7 missile on the brink of cancellation, and work on the Object D satellite had also fallen way behind schedule. Before there could be any thought of a satellite launch, the R-7 first had to prove itself as an intercontinental weapon and OKB-1's hopes now hung on the third, and possibly final, test in late August. Meanwhile in the United States, technical problems and delays were also affecting Project Vanguard's schedule. With hopes for a September launch fast fading, Nelson A. Rockefeller, a trusted advisor within Eisenhower's inner circle, presciently raised concerns regarding the propaganda impact that a Soviet satellite launch might have, warning the president that "the stake of prestige that is involved makes this a race we cannot afford to lose." On 21 August, Korolev's R-7 flew over four thousand miles from its launch site in Kazakhstan to the Kamchatka Peninsula in the far eastern reaches of the Soviet Union. Although the missile's warhead disintegrated on reentry, the flight was deemed a success, and with further weapons tests suspended awaiting the warhead's redesign, Korolev now lobbied hard for permission to use the next available R-7 in an attempt to loft the first Soviet satellite. Having shelved the ambitious Object D in favor of a smaller, simpler design known as the PS-1, the OKB-1 team now swung into action, making their final preparations and calculating a new trajectory to ensure that the R-7's next payload would miss the earth and enter orbit.

On 4 October 1957 a series of unearthly beeps heralded the progress of the highly polished PS-1, or Sputnik, as it arced across the heavens. Korolev's R-7 had successfully placed the two-foot-diameter metallic sphere into orbit around the earth. As the propaganda value of the achievement slowly dawned on Soviet premier Nikita Khrushchev, the White House attempted to downplay Sputnik's significance, but the American media was in no mood to ignore the satellite as it passed over the United States every ninety minutes. President Eisenhower endured strong criticism for apparently allowing American technology to fall behind that of the Soviet Union. Things only got worse for Ike when a second Soviet satellite entered orbit on 3 November. Sputnik 2 weighed in at well over one thousand pounds and, to the amazement of Western scientists, carried the first cosmic traveler—a small dog named Laika. American hopes for a timely response now rested on Project Vanguard, and the nation's eyes turned to Cape Canaveral in Florida as the slim rocket was readied for a December launch attempt. The Vanguard team knew that their vehicle still needed more work and, privately at least, conceded that hopes were not high for the TV-3 launch. On 6 December 1957 the Vanguard rocket slowly lifted away from its launchpad only to collapse back into a blossoming inferno seconds later. America's first satellite had managed to travel only the first four feet of its journey to space, and the press had a field day. *Kaputnik* and *Flopnik* screamed the headlines. Quietly, Eisenhower had already acceded to army pressure, granting von Braun's team permission to make their own launch attempt using a Jupiter C rocket, a modified version of the Redstone missile. On 31 January 1958 the army team succeeded in launching Explorer 1 into orbit—America was back in the race but still some way behind the apparent superiority of the Soviet Union.

Yielding to the intense public and political disquiet that had been building since Sputnik's launch, Eisenhower and his advisors now sought a means to formalize American space policy in order to formulate a longer-term response. The president was faced with a difficult choice: Should he entrust the nascent American space program to the military, with its interservice rivalries, or instead seek to create a new civilian organization? In the wake of Sputnik, NACA director Hugh Dryden had established a Special Committee on Space Technology, composed of representatives from the military, industry, academia, and each of the NACA laboratories. With the organization's existing involvement in space research programs, including the x-15, in mind, Dryden circulated a report in January 1958 stating that "the NACA is capable, by rapid extension and expansion of its effort, of providing leadership in space technology." When James Killian, chairman of the President's Scientific Advisory Committee (PSAC), endorsed Dryden's views to Eisenhower in early March, the president set in motion a chain of events that led to his signing the National Aeronautics and Space Act into law on 29 July 1958.

After forty-three years at the forefront of American aeronautical research, the NACA was officially dissolved on 1 October to be immediately replaced by the National Aeronautics and Space Administration (NASA). As round one drew to a close in the High Desert with the final flights of the Bell X-1E, a new insignia replaced the NACA wings at Building 4800, as the NASA High-Speed Flight Station turned its gaze toward the new frontier of space, a change of emphasis cemented some two weeks later as a sleek black rocket plane made its first public appearance in Los Angeles.

Part 2

America's First Spaceship

7. The Hypersonic Challenge

John Becker seemed destined to pursue a career in aeronautics. Following his first encounter with a biplane as a six-year-old, Becker's growing fascination with aviation led him into the popular hobby of model aircraft building, which soon developed into a yearning to experience the thrill of flight for himself. Having read an intriguing article about hang gliders, the teenage Becker fixed on a plan to use a hill beside the family home in Albany, New York, as a launching point for his own homemade design. When the resulting glider proved too heavy to carry up the hill, Becker simply fitted skids to his creation and persuaded his father to tow him up to takeoff speed behind the family car.

With his passion for flight undimmed by these early exploits, Becker enrolled at New York University to study aeronautical engineering, and while he continued to fly with the university's gliding club, it quickly became clear that his true vocation lay in the laboratory. Graduating during the Depression years, Becker was fortunate to find immediate employment at the Naval Air Factory in Philadelphia before an offer from the NACA in 1935 attracted him to Langley, where the young engineer was assigned to the laboratory's eight-foot tunnel under the guidance of Russell Robinson. When Robinson subsequently transferred across to the Ames Aeronautical Laboratory in Sunnyvale, California, he was replaced by the dynamic John Stack. And Becker was soon immersed in transonic research as a member of Stack's newly formed Compressibility Division, investigating potential solutions to the choking problem that had rendered transonic wind tunnel data so unreliable (efforts which eventually led to the slotted-throat test section, developed alongside—and verified by—the X-1 flight research program).

In 1945, as assistant chief of the division, Becker began to hear of the exciting discoveries Allied inspection teams were making as they exam-

17. Hypersonics pioneer John Becker. Courtesy NASA.

ined the recently captured German aeronautical institutes. Having reached the Bavarian town of Kochel, investigators found two advanced wind tunnels, the first of which had been relocated from the rocket development site at Peenemünde on Germany's Baltic coast and was capable of generating sustained speeds of Mach 5, while a second partially constructed tunnel would have allowed research to take place at a then incredible Mach 10. Used by von Braun's rocket team during the development of the A4 and its proposed successors, the Kochel facilities went far beyond anything the

Allies had on their drawing boards at the time. Suitably inspired by these discoveries, John Becker believed that the NACA possessed the necessary technology to construct a similar facility and consequently proposed that a new experimental hypersonic tunnel should be built at Langley.

Becker began work on the eleven-inch hypersonic tunnel in 1946, and following much trial and error regarding air liquefaction and diffuser design, the new facility made its first test runs during November 1947, producing airflows approaching Mach 7. Recalling these early steps, Becker later recounted, "When we started the hypersonic business, there was no demand whatever for it. It was just a matter of it would be fun—we felt it would be fun to get out, far out, and explore what happens," but hypersonic research did not remain a scientific curiosity for long.

Having witnessed the vulnerability of its bombers to the new generation of Soviet jet fighters during the Korean War, the American military soon began to reconsider long-range missiles as a means to deliver the country's nuclear arsenal. As work began on the first generation of intercontinental ballistic missiles (ICBMs), the challenge of designing warheads capable of surviving reentry through the earth's atmosphere assumed paramount importance. Traditionally, missile designers had favored pointed shapes to minimize drag within the atmosphere, but it quickly became apparent that these shapes could not endure the intense kinetic heating experienced during reentry at hypersonic velocities. As the demand for hypersonic facilities to test new reentry shapes increased, new tunnels were constructed at Ames and the Naval Ordnance Facility, Maryland (which actually reconstructed the recovered Kochel Mach 10 tunnel). By 1952 NACA laboratories were spending an increasing amount of time supporting missile development, and as the rocket research aircraft rapidly closed in on Mach 2, thoughts naturally turned to the possibility of a piloted hypersonic vehicle. Perhaps unsurprisingly, given its heavy involvement in the high-speed research aircraft programs, Bell Aircraft were among the first to raise this idea as Robert Woods addressed the NACA Committee for Aerodynamics in late 1951.

Following the war, Bell had managed to secure the services of von Braun's former commander at Peenemünde, Walter Dornberger. Dornberger had long been fascinated by the studies that Austrian engineer Eugen Sänger and his mathematician partner (later wife) Irene Bredt had carried out into an intercontinental space plane, capable of traversing huge distances by skipping across the upper atmosphere. Although the concept had garnered some support from the German military in the hopes that it might lead to an operational antipodal bomber, Sänger's so-called Silbervogel (Silver Bird) never progressed beyond these conceptual studies. As these plans fell into Allied hands following Germany's collapse, engineers on both sides of the fast-emerging iron curtain began to investigate how plausible the idea really was. When the influential RAND corporation gave the Sänger-Bredt concept a favorable assessment, Dornberger and Bell Aircraft began to study recent advances in both high-temperature materials and rocket propulsion with a view to designing a Silbervogel-type vehicle. Once Dornberger was joined by former Peenemünde colleague Krafft Ehricke, the pair began work on an ambitious winged bomber missile (BoMi) concept, but Bell still needed accurate data regarding the thermal and structural loads that a vehicle of this type was likely to encounter during hypersonic flight.

Although the Committee for Aerodynamics initially declined to follow up on Woods's proposal, the Bell engineer persisted, and in June 1952 a resolution was passed stating that the NACA would "increase its program dealing with the problems of unmanned and manned flight in the upper stratosphere at altitudes between twelve and fifty miles, and at Mach numbers between 4 and 10." The resolution also called for "a modest effort" to examine the problems of flight above fifty miles and at speeds in excess of Mach 10—the NACA's first official step toward research into what would soon be recognized as spaceflight.

By this time, two unsolicited high-speed research aircraft proposals had already emerged from within the NACA itself. The first of these originated at the HSFRS, where two of Walt Williams's most talented research engineers, Hubert Drake and Robert Carman, had devised a two-stage system using a large V-tailed carrier aircraft to launch a smaller hypersonic rocket plane. The two engineers calculated that if the smaller craft were released at Mach 3 and an altitude of 50,000 ft., it should be capable of attaining speeds approaching Mach 10 at altitudes in excess of 180 miles. A second, somewhat less ambitious proposal by PARD engineer David Stone involved fitting the existing Bell X-2 rocket plane with two large Sergeant solid-fueled rockets. Stone believed that the additional thrust generated by these boosters would allow the modified X-2 (featuring an RCS for exo-atmospheric flight) to reach Mach 4.5 at an altitude of 300,000 ft.

When the air force's Scientific Advisory Board (successor to von Kármán's earlier Scientific Advisory Group) concluded during an October 1953 meeting that a hypersonic research aircraft was both desirable and feasible, they also agreed that the cooperative arrangements that had existed between the air force and the NACA since the days of the x-1 should now be extended to cover such a vehicle. The following February, Hartley Soulé chaired a meeting of the RAPP during which it was agreed that rather than modifying an existing research aircraft, as Stone had proposed, the NACA should begin studies toward an entirely new hypersonic rocket plane. With this goal in mind, each NACA laboratory was asked to consider its research requirements for the new aircraft, and in response to this request, John Becker—now chief of Langley's Aerophysics Division—convened a five-person group (which included a young engineer named Max Faget, later a leading proponent for ballistic, rather than lifting, reentry vehicles) to examine the design considerations that would shape the proposed hypersonic aircraft. Although the group chose to focus on the formidable stability and control, propulsion, and thermodynamic challenges that lay ahead, they also examined the potential that such a vehicle would offer for making brief space leaps beyond the atmosphere. According to Becker, "Up until 1952 or '53, there was almost no realization that we were on the verge of the space age. Then, suddenly, we realized we had the propulsion to get up to hypersonic speeds and also to get out of the atmosphere—at least for a little while—and out into space. When that began to sink in, it became a very exciting period."

By April 1954 the group had produced a detailed design study for a hypersonic aircraft capable of achieving Mach 7 and altitudes in excess of 250,000 ft. Echoing Kotcher and Stack's aims for their earlier transonic research aircraft, Becker stressed that the group's design was not intended to be an optimum configuration; rather, it represented the most basic vehicle capable of attaining the desired set of flight conditions. The aim of the exercise was to answer the questions posed by hypersonic and exo-atmospheric flight; after all, he reasoned, if we already had the answers, we wouldn't need the research aircraft. With their 1954 study, Becker and his colleagues had created the blueprint for an aircraft capable of shattering the existing frontiers of flight; now it was time to rally the support needed to build it.

The Langley study addressed many of the key concerns regarding hypersonic flight, foremost of which was how to deal with the intense heat that the aircraft's structure would experience while traveling above Mach 5. Becker's team identified two possible approaches to this problem. The first was a double-wall concept, in which an outer heat-resistant or shielded structure would be separated from a more conventional, inner load-bearing structure by a layer of insulation. Although the Langley team had little practical experience in the construction or use of double-wall structures to draw from, Bell Aircraft had begun to investigate the technique as part of its BoMi studies, and the company's limited research indicated that the approach held promise.

The double wall did, however, come with drawbacks. Complex to develop and construct, new methods would be needed to connect the two structures while still allowing for thermal expansion without buckling. There were also some parts of the aircraft, such as the wing leading edges, that could not be protected in this way. Here, exotic metals or liquid cooling would be used to withstand the heat, but these only added to the system's overall complexity. Becker's group were also concerned that any failure of the outer layer might allow hot gases to impinge on the vulnerable inner structure, potentially leading to the loss of the aircraft. Finally, while the approach made sense for an operational vehicle such as the proposed BoMi, one of the hypersonic research aircraft's key tasks was to investigate heat transfer through the vehicle's structure—something that a double wall would obscure.

The second, and preferred, option was the hot structure, using a single load-bearing skin fashioned from temperature-resistant material to absorb the heat and radiate it back into the atmosphere. Once again, the team had little practical data at their disposal, but the hot structure seemed a far more achievable prospect. Initial concerns that skin panels might need to be excessively thick (and consequently heavy) to absorb the heat were alleviated when tests on Langley's preferred material, a nickel-steel alloy named Inconel-X, revealed that the optimum gauge of metal needed to withstand the expected aerodynamic loads just happened to coincide with the thickness required to absorb the projected temperatures. Although the structure would still require some innovative solutions to cope with thermal expansion, weight was no longer a limiting factor.

By 1954 researchers at Langley were well aware of the problems of high-speed, high-altitude instability. In the wake of Chuck Yeager's battle with inertial coupling in the x-1A, investigations had revealed that the aircraft's vertical fin was too small to dampen its tendency to roll in the low dynamic pressures above 60,000 ft. Although engineers felt confident that inertial coupling could be avoided on future aircraft flying at similar speeds and altitudes by increasing the area of the fin, higher-flying aircraft were expected to face more serious problems. When initial calculations indicated that the hypersonic aircraft might need a fin with a surface area equivalent to the x-1's wing to ensure stability, another Langley researcher, Charles McLellan, suggested a more practical approach. Rather than using an excessively large vertical fin with a thin airfoil section, McLellan's research had shown that a smaller surface featuring a thicker 10° wedge-shaped profile would actually provide better stability at hypersonic speeds. While using a thicker profile initially seemed counterintuitive, McLellan's findings were born out by additional research that revealed that the wedge profile would create less drag than a large thin surface.

One aspect of the design study that proved somewhat controversial was the proposed space leap. In the mid-1950s, human spaceflight still seemed a distant prospect, leading some within the NACA to question whether research into the biomedical effects of weightlessness was really necessary at this stage. Less controversial was the intention to test a new attitude control system in the low dynamic pressures beyond 200,000 ft., but there was still some concern surrounding the vehicle's ability to reenter the atmosphere safely from these extreme altitudes. Becker suggested that a high angle of attack should be maintained during reentry but acknowledged that more research was needed. After some discussion, the exo-atmospheric space-leap was retained—a decision that would prove fortuitous for the United States in years to come.

Propulsion was another area where Becker recognized that more detailed research was needed. The design study had used a cluster of three rocket engines, but it was hoped that contemporary missile development might provide a more suitable power plant in the coming years. This issue would become one of the program's major developmental challenges, bringing much delay and budgetary chaos.

In June 1954, NACA administrator Hugh Dryden sent letters inviting the air force and navy to participate in joint discussions toward a hypersonic research aircraft. When the three parties subsequently met at NACA Headquarters on 9 July, Dryden presented Becker's study as a starting point, but it quickly emerged that both services had already given the matter independent consideration. In the navy's case, Douglas aircraft had been commissioned to produce a design study for a hypersonic successor to the D-558-II. The resulting Douglas Model 671 study (often referred to as the D-558-III) echoed many of the Langley team's recommendations, but it also covered some interesting ground that the NACA study had not, including the suggestion of two discrete flight profiles: one for high speed and the other for high altitude. Using a single RMI XLR30 rocket engine, Douglas predicted that their aircraft would be capable of reaching an incredible million-foot maximum altitude. Rather than the Langley team's preferred hot structure, the 671 would feature a double-wall structure, the outer layer of which would be covered in an ablative material designed to char at high temperatures, carrying excess heat away from the aircraft.

The El Segundo team also put a great deal of work into calculating the effects of thrust deviations or misalignments on the aircraft's ground track, as well as suggesting landing patterns for the 671 as it returned to Edwards. Whereas Becker's team had suggested Convair's B-36 bomber (the largest aircraft then in the air force inventory) as launch aircraft, Douglas chose Boeing's prototype B-52 jet bomber due to its considerable advantages in terms of launch speed and altitude. The El Segundo team acknowledged that many aspects of its 671 study would need further refinement; illustrations showed an aircraft with a clear family resemblance to the earlier D-558-II rather than the more radical appearance of the Langley team's design, suggesting much work lay ahead for the Douglas engineers. But as with the Langley study, the overall conclusion was that a hypersonic research aircraft was both feasible and desirable.

Following agreement during the July 1954 discussions that a hypersonic research aircraft program should be initiated, both BuAer and the WADC carried out their own independent evaluations of the Langley study with a view to their specific requirements. In August the WADC forwarded a positive evaluation of the study to the ARDC with the caveat that the Air Force Power Plant Laboratory should be commissioned to perform a study of suitable propulsion options. The WADC evaluation also included an initial program budget estimate of $12 million. On 13 September the ARDC forwarded its endorsement for the new project to Air Force Headquarters, recommending that the program be initiated "without delay." The new research aircraft moved another step closer to reality when the NACA Committee for Aerodynamics passed a resolution in support of the program during its 5 October meeting at the HSFS. There was, however, one dissenting voice on that occasion; Lockheed's formidable chief designer Clarence "Kelly" Johnson argued that the development of another high-speed research aircraft was not warranted, as their unusual configurations and rocket propulsion had little direct relevance to the aviation industry. Fortunately, this was one argument that Johnson did not win.

On 13 October, Hartley Soulé met with both navy and air force representatives to formalize program requirements. The navy's sole notable request was for an optional two-seat configuration, allowing an observer to fly in place of the NACA instrumentation package. The air force made no additional requests at this stage, but a subsequent meeting between Soulé and WADC representatives on 22 October did confirm one significant break from previous practice. Given the ongoing uncertainties about a suitable rocket engine, the WADC suggested that a separate specification and procurement process should be followed for this item. On that very same day, Hugh Dryden met with senior navy and air force representatives to draw up a memorandum of understanding, outlining each party's contributions. As with previous programs, the NACA would act as technical lead with responsibility for research instrumentation, but a new Research Airplane Committee (RAC) comprising of one representative from each of the participating parties would be formed to provide advice and guidance throughout the program. The air force would assume responsibility for the vehicle's development and construction, covering airframe and engine procurement, while also providing the majority of program funding, just as it had for the X-1 and X-2. Although it would only provide a small percentage of the overall costs, the navy would contribute biomedical support during pilot training and the flight program.

With roles agreed, the air force sent letters to Bell, Boeing, Chance-Vought, Convair, Douglas, Grumman, Lockheed, Martin, McDonnell, Northrop, North American, and Republic, on 30 December 1954, inviting bids for the airframe contract. Interested parties were asked to contact Wright Field by 10 January 1955, with a view to attending a bidder's conference on 18 January. Just nine of the prospective contractors chose to attend the conference, and of these, only Bell, Douglas, North American, and Republic submitted full bids ahead of the 9 May deadline.

On 17 January 1955 the air force informed its partners that the hypersonic research aircraft was now identified as Project 1226, with the resulting vehicle to be designated the X-15. Having received the four bids, the process of selecting a company to build the new rocket plane could now begin. All four contractors had been supplied with the Langley design study to act as a starting point for their proposals, but as the results were circulated among the WADC, BuAer, and the NACA for evaluation, it soon became clear that each had interpreted the requirement in slightly different ways. Alongside their airframe proposal, each bidder had also specified its preferred engine choice from the shortlist of four rocket motors provided by the Air Force Powerplant Laboratory. Both Bell and North American had engines on this list, with the other two candidates coming from RMI and Aerojet.

Given the company's prior experience of building research aircraft, it seemed reasonable to expect that Bell Aircraft would be among the front-runners for the X-15 contract. Unfortunately, its Model D171 showed some unusual design choices that seemed to run against the Langley study's guidance. While a hot structure had not been a compulsory stipulation, the detailed work by Becker's group had certainly made the NACA's preference clear. Bell, however, chose a double-wall construction, with an outer skin of Inconel-X strips separated from the inner load-bearing structure by corrugated sheets of Inconel-X and the lightest insulating material available— air. In adopting this approach, it was unclear whether Bell was designing the D171 to fulfill the X-15 brief or as a technology test bed for its proposed BoMi vehicle.

Unsurprisingly, the company favored its own XLR81 engine, with three units mounted independently in a triangular formation, allowing for a range of thrust increments ranging from 8,000 lbf. up to a maximum of 43,000 lbf. (Oddly, Bell believed that many X-15 flights would be conducted at lower thrust levels, although this was clearly at odds with the program's stated research goals). The XLR81 used a combination of red fuming nitric acid and JP-4 jet fuel as its propellants. As well as being hazardous to handle, the red fuming nitric acid oxidizer was less dense than LO2 and therefore required two nonintegral tanks, complicating the D171's internal layout and reducing the amount of volume available within the fuselage. The wings and tail used wedge-shaped airfoils, as recommended by Charles McLellan, with the wings' leading edges using either lithium or magnesium as a heat sink. A ventral stabilizer provided additional vertical tail area to reduce the risk of inertial coupling but would need to be jettisoned as the aircraft approached the lake bed to provide sufficient clearance on landing.

The aircraft featured a hydrogen peroxide–fueled RCS for exo-atmospheric attitude control and a set of speed brakes to help manage energy during flight. Rather than an escape capsule like the one in the X-2, the D171 would carry a downward-firing ejection seat (an approach used on Douglas's X-3 and then under consideration for Lockheed's XF-104 and the Republic XF-103). If required, a small rear cockpit could be created for an observer by removing research instrumentation and reorganizing the aircraft's environmental systems. Bell believed that the D171 would be capable of exceeding 6,600 feet per second and reaching a maximum altitude of 400,000 ft., using the Convair B-36 as its launch aircraft.

Having already worked on the Model 671 study for the navy, Douglas came to the X-15 bidding process with arguably the best insight of the four bidders into the challenges hypersonic flight would pose. The company's submission—the Model 684—differed from the 671 in many respects. The earlier design's familial resemblance to the elegant Skyrocket had given way to a squatter, bullet-shaped configuration with a broad circular-section fuselage narrowing to a long, pointed nose. Short trapezoid wings were mounted midway up the body, with the horizontal tail set slightly higher. Douglas adopted a hot structure for the 684, but rather than Inconel-X, it chose HK31, a magnesium alloy that would save weight even though thicker-gauge panels would be needed. For propulsion, Douglas chose a single RMI XLR30 Super Viking engine featuring RMI's innovative spaghetti construction, in which hollow stainless steel tubes were brazed together to form the combustion chamber and engine bell walls, making it far lighter than previous models. With LO2 as its oxidizer, the XLR30 used anhydrous ammonia as fuel, with this being circulated through the engine tubing to provide regenerative cooling.

The 684 featured two independent hydrogen peroxide–fueled RCSs, offering redundancy should the primary system fail, along with small speed brakes located on the flared section about the base of the engine compartment. Douglas kept faith with the escape capsule concept used on its D-558 models, believing that "no other system [would] be as good" at the speeds and dynamic pressures the aircraft would encounter. Unlike on the X-2, the 684's capsule would feature a large parachute capable of lowering both capsule and pilot to the ground at a survivable speed. Douglas's provision for a second crewmember saw the unfortunate observer relegated to a cramped, windowless space within the aircraft's nose, in place of the research instrumentation. Surprisingly, rather than recommending the higher-performance B-52, as it had in the earlier 671 study, Douglas chose the increasingly outdated B-50 as its carrier aircraft, hoping to inherit the example then being used by the X-2 program. Even though this incurred a considerable performance penalty, the 684 was still expected to reach a maximum velocity of 6,655 feet per second or a peak altitude of 375,000 ft., depending on the profile flown.

North American Aviation's ESO-7487 proposal came closest to the Langley study's configuration of the four bids. Featuring an Inconel-X hot structure and an uncomplicated configuration, the 7487 put research requirements first, providing a simple, robust response to the NACA's needs. Provision was made for a large accessible instrumentation payload, which could be removed to provide reasonable accommodation for an observer if required. A large all-moving vertical fin featuring a double-wedge profile and a smaller ventral fin would keep the aircraft pointing in the right direction, while the trailing edge of the upper fin could split to act as a speed brake during both powered and unpowered flight. Whereas its competitors had all included ailerons for roll control, North American chose a different approach, with the horizontal tail surfaces acting independently as elevons, controlling both the aircraft's pitch and roll. Having used a similar system on its recent F-107 fighter-bomber prototype, North American felt that elevons would reduce the complexity of the X-15's wings, removing protruding aileron hinges that might cause local heating problems. North American also chose the RMI XLR30 with its LO2 and ammonia propellants housed in two large integral tanks. These tanks would be toroidal, creating a long channel through the center of the fuselage, partially filled by two smaller cylindrical helium tanks for pressurization. This layout led to another unusual element of the ESO-7487—the fuselage tunnels. Fabricated from Inconel-X panels, these chine-like structures ran laterally down the aircraft's flanks from a point level with the front of the cockpit to the rear of the tail, giving the 7487's fuselage its characteristic flattened appearance. As well as housing fuel lines and control cables, the tunnels offered a mounting point for the wings, horizontal tail, and the two extendable skids that constituted the aircraft's main gear. The 7487 would carry an upward-firing ejection seat, with the pilot wearing a full pressure suit for additional protection. Using a B-36 carrier aircraft for launch at around 38,000 ft., the ESO-7487 was expected to attain a maximum velocity of 6,950 feet per second or a peak altitude in the region of 800,000 ft. Interestingly, North American's bid was the most expensive by some margin, coming in at just over $56 million.

The final proposal came from Republic Aircraft, and although this was the company's first venture into rocket-powered research aircraft, it had previously developed the unusual XF-91 jet-rocket hybrid interceptor. Republic could also boast the design genius of Alexander Kartvelli, whose innovative XF-103 Mach 3 interceptor was then under consideration by the air force. Unfortunately, Republic's AP-76 was not only the largest and heaviest of the four submissions, it also offered the lowest performance, especially in terms of altitude. The AP-76's long tubular fuselage housed four Bell XLR81 engines and used a double-wall construction with an inner titanium structure protected by an outer layer of Inconel-X tiles. The aircraft featured relatively large trapezoidal wings and a large vertical tail that, like the North American design, featured a split trailing-edge speed brake (the AP-76's horizontal tail surfaces could also split, adding to the rocket plane's braking capabilities). For unstated reasons, Republic concentrated on the X-15's high-speed profile at the expense of the high-altitude mission, resulting in a disappointing proposed ceiling of 220,000 ft. Cockpit arrangements on the AP-76 were interesting—in an attempt to reduce drag, the pilot was to be housed within the fuselage, with no forward vision (Kartvelli's XF-103 used a similar design). Three small rectangular windows were situated to each side of the pilot, but during approach and landing, the hatch above the cockpit would raise slightly to give forward vision via a periscope system. In the event of emergencies, the pilot would use an upward-firing ejection seat, a surprising inclusion given the amount of work Republic had done on enclosed ejection capsules for the XF-103. The AP-76 offered the best provision for a second crewmember, with the fuselage being stretched slightly to accommodate a second full-size cockpit, while still offering room for a modest instrumentation payload.

Much of the summer of 1955 was spent evaluating the four bids and their respective engine choices, with the designs being ranked on their performance, technical design, suitability for the required research tasks, potential for development or modification, and, of course, projected cost. Each of the bidders was briefed on the outcome of their evaluation during a 12 August meeting at NACA Headquarters. Neither Bell nor Republic had impressed with their designs, but the final decision between the two remaining proposals was close. In the end, North American just edged ahead of Douglas in spite of the higher costs associated with its proposal. RMI was awarded the engine contract for its XLR30, although it was acknowledged that a major man-rating effort would be needed to make the engine suitable for the x-15.

However, things took an unexpected turn when North American informed the x-15 project office at Wright Field that it wished to withdraw its bid. Given its recent selection for the air force's WS-110A bomber project (which led to the B-70 design) and the related LRI-X high-speed interceptor study (leading to the F-108), North American's management felt unable to commit to the x-15's proposed thirty-month production schedule. Although the selection panel briefly considered awarding the contract to second-place Douglas Aircraft, the dialogue with North American continued, and a compromise was eventually reached. With the manufacturing schedule extended and its withdrawal letter duly retracted, North American Aviation was officially informed that it had won the x-15 contract on 30 September. Design and construction the new rocket plane would take place at the company's Inglewood plant in Los Angeles, but first North American needed to put together a team capable of turning the winning bid into reality.

When Robert Woods outlined the case for a new hypersonic research vehicle during a 1952 meeting of the NACA Committee for Aeronautics, Scott Crossfield listened intently from the sidelines. The research pilot was enthralled by the idea of hypersonic flight; if this was going to happen (and Crossfield saw no reason why that should not be the case), then he intended to be at the very heart of the action.

Crossfield had watched with interest as Drake and Carman devised their two-stage hypersonic aircraft proposal, and while their ideas failed to receive official NACA backing, they had also sparked the imagination of station director Walt Williams. Consequently, when the Becker study reached his desk, Williams was quick to offer the station's assistance. Although the Langley engineers clearly had the vehicle's configuration well in hand, Williams recognized that his staff could offer valuable insights into the operational side of the program. The shrewd Louisianan sought Crossfield's input as he began to consider some fundamental questions: How far would this aircraft fly? Where might it launch from? Where could it land if there were an in-flight emergency? How could it be tracked? Williams asked Crossfield to prepare a preliminary report that the pair would later present at the October 1954 meeting of the NACA Committee for Aerodynamics, and as support for the hypersonic research aircraft solidified during the final months of the year, Scott Crossfield approached Williams with a proposition.

Having seen the Bell x-2 become mired in technical problems that still threatened the program's continuation, Crossfield was not prepared to sit by and watch history repeat itself. If the new rocket plane was to achieve its goals in a reasonable time, he argued, someone from the NACA (preferably a pilot with a solid engineering background) needed to be involved from the outset right through to the flight tests, ensuring things remained on track. When Crossfield had previously suggested a similar course of action to bring the x-2 back on schedule, Williams had denied his request, fearing that such direct interventions might strain the already delicate relationships between program partners. Now the station director rebuffed the suggestion for a second time, leaving Crossfield with a dilemma. As a pilot, he dearly wanted to fly the x-15 to its limits, but as an engineer, he saw a golden opportunity to guide the x-15 from drafting table to demonstration—the very role that his old mentor Charles Leinesch had described as they had watched Eddie Allen test the Boeing Clipper over Puget Sound in 1938. After weighing up his options, Scott Crossfield concluded that he should dedicate his efforts to ensuring that the x-15 fulfilled its potential, and as that meant making sure that the engineering was right, he would need to be on hand every step of the way. Consequently, he set about making discreet inquiries to ensure that wherever the x-15 contract went, he would follow.

North American's sprawling Inglewood plant was located at the southeastern boundary of Los Angeles International Airport. Having forged an excellent reputation through such wartime designs as the B-25 Mitchell and the P-51 Mustang, the company was at the forefront of the high-speed revolution by the mid-1950s. Now North American was building on its more recent successes, the F-86 Sabre and the F-100 Super Sabre, by creating innovative new designs such as the F-107 Ultra Sabre fighter-bomber and a supersonic carrier-borne bomber (soon to become the A-5 Vigilante). By 1955 North American was also heavily involved in the air force's push toward triple-sonic service aircraft, producing design studies for the WS-110A Mach 3 bomber specifications, as well as the Long-Range Interceptor, Experimental (LRI-X).

Although the X-15 promised to break new boundaries, North American's senior management were initially lukewarm in their support, fearing the three-aircraft order might divert engineering talent away from larger military contracts. However, the company's recently promoted chief engineer, Harrison Storms, remained a resolute supporter of the X-15, and though his numerous commitments limited his day-to-day involvement with the rocket plane project, he remained as closely involved as possible, selecting the highly capable Charlie Feltz to manage the aircraft's production design. Having joined the company in 1940, Feltz had previously worked on the F-82, the FJ-2, and the F-86, but as project engineer for the X-15, he would need to find solutions to a daunting range of challenges if he was to meet the first of the aircraft's scheduled delivery dates, on 31 October 1958. Taking up residence in a small space near the plant's busy staff cafeteria, Feltz assembled a dozen-strong team, in many ways a mirror to Becker's group; where the Langley team had proposed general solutions to the problems of hypersonic flight in their design study, Feltz and his colleagues would now wrestle these requirements into a final flightworthy form. Joining the team with the nebulous title of specialist consultant, Scott Crossfield quickly surveyed the challenges that lay ahead and arrived at a less formal but more accurate description of his role: "I would be the X-15's chief son-of-a-bitch. Anyone who wanted Charlie Feltz to capriciously change anything or add anything in the cockpit, or in the whole X-15 for that matter, would first have to fight Crossfield."

The former NACA man was soon called into action when a new air force directive arrived at Inglewood stating that all future high-speed aircraft must feature encapsulated escape systems for their crew. Much of the air force's concern stemmed from a February 1955 incident in which North

American test pilot George Smith had become the first human to eject at supersonic speeds. Having left his malfunctioning F-100 at Mach 1.05, Smith sustained serious injuries that saw him hospitalized for seven months. With double- and even triple-sonic aircraft now on the drawing boards, the air force felt compelled to provide a more effective escape system than the existing ejection seat. For Crossfield, however, escape capsules meant two things—complexity and weight.

Given his firsthand knowledge of both the D-558-II and X-2 escape systems, Crossfield remained skeptical that capsules would improve a pilot's prospects for survival, but he knew that adding one to the X-15 would almost certainly delay development, increase costs, and degrade performance through increased weight. Crossfield argued that the additional protection afforded by a full pressure suit, coupled to the aircraft's inherent strength, offered the pilot a better chance of survival than an as-yet-unproven capsule. To bolster his case, North American produced a study showing that over 90 percent of incidents requiring pilot ejection from the X-15 were likely to occur below Mach 3 and 80,000 ft. To ensure safe escape within these limits, the company designed a unique ejection seat that would safely restrain the pilot, while long telescopic booms extended behind him to ensure stability. Although still unconvinced, the air force allowed North American to proceed with this approach, much to the relief of Crossfield, Feltz, and Storms.

Crossfield also drew on his experience to ensure that the X-15 pilots would be able to control the aircraft as they probed new speeds and altitudes. Although the cockpit would feature a standard center stick, he knew that this would be difficult to use during high-g phases of the flight, such as climb out and reentry. A right-hand side stick (mechanically linked to the center stick) would allow the pilot to operate the aircraft's aerodynamic controls without having to raise an arm against the g-forces. The side stick was pivoted in such a way that it prevented sudden arm movements resulting from abrupt g-force changes (such as at engine shutoff) from being transmitted through the controller—a problem that had been noted on earlier rocket planes using only the standard center stick. A left-hand stick would be used to control the RCS thrusters during flight beyond the atmosphere.

The aircraft's original configuration underwent some significant changes during 1956. The unusual side tunnels had raised concern during the bid-

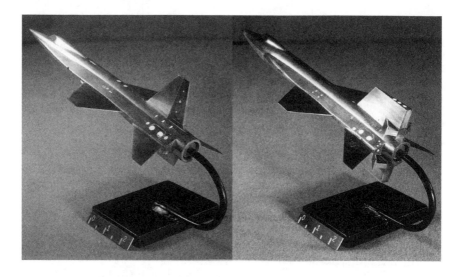

18. Wind tunnel models of the x-15 showing North American
Aviation's original ESO-7487 configuration (left) alongside the final
configuration (right). Note the shortened side tunnels and the changes
to the vertical tail surfaces. Courtesy NASA.

ding process, and further wind tunnel testing now confirmed that they
could contribute to pitch-up under certain conditions. To alleviate this,
they were shortened to start aft of (rather than level with) the cockpit. The
original dorsal and ventral tail fins were also redesigned to increase their
surface area, addressing concerns over the aircraft's high-speed stability. The
new fins were almost equal in size and featured a thick 10° wedge profile
with a blunt trailing edge. The fixed inboard sections each carried a pair
of extendable speed brakes, while the outboard sections were fully moving, acting as large rudders. In order to provide enough ground clearance
for landing, the lower section of the ventral fin would be explosively jettisoned while the x-15 was in the landing pattern, descending under its own
parachute. In North American's 1955 proposal, the extendable main gear
skids had been located beneath the wings, but these were now moved to
the rear of the aircraft where they could share the engine's support structure. In this position, the skids promised far better stability during slide-
out, but this came at a price, as the aircraft would now slam down heavily
onto its front gear after the skids hit the lake bed.

Like most aerospace projects, the x-15's weight increased throughout development. After Hartley Soulé had advised Harrison Storms that if the company was to make any errors during construction, it should make them on the strong side, Storms was determined that the x-15 would not be found wanting on that score. When NACA research suggested the lower sections of the airframe were likely to encounter higher temperatures than originally predicted, the Inconel-X skin in these areas was thickened, while proposed materials for the speed brakes and wing leading edges were changed to more durable, but heavier, alternatives. Dampers were added to the control system after tests in North American's fixed-base simulator suggested stability problems during high-altitude flights. These added yet more weight, but while Charlie Feltz fought to contain the change requests at Inglewood, events on the other side of the continent posed a far greater threat to the x-15's prospects.

Having been awarded the engine contract on 26 October 1955, RMI now faced a development task every bit as daunting as North American's. The selected XLR30 Super Viking power plant was itself a development of the earlier XLR10 engine used on the navy's Viking sounding rocket. Development of the XLR30, rated at 50,000 lbf., for planned missile projects had slowed during 1955, and its selection for the x-15 came with major caveats, suggesting a complex modification program. The final engine would need to be freely throttleable between 30 and 100 percent thrust, while also being restartable—both would be firsts on an engine of this size. More importantly, the contract stated that no single component failure should threaten pilot safety. This man-rating requirement meant that numerous safety systems would need to be developed and tested, with each adding further weight and complexity. The NACA was also concerned that the XLR30's ammonia propellant might lead to ground handling and corrosion issues and possibly even interfere with its research instrumentation. RMI was eventually able to dispel these fears, but only after additional testing and delay.

Contracts for the highly modified engine, designated XLR99 by the air force, were finally signed in September 1956, with RMI optimistically predicting delivery of the first units within thirty months at a cost of $10 million. Development of the XLR99, or Pioneer as it was known within RMI, would turn out to be far more complex than the company expected, leading to long delays and huge cost overruns. Problems with combustion instability forced a lengthy redesign of the engine's injector, while the powerful turbopump also proved temperamental. Two major explosions during testing led to further delays and modifications, and by January 1958 air force patience was beginning to wear thin. The xlr99 was overweight, over budget, and overdue. In an attempt to bring the situation back under control, the wadc asked rmi to provide a revised production schedule, but the contractor's response brought more unwelcome news. Incredibly the xlr99's weight increases and schedule overruns were now accompanied by a new budget estimate of nearly $34.5 million—more than triple the amount quoted less than two years earlier. The air force placed rmi under increased pressure to demonstrate progress or face cancellation, and while the xlr99 situation improved steadily throughout 1958, delivery of working units before mid-1959 now looked unlikely.

Meanwhile, Charlie Feltz's team at North American had unveiled a full-scale engineering mock-up of the x-15 during December 1956. Following two days of detailed inspections by air force and naca representatives, the design was accepted subject to minor changes, allowing Feltz to begin work on the final drawings required for construction. Unfortunately, almost as soon as this process began, news of a major change with far-reaching implications reached Inglewood. North American had always intended to launch the x-15 from the b-36 bomber; however, by January 1957, sac had already begun withdrawing the aircraft from service. Maintenance crews at Edwards had fought a running battle to keep their b-29 and b-50 launch aircraft in serviceable condition as parts had become increasingly scarce; now they foresaw similar problems maintaining what might well become the last flightworthy b-36, as the x-15 program progressed. Meanwhile, the x-15's increased weight and consequent drop in performance led the naca to investigate faster, higher-flying launch aircraft that might reclaim some of the lost ground.

Initially, the hsfs studied three alternatives: Boeing's b-52 Stratofortress and kc-135 Stratotanker and Convair's Mach 2–capable b-58 Hustler bomber. As attractive as the b-58's speed was, the aircraft would require significant structural modification in order to carry the x-15, while the kc-135's low wing and ground clearance also ruled that aircraft out. Although the b-52 was unable to carry the rocket plane in its bomb bay as previous carrier aircraft had, it could carry the x-15 underwing at a point between the fuselage and inboard engine pod. This would, however, require the pilot to enter the x-15 before takeoff, rather than during the climb to the launch point. The rocket plane's tail would also need to be strengthened due to its physical proximity to the B-52's engine pod—sending an unhappy Feltz back to the drawing board. But despite these drawbacks, the B-52 offered considerable altitude and speed advantages over the B-36, and as it had only recently entered service, spares would not be a problem for the foreseeable future. With the decision made, the air force obtained two early production models (003 and 008) from Boeing and delivered them to North American for the required modifications, which saw them redesignated as NB-52s.

Although the x-15's manufacture brought its own problems, with the Inconel-X structure demanding new welding and high-temperature testing techniques, North American made steady progress toward the first aircraft's October 1958 delivery date. Unfortunately, with no engine in sight, it seemed that the x-15 would be going nowhere fast. Following the XLR30's selection in 1955, Walt Williams had suggested that a less powerful interim engine could be fitted, allowing initial flight-testing to take place while RMI perfected its big engine. Although rejected at the time, delays with the XLR99 now led to this idea being reconsidered. Former HSFS engineer Robert Carman, now a member of Feltz's team, calculated that the x-15's existing XLR99 support structure would be able to house two XLR11 motors. This would mean switching to water-alcohol fuel, rather than ammonia, but the tanks would require no modification. Although both Feltz and Storms supported the plan, Scott Crossfield was less convinced, fearing that the x-15 might be stuck with this less powerful substitute arrangement once it was installed. However, even with dual XLR11s, the x-15 still promised Mach 3 performance, allowing North American to proceed with initial demonstration flights in aircraft one and two, while the third x-15 awaited its XLR99. By the time the air force independently arrived at a similar conclusion, Feltz was ready with the necessary engineering fixes, allowing rocket technicians at the AFFTC and HSFS to swing into action, refurbishing XLR11 and LR8 motors from the x-1 and D-558-11 programs for a new life in the x-15.

As production of the three x-15 airframes progressed during 1958, Scott Crossfield turned his attention to the physiological demands of hypersonic flight. Following Marion Carl's 1953 high-altitude flights in the D-558-11, Crossfield had remained in close contact with David Clark. Having developed Carl's pioneering full pressure suit used for those flights, Clark had continued to evolve his ideas using a combination of navy funding and his own resources. For his part, Crossfield became a willing collaborator, testing the improved suits and offering practical suggestions from a pilot's point of view. One of Clark's major breakthroughs was his Link-Net restraining layer, which prevented the suit from ballooning when inflated. This lightweight, flexible material alleviated many of the comfort and mobility problems associated with earlier pressure suit designs, which used bellows or hinges at their joints. In 1955 the David Clark Company received a development contract from the Air Force Aero Medical Laboratory (AFAML) for the MC-2, a full pressure suit that could be worn comfortably for extended periods by the crews of high-altitude aircraft. Having already advised North American to adopt a full pressure suit for their X-15 bid, Crossfield felt sure that Clark's MC-2 could fulfill the requirement. With North American responsible for furnishing the suit under its contract, a subcontract was issued to the David Clark Company in April 1956. Surprisingly, although they were already working with Clark, the AFAML took some convincing that the MC-2 was suitable for the X-15 program. But Crossfield persisted, and by July the air force assumed responsibility for the suit, relieving a grateful North American of one of its many tasks.

Crossfield became a regular visitor to Clark's Worcester, Massachusetts, factory during development of the MC-2. Spying a scrap of aluminized fabric on a workbench during one such visit, Crossfield suggested to Clark that he use the silver material for the MC-2's protective outer layer, in place of the current khaki fabric. Although the aluminized material would offer pilots extra thermal protection, the savvy Crossfield recognized that the image of a silver-suited pioneer venturing into the unknown would likely prove popular with press and public alike. Once a prototype MC-2 became available, Crossfield embarked on a grueling series of tests to prove that a suited pilot could operate under the full range of conditions they were likely to experience while flying the X-15. He endured run after run in the navy's powerful centrifuge at Johnsville, Pennsylvania, where engineers had equipped the gondola with an approximation of the X-15's cockpit, allowing Crossfield to test flight profiles under realistic g-loads. By subjecting himself to this repeated experimental punishment, Scott Crossfield helped establish the physical standards that other X-15 pilots (and later, astronauts) would be required to meet.

19. Neil Armstrong egressing the Johnsville Centrifuge gondola during training for the x-15 program. Courtesy Southeastern Pennsylvania Cold War Historical Society.

During 1956, engineers at North American had constructed a fixed-base x-15 simulator, allowing them to test the x-15's flight responses well before the first metal was cut. Like Dick Day's earlier GEDA setup for the x-2, the x-15 simulator was powered by large analogue computers, but unlike its rudimentary forbear, the pilot was now able to sit in a full cockpit with a rig stretching out behind him to replicate the aircraft's control system. Known as an "iron bird," the simulator became a vital tool for flight planners and pilots alike as the program progressed.

While the x-15 would glide back to Edwards like its rocket plane predecessors, it had a far lower lift-drag ratio (L/D), meaning it would descend very rapidly. In order to test the x-15's landing characteristics, Crossfield devised a technique to produce similar high sink rates in an F-100 by idling the engine, lowering the gear, and releasing the drag chute while in flight. Although the experience proved useful, the F-100 was soon replaced by the F-104, as this could perform low L/D approaches without the risky drag chute deployment.

As his preparations continued, Scott Crossfield began to work with other prospective x-15 pilots, including his North American backup, Al White, and the NACA's Joe Walker, Jack McKay, and Neil Armstrong. Although the memorandum of understanding signed by the NACA, air force, and navy in 1954 had stated, "Upon acceptance of the airplane and its related equipment from the contractor, it will be turned over to the NACA, who shall conduct the flight tests and report the results of same," it contained no explicit mention of any military pilots taking part in either the envelope expansion or research flights. While it was expected that Walt Williams would extend an invitation to the AFFTC to nominate project pilots, Iven Kincheloe was not about to take any chances regarding his seat in the x-15. When General Boyd (now commander at the WADC) visited Edwards to fly the new Convair TF-102 trainer, Kinch seized the opportunity to state his case. Soon after, the x-15 development office at Wright Field confirmed that while the flight test program would be conducted under NACA supervision, the test team would indeed include AFFTC pilots. Kincheloe was duly assigned as project pilot, with Captain Bob White acting as his backup. Although its financial commitment to the program had been modest, the navy was also invited to contribute a pilot and nominated Forrest Petersen, a graduate of the United States Naval Test Pilot School (USNTPS) at Patuxent River, Maryland.

Elsewhere at Edwards, construction of the facilities that would be required to support the x-15 program began in earnest. A new Rocket Engine Test Facility featuring static test stands capable of withstanding the million-horsepower thrust of the XLR99 was built along with a reinforced blockhouse and bunkers to protect the ground crews during test runs. New hydraulic lifts were sunk into the ramp near North American's hangar on Contractors' Row, allowing the x-15 to be raised beneath the wing of its NB-52 carrier aircraft. The base's experimental high-speed track was lengthened from ten thousand to twenty thousand feet in 1957, and the facility's rocket sled was soon pressed into service to qualify the x-15's stabilized ejection seat and investigate how the MC-2 suit would cope with supersonic windblast. Meanwhile, at the HSFS, Building 4800's third floor was extended to accommodate an x-15 control room, one of three sites that would now make up the High Altitude Continuous Tracking Range (more commonly known as the High Range).

Since Crossfield and Williams had first shared their thoughts on operational requirements for Becker's hypersonic design back in 1953, work had begun on a suitable flight corridor for the x-15. Whereas all its rocket-powered predecessors had operated within the vicinity of Rogers Dry Lake, the x-15 would need far greater distances in order to truly stretch its legs. Given that much of a flight was likely to occur outside the atmosphere or at hypersonic velocities, a straight flight path was required, with alternate landing spots needed at both the launch point and at intermediate locations along the route, should engine failure or other emergencies demand an immediate landing. Flights also needed to avoid commercial air routes and highly populated areas, but fortunately, the barren expanses of the American West offered a suitable stretch of rugged terrain running on a southwesterly course from Wendover, near the Bonneville Salt Flats of Utah, all the way back to Edwards, a distance of some 480 miles. This route was peppered with dry lakes large enough to act as alternate landing sites within the x-15's glide range, and with much of the flight path traversing airspace already reserved for military activities, interference with airline routes would be kept to a minimum. New tracking stations were constructed at Ely and Beatty in Nevada, and together with the control room at Edwards, they would provide continuous tracking and communications across the entire range. The three stations were linked to allow for the uninterrupted transfer of data across the network, giving controllers back at the HSFS the ability to monitor telemetry from the x-15 in real time. One of the masterminds behind the High Range was the HSFS's head of instrumentation, Gerald Truszynski—the same engineer responsible for tracking the very first xs-1 flights at Pinecastle Field using a borrowed radar set back in 1946.

Recognizing that the new hypersonic program would mean a significant increase in workload at all levels, the AFFTC and HSFS both began welcoming an influx of new personnel, while also making changes to their operating structures. Paul Bikle was a veteran flight test engineer who had first arrived at Edwards from Wright Field in 1951 as General Boyd relocated air force flight test activities to the desert base. As the AFFTC's technical director, Bikle liked to spend his days circulating around the engineering offices and workshops, keeping a close eye on things and spotting problems before they escalated. Always well informed, Bikle had a knack for identifying practical solutions to complex engineering problems. In the words of Bob Hoey, an AFFTC flight planner during the X-15 program, "He would listen to the briefings from all sides, and then say, 'strikes me that you should do this—,' then put forth a plan that we would each say *why didn't I think of that!*"

In preparation for the X-15 (and its expected round three orbital follow-on), Bikle established a Manned Spacecraft Engineering Office at the AFFTC featuring some of the center's most talented test engineers and flight planners, individuals who would go on to play a crucial role in the success of the center's research programs for years to come. Bikle also worked closely with his HSFS counterpart Walt Williams, helping rebuild the relationship between the centers that had become strained during the troubled X-2 program. For his part, Williams listened to AFFTC concerns regarding potential operational challenges and set about creating the X-15 Flight Research Steering Committee at Edwards to ensure that all parties had an equal voice when it came to the planning and conduct of the flight research phase.

Although the X-15 had already generated a groundswell of interest among the aviation press by late 1957, the launch of Sputnik on 4 October proved to be a watershed moment in America's attitude toward space exploration. The popular press—having initially reacted to the Soviet space spectacular by questioning American educational standards and the country's perceived technological lead over its Cold War foe—now picked up on the new rocket plane, specifically its once-controversial space-leap capability. Suddenly the X-15 was thrust into the national spotlight as a response to Korolev's increasingly complex Sputniks. Through breathless editorials, America at large learned that the dart-like X-15 represented a huge technological leap into the future, capable of doubling the performance of previous rocket planes while carrying the first brave Americans beyond the atmosphere and out into the new frontier of space. This wave of increasingly enthusiastic coverage was often accompanied by pictures of Scott Crossfield in his futuristic-looking silver MC-2 suit—a modern-day knight in shining armor ready to do battle with the unknowns that awaited him beyond the sky.

One particularly lurid article in the *Washington Daily News* explained to its readers the routine awaiting x-15 pilots following high-altitude flights, stating that "the minute [the pilot] lands, he'll be hustled into a specially lined plane and flown to [the] AEC laboratory at Los Alamos N. M. There he'll be rushed into a cell where a machine will examine him from head to toe to see how much radiation he has absorbed." The source for this particular story is unclear, but such urgent flights in "specially lined" planes never materialized, no doubt to the relief of the x-15 pilots. The doubts voiced by North American's management regarding the x-15 had long since evaporated, and now the company fully embraced the new mood, proudly erecting large illuminated letters above the Inglewood plant, declaring it the "Home of x-15."

Finally, on 15 October 1958, sixteen days ahead of the date agreed some three years earlier, x-15-1 rolled out from the Inglewood plant before a large assembly of VIPs and North American employees. Leading the praise, Vice President Richard M. Nixon felt moved to declare, "Americans can proudly say that the United States has recaptured the lead, that we have moved into first place in the race to outer space." (Nixon's first reaction to the sight of the sleek black rocket plane had reportedly been the slightly less stirring: "That's a funny looking thing.") Harrison Storms took the opportunity to extoll the new aircraft's seemingly fantastic capabilities, telling those assembled, "The performance of the x-15 is hard to comprehend: It can outfly the fastest fighters by a factor of three, a high-speed rifle bullet by a factor of two, and easily exceed the world altitude record by many times." The man tasked with proving that the aircraft could actually fly was more succinct; when asked for his feelings, Scott Crossfield simply added, "It's a good simple plane, easy to fly, and I'm looking forward to trying it out." Also present that day were five other pilots who, in time, would fly the x-15. The newly formed NASA was represented by HSFS research pilots Joe Walker, Jack McKay, and Neil Armstrong, while air force captains Robert White and Robert Rushworth represented the AFFTC. Tragically, Iven Kincheloe hadn't lived to see the X-15 completed; Kinch had been killed on 26 July 1958, attempting to eject from his F-104A when it lost power soon after takeoff from Edwards. White would now take his place, sharing the initial government flights with Walker.

Deputy commander of the ARDC, Major General Victor Haugen estimated the air force investment in the X-15 program at the time of the rollout ceremony to be in the region of $120 million, some ten times the estimate provided by the WADC in August 1954. It was time for the X-15 to prove that this was money well spent. The very next day, X-15-1 embarked on one of its slower journeys, following the winding route up through the Los Angeles hills and on toward the High Desert where a new home awaited it.

8. Higher and Faster

As X-15-1 arrived at arrived at Edwards AFB on 17 October 1958, it certainly didn't look like the fastest aircraft ever created. Whereas many of its predecessors had descended on the base tucked beneath their launch aircraft, the first X-15 rolled into the Mojave Desert strapped to a flatbed truck and wrapped in thick protective paper. Reaching North American's hangar situated on Contractors' Row to the north of the base's flight line, the partially disassembled research aircraft was unwrapped like some huge, unwieldy gift and readied for the task ahead.

North American planned to conduct four test phases while waiting for the delayed XLR99 engine. First would come the lightweight captive-flight evaluation, testing many of the X-15's systems while also verifying the flight characteristics of the NB-52 and X-15 combination. This would be followed by the lightweight glide-flight evaluation, when Scott Crossfield would make short glides to test the X-15's low-speed handling and landing characteristics. Once these phases were complete, the heavyweight captive-flight evaluation would test both the NB-52's LO2 top-off system and the X-15's propellant jettison system, with the rocket plane remaining attached to the launch aircraft. The final test would be the powered-flight evaluation—the X-15's first flight using the interim XLR11 motors. North American needed an experienced test director on the ground to keep things running smoothly during Crossfield's demonstration flights, and this responsibility fell to Quinton C. Harvey, more commonly known to his colleagues as QC.

Harvey was a flight test veteran, having first arrived in the High Desert a decade earlier with the McDonnell XF-85 parasite fighter program. After joining Bell Aircraft, he had gained rocket plane experience as an engineer on the ill-fated X-1-3 and the record-breaking X-1A. Following his move to North American in 1953, Harvey earned an excellent reputation through his work as a test engineer on the F-86, F-100, and F-107 programs, making him a natural choice to take on the x-15. But before Harvey could begin his four-phase flight test plan, the x-15 needed to undergo a thorough series of ground tests, and one system in particular would cause the test director no end of headaches during this period.

Whereas conventionally powered aircraft relied on their constantly running engines to provide any power required for flight controls, instrumentation, and radios, rocket planes had generally used heavy batteries to supply auxiliary power during their short, partly unpowered flights. For the x-15, North American dispensed with batteries, instead opting to use auxiliary power units (APUs). These small, hydrogen peroxide–fueled, steam-driven turbine generators were able to power the aircraft's electrical systems from launch until landing, and the x-15 carried two forty-horsepower General Electric APUs in a small compartment behind the instrumentation bay. Unfortunately, as the x-15's ground tests got underway, these APUs began to fail with alarming regularity. Given that a double APU failure during flight would leave the pilot with no means to control the aircraft, this was clearly an issue that needed to be rectified quickly.

Although the APUs operated independently, it was common for both units to fail in rapid succession—a clue that helped North American's engineers to identify that vibrations from the first failed APU were being transmitted to the second unit via a common bulkhead on which they were mounted, hastening its failure. Having remedied this issue, Storms summoned a team from General Electric to thoroughly debug the units until, following much trial and error, the temperamental APUs were largely tamed (although they would continue to cause intermittent problems throughout the program). By early March 1959, x-15-1 was finally ready to take to the air for its first captive test flight.

Unlike previous research aircraft, the x-15 operated in the full glare of publicity throughout its early years, with journalists and photographers jostling for access to every facet of the program. The start of NASA operations on 1 October 1958 only heightened the public's growing fascination with space, and while the new agency soon announced Project Mercury, a capsule-based effort to place an American in orbit, the x-15 still looked likely to be America's first piloted salvo in what journalist Richard Tregaskis dramatically called "the War Against Space." As Scott Crossfield made his way to the Edwards flight line on the morning of 10 March, clad in regulation shirt and tie as if this were just another day at the office, the *Los Angeles Mirror News* was informing its readers that he was "poised today to take man's first step toward space—inside rocket ship x-15, the world's first manned space craft."

Crossfield soon became expert at fielding the question of whether he would become the first man in space. A year earlier he had been one of nine test pilots named in an astronaut availability study for the Man in Space Soonest (MISS) program, an air force forerunner of Mercury (the candidates had all been drawn from the x-15 program, with the exception of Bill Bridgeman). Although MISS went no further, Crossfield had also served as a consultant to the NACA's Stever committee, advising on human factors relating to piloted spaceflight. Through training for the x-15, Crossfield could offer practical insights into some of the challenges awaiting future astronauts, although he later claimed that Brigadier General Don Flickinger and Dr. Randy Lovelace—both colleagues from the NACA human factors panel—had urged him not to apply for the Mercury program, feeling he was "too independent" for NASA's needs. (Interestingly, Crossfield was later identified as an astronaut by Congress, when the House Science and Technology Committee began to issue a series of reports identifying astronauts and cosmonauts during 1969.) Now, as flight-testing of the x-15 grew near, he offered journalists the well-rehearsed line that his job was limited to proving that the x-15 could fly; the maximum-speed and maximum-altitude flights belonged to the government pilots who would follow him.

Back on the Edwards ramp, Crossfield reviewed the x-15's readiness and signed for the aircraft before entering the Sixteenth Physiological Training Flight's van, where he donned his MC-2 suit. Once suitably attired, he was then helped across to the rocket plane suspended beneath NB-52 003's starboard wing, where he ran through preflight checks while air force captains Charlie Bock and Jack Allavie and North American's launch panel operator, Bill Berkowitz, did the same in the launch aircraft. At around 10:00 a.m., the huge eight-engine bomber taxied out to the main runway, pausing briefly as its engines built up to military power, before making its long takeoff run and soaring away into the blue desert sky. Q. C. Harvey, situated in North American's control room, stayed in constant contact with Crossfield, while Storms and Feltz listened on from the company's dark-green communications van out on the lake bed. During the climb to altitude, Crossfield opened the cockpit ram-air door to test his MC-2 suit, which inflated right on cue as he passed through 35,000 ft., leaving the partially immobile pilot straining to close the ram-air door before repressurizing the cockpit with nitrogen.

During its initial test phases, the x-15's nose bore a striped flight-data boom, with readings from its captive flights being used to calibrate the aircraft's instruments before glide flights began. Moving through his test plan, Crossfield first checked the aircraft's controls with the corresponding movements of each surface being confirmed by his backup, Al White, flying chase in an F-100. Next came a mock launch, testing the x-15's ability to power up as it transitioned from the NB-52's power supply to its own internal generators. Crossfield's hopes that the temperamental twin APUs would ace their in-flight debut were quickly dashed when the first unit failed instantly as he attempted to bring it on line. After similar problems with the second unit, disappointment was quickly replaced with concern as smoke began pouring into the cockpit. Although his propellant tanks were empty today, Crossfield feared that even a minor fire might spread to the NB-52's fuel-filled wing, so he wisely called an abort. In the thickening smoke, he performed one final test, releasing the x-15's landing gear as the mated aircraft descended back to Edwards. Once on the ground, technicians confirmed Crossfield's suspicions that the smoke had originated from the now-charred APUs.

Despite hopes that these teething troubles would be quickly rectified, Crossfield's next four flights all suffered APU, damper or radio failures. North American had originally hoped that x-15-1 would make its first glide flight on 1 April, but it was early June before this became a realistic prospect. An attempt on 5 June never left the ground, with Crossfield again reporting smoke in the cockpit as the NB-52 taxied toward the runway. Demoralized by this run of failures, the North American team's spirits began to lag, just as those of the Bell and Douglas crews had under similar circumstances. Feeling personally responsible for the team's morale, Crossfield made a point of suiting up and waiting out preflight delays in the x-15's cockpit, figuring that as long as the team saw that he was ready, they wouldn't let their efforts drop.

On 8 June 1959 things finally came good. As Bock and Allavie brought the NB-52 onto its launch heading high above Rosamond Dry Lake, Crossfield briskly moved through the prelaunch checklist, confirming each item with Harvey back in the tower. The x-15 was finally running on its own power, with both APUs on line. The only glitch so far had been a failed pitch damper, but Crossfield felt confident that he could fly without it. So following a short countdown, the x-15 dropped away from its pylon over Rosamond Dry Lake to make the short glide back to Edwards. With its low L/D, the x-15 didn't offer Crossfield much time to unlock its secrets, but as he put the black rocket plane through a series of control pulses and stall tests, the test pilot reported that it "handled like a champion."

Approaching the lake bed, Crossfield armed and jettisoned the stubby lower ventral fin before lining up with the black runway lines marked along the hard clay surface. Falling fast, with little altitude left to spare and no option to go around, the x-15 suddenly began to pitch wildly. With Harvey, Feltz, and Storms looking on powerlessly, Crossfield fought the porpoising aircraft, finally bringing it down safely at the bottom of an oscillation. As the x-15 slid across the lake bed, trailing rooster tails of dust, Crossfield's relief gave way to a creeping fear that despite years of research and testing, he and North American had somehow managed to build an unstable aircraft.

Harrison Storms reached a different conclusion, however. The chief engineer was convinced that the problem lay in a power boosting system on the x-15's side-stick controller (Crossfield's preferred means of flying the aircraft). During development, engineers had used their best estimates to judge how much power assistance the small movements of the side stick would require and how quickly these forces should be applied. Although the system had worked well in the simulator, Storms believed that it was now reacting too slowly to Crossfield's inputs, meaning the pilot and controls had become out of sync, with every new input worsening, rather than improving, the situation. When postflight analysis validated this hypothesis, the problem was solved via a simple valve adjustment.

With the first glide flight out of the way, ship one was withdrawn for maintenance and engine tests prior to beginning its powered flights. Attention now turned to x-15-2, which, having arrived at Edwards during April, had already undergone static tests with the dual XLR11 engines. It would now fall to this aircraft to perform the next phases of testing—the heavyweight captive and powered flights.

The decision to use dual XLR11 engines to power the X-15 had appeared to be a relatively simple proposition on paper, but the reality proved to be somewhat different. Many of the XLR11 and LR8 motors had been in storage since the end of the X-1 or D-558-11 programs, while some technicians who worked on these programs had since moved on or been transferred to other duties. As a result, progress was slow as a new band of technicians discovered the multichambered engine's idiosyncrasies for themselves.

The X-15's propellant system also caused its share of problems, with sensitive valves failing on a regular basis. Consequently, it took months of trial and error before the Rocket Engine Test Facility finally shook to the combined 11,800 lbf. thrust of the dual XLR11s. But just as things seemed to be getting on track, disaster struck. Following every engine run, nitrogen gas was forced through the aircraft's hydrogen peroxide lines to remove any traces of the highly volatile liquid. Unfortunately, on one occasion, a new hose was used for this procedure, and unbeknownst to the ground crew, it still carried a film of oily residue from the manufacturing process on its inner surface. As nitrogen gas raced through the hose, small amounts of oil came into contact with the highly reactive peroxide, leading to an explosion. Although the resulting fire was quickly extinguished, damage to X-15-2's engine bay meant a trip back down to Inglewood for repairs.

When the aircraft returned to Edwards, it headed straight back to the test stand before finally taking to the air on 24 July for the first heavyweight captive flight. Having reached launch altitude, the team began with a test of the NB-52's LO2 top-off system. By using sensors buried deep within the rocket plane's oxidizer tank, a supply of LO2 in the NB-52's bomb bay could replace any oxidizer that had boiled off during the flight to the launch point, via pipes running down through the launch pylon. However, during its first in-flight test, the system stubbornly refused to work despite launch panel operator Berkowitz's best efforts. Moving on, Crossfield pressurized the X-15's tanks for a full launch rehearsal before jettisoning their contents. Excepting the uncooperative top-off system, the test had been a complete success, allowing Q. C. Harvey to declare that the X-15 was now ready to make its inaugural powered flight.

As preparations continued at the North American hangar, changes were afoot elsewhere on base. With Project Mercury beginning to gather pace back at Langley, Walt Williams was tempted away from the High Desert to

20. Paul Bikle, who replaced Walt Williams as director of the NASA High-Speed Flight Station in September 1959. Courtesy NASA.

become associate director of NASA's newly formed Space Task Group. Having arrived as head of the embryonic NACA Muroc Flight Test Unit in late 1946, Williams had overseen the growth of the now three-hundred-strong research facility about to embark on its most ambitious flight research project to date. NASA's nascent manned spaceflight program would now benefit from his wealth of flight test experience, and he would be joined in this new venture by some of the station's most experienced managers, including influential head of instrumentation Gerald Truszynski.

Williams did not look far when it came to recommending a suitable successor. The outgoing director believed that his AFFTC counterpart, Paul Bikle, was the right man for the job, and so on 15 September 1959 Bikle was officially named as the station's second director. Like Williams, Bikle was a resolutely practical manager who believed in working with the minimum of fuss or paperwork. Admired by air force and NASA colleagues alike, Bikle's affiliations would help him to strengthen the relationship between the AFFTC and the HSFS which was so vital to the X-15 program.

Having operated largely autonomously from Langley for some time, the organizational ties were finally cut on 27 September as NASA Headquarters announced that the HSFS would now receive full center status, becoming the NASA Flight Research Center (FRC).

Although North American had hoped to hand over X-15-1 in August 1959, the various problems that Harvey and his test team had encountered pushed that date back, placing pressure on Harrison Storms to get the X-15 flying under power and into the customer's hands as soon as possible. Never one to be intimidated, Storms stood his ground, insisting that haste at this stage might prove disastrous in the longer term. North American were moving as fast as circumstances allowed, and the X-15 would be ready when it was ready.

The first powered flight finally came on 17 September. Flying X-15-2, Crossfield dropped away from the NB-52 over Rosamond Dry Lake and rapidly flicked the eight ignition switches for his dual XLR11 engines. Tracing a rectangular course around the Edwards area, Crossfield reached a maximum speed of Mach 2.1 at 52,000 ft., testing the aircraft's stability and control as he went; the X-15 had exceeded his hard-won 1953 Mach 2 mark on its powered debut and had barely broken a sweat doing it. To celebrate, Crossfield allowed himself an unauthorized aileron roll before returning to the serious business of getting the aircraft back on the ground. As he entered the landing pattern with Bob White's F-104 in close attendance, Crossfield jettisoned the ventral fin as the X-15 sank steadily toward the lake bed, without any of the wild pitching he had experienced during the first glide flight. A second powered checkout flight followed on 17 October, following two aborts due to continuing problems with the LO2 top-off system. Despite a failed damper that made the aircraft extremely sensitive to roll inputs, the otherwise-successful flight was marred by a fire in X-15-2's engine bay after landing. Although the blaze was quickly extinguished, the aircraft had sustained significant damage and remained grounded while repairs were carried out.

Two more aborts followed before the third powered flight on 5 November, but within moments of launch Crossfield was facing the first major in-flight emergency of the program. Upon igniting the two XLR11s, he felt a muffled explosion emanating from the rear of the aircraft, and a fire warning light quickly confirmed that he had a problem. Within seconds,

21. X-15-2 sits broken backed on Rosamond Dry Lake following Scott Crossfield's emergency landing on 5 November 1959. Courtesy NASA.

chase pilot Bob White reported that a chamber in the lower engine had exploded, rocking his F-104 and causing a fire in the rocket plane's engine bay. Instinctively, Crossfield cut the remaining chambers and began to jettison his heavy propellant load in readiness for an emergency landing on Rosamond Dry Lake. Unfortunately, to reach this alternate location, he had to make a steeper-than-normal approach, slowing the speed at which the ammonia and LO2 could be expelled from the rocket plane's tanks.

The aircraft was still carrying a significant amount of propellants as Crossfield touched down, causing the X-15's nosewheel to slam down onto the hardened clay with even greater force than usual. As he slid across the dry lake, Crossfield decelerated rapidly. The reason for this became abundantly clear once the dust settled; as the overweight X-15 had slammed forward, its fuselage failed near the LO2 tank's forward bulkhead, leaving the fractured body to gouge into the lake bed like an oversized brake. Once again, X-15-2 was loaded up for the journey to Inglewood.

The X-15's nose gear was stored in a compressed state during flight to save space. Investigations into the landing accident revealed that the gear had extended so rapidly on release that oil in its cylinder had foamed. In this state, it had been unable to absorb the shock of the heavy landing, causing the fuselage to fail at a previously undetected weak spot. North American modified the gear strut to prevent any reoccurrence and strengthened all three aircraft at the point where x-15-2's fuselage had failed, but some wondered whether Crossfield's piloting technique might also have played a part in the failure. Based on his experiences in the x-15 fixed-base simulator, AFFTC engineer Bob Hoey had previously raised concerns that Crossfield's recommended 276 mph approach speed would produce a high sink rate, potentially placing excessive loads on the aircraft as its nose gear slammed down. After calculating that a higher approach speed would reduce these loads, Hoey learned that FRC engineers had reached a similar conclusion. Following the 5 November incident, Hoey's recommendation to increase the approach speed to 345 mph was implemented and used for the rest of the program.

Given the extreme speeds and altitudes at which the x-15 would fly, some novel systems were needed to provide flight data. Wisely, North American chose to stagger the introduction of these systems during the demonstration phase, allowing each one to be thoroughly tested before the team progressed. The new decade's first flight took place on 23 January 1960 and marked the debut of the Sperry-manufactured stable platform—an early inertial guidance system designed to provide data on the aircraft's attitude, velocity, and altitude under flight conditions where standard instrumentation would be rendered ineffective. The platform was aligned using data from the NB-52 prior to launch and would be used until the final stages of a flight when the pilot would revert to standard flight-data instruments. Although not required at the modest speeds and altitudes of the demonstration phase, North American was eager to fly the stable platform in order to gain experience and demonstrate that it could function as required.

Fortunately, all of x-15-1's systems performed well during the flight as Crossfield reached Mach 2.53 at 66,000 ft., and following this textbook performance, the air force and NASA agreed that x-15-1 had met its acceptance criteria. On 3 February the aircraft was officially signed into air force ownership, before being immediately transferred to NASA on long-term loan. Having delivered their first x-15, Q. C. Harvey and the North American team now turned their attention to the recently repaired x-15-2, as well as ground tests of the long-awaited XLR99 big engine in x-15-3.

RMI had delivered its first XLR99 to Edwards in June 1959, the same month that the final x-15 had arrived from Inglewood. Although engine test-ing began that August, it would be June 1960 before an XLR99 was finally installed in x-15-3 and the entire propulsion system could be fired on the test stand. On 8 June, Scott Crossfield climbed into the aircraft to make the XLR99's third ground run, following two earlier brief but largely successful tests. Having no means to operate the engine remotely, these tests required a pilot to be in the cockpit, but as the aircraft would stay firmly anchored to the test stand, Crossfield preferred to wear regular clothes rather than his bulky MC-2 suit. With the ground crew having retreated to the reinforced concrete blockhouse or other protected viewing points, Crossfield powered up the aircraft and primed its million-horsepower engine before advancing the throttle to 50 percent thrust for a few seconds. With the thrust steady, he advanced the XLR99's throttle to the 100 percent mark, unleashing the engine's full 57,000 lbf. of screaming rocket power.

With the first section of the test successfully completed, he brought the throttle back, shutting the engine off. Crossfield would now demonstrate the XLR99's in-flight restart capability. To do this, he brought the throt-tle out of idle, advancing it back up to the 50 percent mark, but at this point, the engine unexpectedly quit. Under these circumstances, the pilot was advised to bring the throttle back to idle before resetting the engine's safety devices, allowing a restart to be attempted. Following this proce-dure, Crossfield reached forward to reset the engine, noting later that "the reset button was depressed at which time the aircraft blew up."

A violent explosion tore through the x-15, throwing its forward fuselage and wings some thirty feet down the test stand, while subjecting Cross-field to an acceleration he described as "beyond the experience of this pilot." After calmly shutting the aircraft's systems down and switching to an emergency oxygen supply, Crossfield was soon rescued by the advanc-ing fire crew. Through his earlier insistence that the x-15's cockpit should be strong enough to keep the pilot safe under extreme conditions, he had unwittingly ensured his own survival on the test stand; unfortunately, x-15-3 had not fared so well.

After its XLR99 had been removed and taken to the AFFTC rocket shop, the twisted remains of x-15-3 were loaded up and returned to Inglewood, where a full investigation and damage assessment was carried out. The sub-sequent inquiry revealed that the problem had originated in the aircraft's ammonia tank, after a faulty regulator combined with back pressure from the test stand's ammonia vapor disposal system, subjecting it to excessive pressure. The tank's rear bulkhead had then ruptured the aircraft's peroxide tank, causing the explosion. The good news was that the incident had not been caused by design flaws in either the x-15 or the xlr99. Unfortunately, for the already massively over-budget program, x-15-3 now needed to be rebuilt using spares. As the air force considered the shattered aircraft's fate, xlr99 testing moved to aircraft two.

Following its backbreaking emergency landing, the repaired x-15-2 was now back in the air, performing stability and control tests across a range of flight conditions. In the weeks prior to his explosive test stand experience, Crossfield had begun to test the aircraft's ballistic controls—its system of peroxide-fueled rcs jets. After being withdrawn to have its big engine installed, x-15-2 was aloft again on 13 October 1960, when a succession of technical problems caused an abort. A second launch attempt on 4 November suffered a similar fate, so it was a case of third-time lucky as everything came together perfectly during the next attempt on 15 November. Launching from the nb-52 at 46,000 ft., Scott Crossfield advanced the xlr99's throttle to 50 percent and streaked away from his chase aircraft, achieving a maximum speed of Mach 2.97 at 81,297 ft. The terms of North American's contract had originally prohibited Crossfield from exceeding Mach 2. Even using the less powerful xlr11s that had proven impractical in the x-15. With the xlr99, it was all the pilot could do to keep the aircraft below Mach 3.

On 22 November he successfully demonstrated the engine's throttling and restart capabilities, but postflight inspections revealed that the thrust chamber's ceramic Rokide coating had begun to degrade, prompting concern among afftc and frc engineers. After North American and rmi judged the damage to be within acceptable limits, Crossfield elected to push on with the third and final xlr99 demonstration flight on 6 December 1960. Launching at a point just to the west of Lancaster, Crossfield set to work performing multiple engine restarts and testing the aircraft's ballistic controls before touching down on Rogers Dry Lake eight minutes later, bringing his fourteenth and final flight in the x-15 to a close.

In the decade that had passed since John Griffith met Scott Crossfield at Mojave station, the ever-enthusiastic Crossfield had become a legend in flight test circles. Although he had occasionally gained notoriety for the wrong reasons, few could deny Crossfield's numerous contributions to high-speed flight research. During five years with the NACA, he had made eighty-seven rocket plane flights; the majority of these (including his November 1953 Mach 2 flight) were in the D-558-II. By leaving the NACA for North American, Crossfield had followed his engineering instincts, but in doing so, he had also given up the chance to take the X-15 to its ultimate speed and altitude marks. Publicly, he never expressed regret for this decision, declaring, "I got just about all out of it any man could get out of a single program—and I can't ask for it all you know—I got all eight yards, didn't get to nine yards that's all."

Privately, though, Crossfield had hoped to keep his place in the X-15, approaching Paul Bikle to seek a return to the FRC as prime X-15 pilot. Unfortunately for Crossfield, Walt Williams and Joe Vensel had already assigned pilots to the program, and the new director was not about to overturn those decisions. Besides, Crossfield's headstrong manner and proprietorial attitude toward the X-15 had done little to endear him to the FRC and AFFTC pilots, meaning there would have been little support for his return, especially from chief pilot Joe Walker.

With the door to the X-15 firmly closed, Scott Crossfield stayed with Harrison Storms, as the chief engineer revitalized North American's Missile Division in Downey with key Apollo contracts. Although he broke no records in the X-15, Crossfield did receive one unusual award, when the Southern California Soaring Society presented him with a mahogany-mounted brick in recognition of the quickest gliding descent from 38,000 ft. to landing.

As Q. C. Harvey and the North American test team wrapped up their contractor demonstration phase at Edwards, they left behind a solid and resilient aircraft that, true to Soulé's request, had proven itself to be "on the strong side." Perhaps Scott Crossfield's true legacy was the faith that successive X-15 pilots showed in the aircraft's ability to bring them home safely from the farthest corners of the flight envelope—a true pilot's aircraft.

With X-15-1 now available, the FRC was eager to begin pushing the new research plane toward its design goals of Mach 6 and 250,000 ft., with piloting duties for this envelope-expansion phase to be split equally between NASA's Joe Walker and the AFFTC's Captain Robert White. While Walker brought many years of rocket-powered flight experience to the role, White was a rocket plane rookie, having only assumed the prime spot after his friend Iven Kincheloe's tragic death. Hailing from New York City, White joined the USAAF in 1942, seeing active service in Europe during the later years of the war. His tangles with the Luftwaffe were curtailed when his P-51 was downed in February 1945, leaving the young airman as a prisoner of war until hostilities ended. Once back in the United States, he enrolled at New York University, earning a degree in electrical engineering before returning to active duty and flying F-80s in Korea. White attended the USAF TPS at Edwards in 1954, graduating top of his class before joining the AFFTC's Fighter Ops Division, where he served as the project pilot for the Republic F-105B's phase two evaluations. Originally named as Kinch's backup for the X-15, White was well prepared when he inherited the lead role in 1958, having made regular visits to Inglewood to study the X-15's systems and log simulator time during the aircraft's development. In the wake of Kincheloe's untimely death, White also took ownership of the legendary Model A Ford—a mark of his elevation to the chief test pilot role. (White was the last of the rocket research aircraft pilots to own the vintage automobile, although it was later driven by future Gemini and Apollo astronaut Michael Collins during his time at Edwards.)

Both White and Walker were excellent pilots and highly regarded by their peers, but while Walker's huge grin and Pennsylvania drawl made him instantly popular with the X-15 press corps, White cut a more reserved figure. With his clipped and accurate statements, White remained every inch the air force test pilot—a point further reinforced when he politely declined a desk in the FRC pilots' office, preferring to remain among his fellow servicemen at the AFFTC. As the new decade dawned, both men logged countless simulator hours (assisted by flight planners Bob Hoey and his NASA counterpart Dick Day) and made numerous low L/D landings in F-104s, in preparation for their upcoming flights. But one question remained—Who would fly first?

The original flight schedule listed Walker for the first flight, but in October 1959 the AFFTC's new commander, General John W. Carpenter III, raised the subject with Paul Bikle. Given that the air force was shouldering the majority of the program's considerable financial burden, Carpenter felt that White should fly first, allowing the service to reap the benefits of any

22. The dual-XLR11-powered X-15-1 falling away from its NB-52 launch aircraft during an early flight. Courtesy NASA.

associated publicity. Perhaps Carpenter expected the former AFFTC technical director to acquiesce, but Bikle stood his ground, pointing out that the X-15 concept had originated with the NACA, making it entirely appropriate that Walker should go first. It is not known whether Carpenter took the matter further, but when the X-15 made its first government flight on 25 March 1960, Joe Walker was in the pilot's seat.

For his first flight, Walker would stay well within the X-15's limits, giving him a chance to become acquainted with the aircraft before moving on to the more demanding envelope-expansion flights. As the NB-52 (piloted by Fitz Fulton and Jack Allavie, with FRC veteran Jack Russell as launch panel operator) made its early afternoon takeoff, Walker's chase pilots took up their positions. Today he was accompanied by Crossfield, White, and fellow FRC research pilot Jack McKay, and following some problems with the X-15's stable platform and concerns about strengthening winds, Walker dropped away and set about lighting the dual XLR11s. Unfortunately, he soon had his hands full, as the upper motor required two restarts before he was able to reach Mach 2 at just less than 50,000 ft. Entering the final leg of his circular landing pattern, a sizable tailwind carried Walker way beyond the smoke canisters that marked his planned touchdown point, much to the chagrin of the waiting press photographers. At the customary postflight press conference, the beaming pilot declared that the x-15 "handled pretty good."

White did not have to wait long for his turn, repeating Walker's flight plan on 13 April. As a matter of service pride, the air force pilot had no intention of landing long, as Walker had, and made a small wager with Dick Day that he could bring the x-15 down within one thousand feet of the smoke canisters. Although he forgot to activate the research instrumentation until after engine burnout, White made an otherwise textbook flight and a pinpoint landing (earning two martinis at Day's expense). With both pilots now checked out, the business of envelope expansion could begin.

The x-15 was heavily instrumented, carrying a far more comprehensive payload than its predecessors. During construction, North American had installed 656 thermocouples, 112 strain gauges, and 140 pressure sensors in the fabric of the aircraft (mainly on the right side of the airframe). Although the x-15 made greater use of telemetry to transmit real-time data than its predecessors, oscillographs were still used to record data on filmstrips for postflight analysis. Whereas data reduction had previously been an entirely manual task for Roxanah Yancey's "human computers," the advent of, first, analogue and, later, digital computers gradually brought some automation to the role. Rather than being computers themselves, some new recruits were versed in the emerging discipline of computer programming, and while data input remained a time-consuming task, it did represent something of an improvement over the noisy Friden calculators that were gradually phased out during the x-15 years. One thing that had not changed, however, was the belief that women were better suited to these tasks, demonstrating greater patience and attention to detail. Beyond the computer department, new opportunities were beginning to appear for female employees at the FRC. Both Yancey and Betty Scott Love moved to engineering positions, and more female author and coauthor credits began to appear on the center's research papers.

In the air, Walker and White alternated flights as they incrementally pushed the interim XLR11 motors toward their performance limits. Joe Walker took the x-15 to Mach 3 for the first time on 12 May, reaching Mach 3.19 at 77,882 ft. in a flight that also marked the first use of a remote launch lake, with the drop taking place over Silver Lake, nearly one hundred miles east of Edwards. The use of remote launch lakes allowed the x-15 to maintain a straight heading for Rogers Dry Lake as it reached greater speeds and altitudes, but it required major logistical support from both the AFFTC and FRC. Given the greater distances involved, weather became increasingly important, as good conditions were required not only at the launch lake and Edwards but also at any intermediate lakes that could serve as emergency landing sites. Before each flight, emergency vehicles were positioned at these locations, meaning an early morning c-130 flight for an Edwards fire truck. Depending on the distance a flight would cover, the c-130 might orbit along the flight path with a second fire truck, ready to descend on an intermediate landing lake in case of an emergency landing. Rescue helicopters were also on hand at the lake beds to provide immediate assistance should the x-15 pilot be injured during landing.

Although most launches would now take place beyond the boundaries of Edwards AFB, the pilot would always be in direct contact with the ground. Where Crossfield had relied on Q. C. Harvey for guidance during his flights, government x-15 pilots would have the reassuring voice of NASA-1 to support them. This role originated as a means to provide the pilot with a single point of contact readily able to understand and convey information during the fast-paced flights. It was decided that NASA-1 should always be another x-15 pilot, ensuring familiarity with the aircraft's systems and the ability to speak the same language as the man in the cockpit. Although the control room was always crowded with engineers monitoring various events during a flight, the x-15 pilot could always request and receive information via NASA-1, rather than having to contend with multiple voices, adding to his already formidable workload. On Walker's 12 May flight, Neil Armstrong acted as NASA-1, relaying information on the x-15's ground track to Walker, along with updates on the closest emergency during every phase of the flight.

One week after Walker's Mach 3 flight, a second launch from Silver Lake saw Bob White reach 100,000 ft. during the program's first altitude-buildup flight. Although the dual XLR11 engines produced less than a quarter of the XLR99's thrust, they still allowed the x-15 to break the x-2's unofficial speed and altitude records set by Apt and Kincheloe, respectively. On 4 August 1960 Joe Walker launched over Silver Lake, running all eight XLRII chambers for 260 seconds to reach a maximum speed of Mach 3.31 at a little over 78,000 ft., besting Apt's mark. Replying to a postflight question regarding the sensation of speed during his flight, Walker explained, "You're so busy in there the sense of speed is when you watch the Mach meter climb to another number." Among the congratulatory messages was a telegram from Pete Everest, asking, "When are you publishing your book entitled 'Fastest Man Alive'?" (the title of Everest's memoir published in 1958, following his flights in the x-2). In the wake of Walker's flight, the *Los Angeles Herald-Express* informed its readers that "Rocket Ship x-15 flies 2,150 mph— Warm-up hop for space assault." Indeed, a new unofficial altitude record followed just eight days later, when White reached 136,500 ft., easily surpassing Kincheloe's September 1956 record of 126,200 ft. Although x-15-1's ballistic controls were not yet operable, the aircraft's wedge-shaped vertical surfaces proved effective, sparing White the directional stability problems that Kinch had encountered on his flight. With Walker and White having taken the x-15 about as fast and as high as it could go while equipped with XLRIIs, further envelope expansion would have to wait until the XLR99 became available. In the meantime, four more pilots were introduced to the aircraft ahead of the upcoming research program.

On 23 September, Lieutenant Commander Forrest Petersen became the first (and only) serving navy pilot to fly the x-15. Although Petersen's inclusion came as a result of the navy's relatively modest financial contributions to the program, he was no mere token during his time in the High Desert. In spite of limited flight opportunities, Petersen threw himself into training and support activities, becoming a popular member of the team. Having graduated from the USNTPS alongside future NASA astronaut Alan Shepard, Petersen had reported for duty at Edwards in August 1958, and now, just over two years later, his opportunity to fly the x-15 had finally arrived. Like Walker and White before him, he would begin with a low and slow familiarization flight to a maximum Mach 2 at 50,000 ft. However, having launched near Palmdale and accelerated on a northeasterly heading toward the town of Boron, the x-15's engines quit prematurely, and following two unsuccessful restart attempts, the naval aviator had no option but to use what energy remained to get back to Edwards. After one of the lowest approaches of the program, Petersen brought the x-15 in for a smooth landing, wryly noting that nothing had surprised him during his debut flight, "with the exception of early engine shutdown." Fortunately, things went more smoothly when he repeated the flight on 20 October, reaching Mach 1.94 at 53,800 ft.

Next up was FRC research pilot Jack McKay. By the time he joined the x-15 program, McKay had made forty-four flights in the D-558-II, x-1B, and x-1E, ranking him second only to Scott Crossfield in terms of rocket plane experience. His checkout on 28 October proceeded without a hitch, save for the ventral fin's parachute failing after it was jettisoned during the landing pattern. The flight was in many ways the epitome of McKay himself— efficiently achieving its goals with the minimum of fuss.

One week later, Robert Rushworth became the second air force pilot to fly the x-15. Although he would fly the same low and slow profile, the Mach 2 flight would still be a far cry from Rushworth's wartime experiences flying c-47 transports over the Hump between India and China. Having studied mechanical engineering at the University of Maine after the war, Rushworth later flew F-80 Shooting Stars for the Air National Guard as war raged in Korea. After earning his second degree, this time in aeronautical engineering from the Air Force Institute of Technology at Wright-Patterson AFB, he headed for Edwards to attend the USAF TPS, following White's path to the Fighter Ops Division after graduation. As White advanced to the prime spot in the x-15 following Kincheloe's death, Rushworth assumed the backup position, a decision that apparently did not please White, who still regarded his colleague as a multiengine cargo pilot. Whatever misgivings White may have initially harbored, Rushworth proved himself equal to the task of flying the x-15 on 4 November, reaching Mach 1.95 at 48,900 ft. in ship one.

One more pilot checked out in the x-15 before the end of 1960, and like McKay and Walker, he also brought rocket plane experience to the role. Neil Armstrong was still a relatively junior pilot at the FRC, but his earlier RCS research flights in the x-1B and NF-104, combined with his recent selection as a NASA consultant to the air force Dyna-Soar program, earned him a place in the x-15. Armstrong made his debut in x-15-1 on 30 November, but the flight was marred by engine problems, restricting his maximum speed to Mach 1.75 at 48,840 ft. Armstrong would make one more

23. The NASA Flight Research Center pilots pose in front of the Bell x-1E situated outside Building 4800 in 1962. Left to right: Neil Armstrong, Joe Walker, Bill Dana, Bruce Peterson, Jack McKay, Milt Thompson, and Stan Butchart. Courtesy NASA.

flight before the year's end, this time performing the first in-flight test of a vital new piece of x-15 hardware—the Q-ball nose.

During the program's first thirty flights, x-15-1 and x-15-2 had carried nose booms fitted with standard sensor vanes to determine their angle of attack and sideslip relative to the airstream. While similar systems had been used on all previous rocket research aircraft, the x-15 would soon be encountering temperatures high enough melt these delicate instruments, as well as dynamic pressures so low that there would be no airstream for them to detect. With this in mind, the NACA had developed a novel air-data nose sensor consisting of an Inconel-X sphere with two pairs of pressure orifices drilled into it, one pair in the vertical plane to detect angle of attack, the other pair in the horizontal plane to detect sideslip (a fifth, central orifice was used to provide a total source pressure). During flight, a set of servos could move the ball both vertically and horizontally until the dynamic pressure detected by all the orifices was equal, meaning the ball was facing directly into the airstream. These relative vertical and horizontal deflections from center were then displayed to the pilot as his angle of attack and sideslip, respectively. The Northrop Aircraft Corporation's Nortronics Division handled the detailed design and manufacture of the Q-ball, but the need to test the nose sensor at temperatures approaching 2,000°F led to a classic example of FRC improvisation. The nose was simply placed behind one of the center's F-100s, with the searing heat of the J-57's afterburner subjecting the now-glowing sensor to the required temperature. On 9 December, Armstrong dropped from the NB-52 near Palmdale to repeat the same basic profile he had flown nine days earlier, allowing data from the Q-ball to be compared with previously recorded nose-boom readings. The demonstration was regarded as a success, clearing the way for all subsequent X-15 flights to use the Q-ball nose.

As 1960 drew to a close, six pilots had checked out in the X-15, and the XLR99 was now qualified for flight. The program would now adopt a two-pronged approach. Although aerothermal data was recorded on all flights, detailed research programs designed to probe specific flight regimes could now begin. These would predominantly be flown by Petersen, McKay, Rushworth, and Armstrong, while Walker and White would continue their envelope-expansion flights, pushing the X-15 toward Mach 6 and the lower reaches of space.

Throughout 1961 the X-15 began to show what it was truly capable of, as records fell on a regular basis. On 7 February, Bob White made the program's final XLR11 powered flight in X-15-1, reaching Mach 3.5 at just over 78,000 ft., a record speed for the interim engines. Following two aborted attempts, he also made the first government flight in the XLR99-equipped X-15-2 on 7 March, using the engine's 57,000 lbf. thrust to become the first pilot through Mach 4, reaching Mach 4.43 at 77,450 ft. As White investigated the X-15's high-speed stability, the 500°F temperature differential between the aircraft's fuselage and side tunnels caused some of the Inconel-X tunnel panels to buckle, a problem that would return on later flights.

Following an aborted attempt on 21 March, Joe Walker made his first flight of 1961 on 30 March, marking the XLR99's first use on an altitude-buildup flight, with a planned high point of 150,000 ft. Walker would also be testing the David Clark Company's improved A/P22S-2 full pressure suit, which addressed many of the MC-2's shortcomings. Where the earlier suit had used a two-piece pressure garment joined at the waist, the A/P22S-2's single inner garment featured a pressure-sealing zipper, making donning and doffing the suit far quicker and simpler. The A/P22S-2 helmet was also a great improvement on its predecessor, offering a far wider field of view. The new suit also featured detachable gloves and improved biomedical-monitoring connectors, improving the pilot's general comfort.

To reach his planned altitude, Walker was launched from Hidden Hills Lake, near Death Valley on the California-Nevada border. As x-15-2 fell away from the NB-52, Walker's first attempt to light the XLR99 failed, but with the heavily laden rocket plane falling fast, he coolly made a second successful attempt and brought the aircraft back on profile for the long climb. Walker shut the XLR99 down after eighty-two seconds, experiencing around two minutes of weightlessness as the sleek black rocket plane soared upward, eventually peaking at 169,600 ft. Although x-15 flights left little time for sightseeing, he was able to appreciate the unique view of the southwestern United States, commenting, "You have no doubt from external visual cues that you're really high up." Walker took the opportunity to test his ballistic controls in the low dynamic pressure while at altitude. But time outside the atmosphere was brief, and he was soon lining up for reentry, increasing the x-15's angle of attack so that its underbelly would bear the highest temperatures during the plunge down through the atmosphere. Descending through 100,000 ft., the aircraft began to vibrate heavily, an experience Walker likened to "slamming over a corduroy road." The shaking lasted for around fifty seconds, leading the NASA pilot to comment, "It didn't shake my teeth loose, but I was mighty glad I had permanent bridges." Postflight analysis revealed that the vibrations had been caused by the aircraft's stability augmentation system, and changes were made to prevent similar problems in the future.

Although Walker had reached a peak altitude of thirty-two miles, his achievement went largely unheralded. Press and public alike were now transfixed by events on the other side of the continent, as NASA prepared to send America's first astronaut into space aboard a Mercury capsule. To many involved in the x-15 program, the upcoming suborbital Mercury flights atop a Redstone missile lacked the sophistication of the rocket plane's planned space leaps. Indeed, when Wernher von Braun had first proposed similar Redstone-boosted suborbital hops on behalf of the Army Ballistic Missile Agency in 1958, the NACA director at the time, Hugh Dryden, had disparagingly described the plan as having "about the same technical value as the circus stunt of shooting a young lady from a cannon." NASA's Mercury astronauts would have no control of their spacecraft during the boost and reentry phases, and they would need to be fished from the ocean by a flotilla of rescue vessels, following their return. By contrast, the X-15 pilots would fly their vehicle (complete with throttleable engine) into space, before manually performing the reentry and recovery to a pinpoint runway landing. But by 1961 the American public, desperate for space parity following a string of Soviet spectaculars, had elevated the Mercury astronauts to an exalted status far beyond the test pilots toiling in the Mojave Desert.

Unfortunately for NASA, things got worse before they got better. On 12 April 1961 the Soviet Union announced that Yuri Gagarin had become the first human in space, completing a single orbit of the earth in his Vostok spacecraft. A disappointed Alan Shepard finally made his much-delayed flight on 5 May, but the fifteen-minute suborbital lob paled in comparison to Gagarin's orbital adventure. America's dynamic new president, John F. Kennedy, recognized the need to show that the United States could not only equal but surpass its Cold War rival. On 25 May he rose before Congress to issue a dramatic new challenge: "I believe that this nation should commit itself to achieving the goal, before this decade is out, of landing a man on the moon and returning him safely to the earth."

When Hartley Soulé and John Becker had first raised the possibility of brief forays into space in a hypersonic research aircraft some seven years earlier, the idea had seemed fanciful to many within the NACA. Although the concept of spaceflight arising as a natural evolution of high-speed aviation had been accepted, such ventures were still regarded as lying decades in the future; however, in the aftermath of Sputnik, the future arrived more quickly than anyone had anticipated. When the newly formed NASA had announced Project Mercury as America's primary man-in-space program in 1958, many—including Harrison Storms and Scott Crossfield—felt that this flirtation with ballistic capsules would be brief, a quick and dirty means to get America into orbit before attention switched back to winged spacecraft capable of providing regular, economical access to space. But now, faced with a finite goal and deadline, NASA had no time to pursue this seemingly logical progression. Capsules were an existing and proven technology; orbital space planes were not. Consequently, the x-15, once heralded as America's first spaceship, began to fade from view. Out at Edwards, however, few would miss the attention. The program's primary goal remained the pursuit of real-world flight data to verify laboratory predictions, and the x-15 was edging ever closer to the speeds and altitudes where that proof could be found.

On 23 June, Bob White launched over Mud Lake, near Tonopah, Nevada, on the program's first hypersonic flight. After 78.7 seconds, the XLR99 had propelled x-15-2 to Mach 5.27 at an altitude of 107,700 ft., making White the first pilot to break two Mach numbers. The flight wasn't without its problems, with a loss of cabin pressure causing White's suit to inflate, but fortunately this didn't affect his ability to control the x-15 using the side stick until his suit deflated at lower altitudes.

With the x-15 closing in on its design speed, data from the first year of government flights raised fresh concerns regarding the aircraft's ability to reach its design altitude of 250,000 ft. Using the simulator (now located at the FRC), engineers discovered that the aircraft's large ventral fin, thought crucial to preserve high-speed stability, could actually cause the x-15 to become unstable at high angles of attack if the roll damper failed during reentry. The higher the aircraft went, the higher its angle of attack would need to be to ensure a safe reentry—unless a solution could be found, it seemed unlikely that the x-15 would fly much beyond 200,000 ft. The engineers at North American suggested adding an emergency roll damper, providing redundancy should the first unit fail, but back at Edwards, flight planners Hoey and Day believed that there might be a way to prevent the instability from occurring in the first place.

In the ill-fated x-2, thrust misalignments and the lack of adequate vertical tail area had resulted in stability problems at higher altitudes where dynamic pressures were lower. These issues had influenced North American's decision to dramatically increase the size of the x-15's vertical surfaces and rudders in order to guarantee the rocket plane's stability. However, once x-15 ground crews developed techniques for aligning the XLR99 more accurately, largely eliminating thrust misalignments, the need for such large symmetrical surfaces decreased. Hoey and Day now theorized that by removing the x-15's lower ventral, safe reentries could be flown at higher angles of attack, even without dampers. While simulator runs suggested that this approach would work, there was only one way to know for sure. Having won Paul Bikle's approval, Bob Rushworth was to test their theory in the program's first ventral-off flight, on 4 October, but as flight day approached, the planners' apprehension increased. "I suspect that Bikle and Rushworth had more confidence in Hoey and Day than did Hoey and Day," Hoey recalled later. During the flight, Rushworth established that the x-15 did indeed remain stable, especially at higher angles of attack, verifying the simulator predictions. Additional tunnel testing would be needed before high-altitude ventral-off flights took place, but in the meantime, the aircraft would be fitted with the backup roll damper, allowing altitude flights to continue with the ventral on.

One week later Bob White took x-15-2 to 217,000 ft., overshooting his planned altitude by 17,000 ft. and becoming the first pilot to exceed 200,000 ft. As he plunged back through the heat of reentry, White was alarmed to see his left outer windshield glass crack. Each window usually consisted of an outer pane of aluminosilicate glass and two inner panes of soda-lime tempered glass, but on this occasion, the left window had been mistakenly fitted with a soda-lime glass outer pane. Although his visibility was impaired, White brought the x-15 in for a safe landing, with the incident warranting only a brief mention in his flight report; however, this would not be an isolated occurrence. On 9 November, White was to take the x-15 to its design speed of Mach 6 at an altitude of 110,000 ft., and after launching over Mud Lake, the air force pilot streaked back toward Edwards AFB, reaching Mach 6.04 in the process. Unfortunately, as the x-15 decelerated, the right outer windshield glass shattered, leading White to exclaim, "Oh, good Lord! Hope this one holds." He later explained, "You know what I was hoping? That the left windshield wouldn't crack. If it did, I planned to fly by instruments down to 35,000 feet, then jettison the canopy and see if the plane could be controlled in the wind blast with the canopy off. I felt I could get away with it, but if the plane proved uncontrollable, I would have no choice but to eject myself and parachute down."

Fortunately, the left windshield did hold, and although visibility was badly affected, White brought the aircraft home with the assistance of his chase pilot. Postflight inspections revealed that the window's Inconel-X retainer frame had buckled during the flight, creating a hot spot that caused the glass to fail. The frames were replaced with heavier-gauge titanium versions to prevent any reoccurrences, although a better solution would be needed if the aircraft were to fly faster. It had taken just under nine years, from October 1947 until September 1956, for the rocket planes of Rogers Dry Lake to extend the envelope from Mach 1 to Mach 3. Now, Bob White had claimed Machs 4, 5, and 6 in just over nine months, and the x-15 still had more to give.

On 10 August, Forrest Petersen had made the first XLR99-powered flight in x-15-1, reaching Mach 4.11 and adding to the FRC stable a second aircraft equipped with a big engine. Neil Armstrong finally brought the fleet up to full strength, closing out the 1961 flight schedule with the maiden flight of x-15-3. Following its test stand explosion the previous year, the air force had contracted North American to rebuild the shattered rocket plane and took the opportunity to install the advanced Minneapolis-Honeywell MH-96 adaptive flight control system in place of the x-15's standard stability augmentation system. The MH-96 was then under development for the air force's Dyna-Soar space plane and promised to simplify the pilot's task by making the aircraft's response consistent for a given stick movement, regardless of how far the control surfaces needed to move to achieve this. By blending both ballistic and aerodynamic controls, adaptive control systems were expected to offer real benefits for vehicles transitioning between air and space, and in the x-15, the MH-96 allowed the pilot to control the entire flight using a single stick. There were, however, concerns that the system might mask a pilot's feel for the aircraft, making the onset of stalls and other flight conditions more difficult to detect. Having already worked closely with Minneapolis-Honeywell to evaluate the MH-96 through his role as a Dyna-Soar pilot-consultant, Armstrong recommended the system for the x-15 and was duly tasked with performing a series of evaluation flights. The first of these took place on 20 December, when Armstrong flew x-15-3 to a relatively modest Mach 3.76 at 81,000 ft., and although the MH-96 dropped out and had to be reset immediately after launch, the results were encouraging enough to warrant further investigations.

The later months of 1961 also saw the first x-15 industry conference since the flights began, giving representatives from North American, NASA, the air force, and the navy an opportunity to share their initial research findings. The data gathered so far broadly confirmed experimental predictions regarding aerodynamic heating, but temperature-sensitive paints also revealed localized heating patterns across the x-15's airframe in detail. After superheated hot spots had caused problems during early high-speed flights, engineers fitted fairings over expansion gaps in the x-15's wing leading edges and side tunnels to prevent buckling. With the x-15 providing the first full-scale demonstration of how the superheated boundary layer might behave at high speeds, these research results were considered vital for any future hypersonic programs. Beyond aerodynamic research, the first forty-five flights had also produced a wealth of aeromedical data, proving that pilots could function effectively under high-g, high-stress, and weightless conditions. The high heart rates that x-15 pilots experienced during their short flights were especially noteworthy. During 1960, NASA flight surgeons had observed elevated heart rates in chimpanzees during Mercury test flights, leading to worries that astronauts might lose consciousness or even suffer cardiac arrests during spaceflights. Data from the x-15 helped dispel these fears, although Walt Williams would later relate a conversation between a Project Mercury doctor and Joe Walker during which the physician asked Walker how he had felt as his heart rate exceeded 150 beats per minute during flight. Responding that he had felt fine, Walker suddenly realized the underlying implication: "Now wait a gosh-darn minute. Are you trying to ask me whether or not I fainted?" When the physician suggested that Walker might have lost consciousness momentarily without realizing, the exasperated pilot exclaimed, "Look, what I did one second depended on what I had done the second before, and I'm here talking to you!"

The 1962 flight campaign began with an emergency and a departure. After almost four years with the program, it was time for Forrest Petersen to move on in order to advance his naval career. The flight plan for Petersen's final x-15 sortie on 10 January called for a series of high angle of attack maneuvers at Mach 5.7 to obtain dynamic heating and stability data, but following a routine launch over Mud Lake, x-15-1's engine shut down after a mere three seconds. It took around twenty seconds to attempt an engine restart in the x-15, but with the fully laden aircraft falling fast and Petersen knowing that he needed at least 26,000 ft. of altitude to jettison his propellants if he were to make an emergency landing at a safe weight, that meant just one restart attempt would be possible. When the recalcitrant XLR99 failed to respond, the naval aviator turned back toward Mud Lake and set up for the first remote emergency landing of the program. All X-15 pilots spent many hours practicing low L/D landings at their alternate lakes, and Petersen had no trouble bringing X-15-1 in for a safe, if premature, landing. Although he hoped to fit in one more flight before his new orders arrived, Petersen's chance never came, and he left the High Desert with five X-15 flights to his name.

The early months of the year also saw Neil Armstrong continue his evaluation of the MH-96, with a series of altitude-buildup flights, testing the system at progressively lower dynamic pressures. On 17 January he took X-15-3 to 133,500 ft., and he followed this with a flight to 180,000 ft. on 5 April. His next flight, on 20 April, called for an altitude of 205,000 ft. but resulted in one of the most infamous incidents of the X-15 program. Following a turbulent ride out to Mud Lake below the NB-52's wing, Armstrong launched safely and headed for altitude, eventually reaching 207,500 ft. With everything working well as he coasted over the top of his arcing flight path, Armstrong began to set up for reentry, during which he planned to test the MH-96's g-limiter. When the g-limiter failed to kick in at the expected point, Armstrong increased his angle of attack slightly in the hope of prompting a response. Distracted by this as he pulled out of his descent, the young test pilot failed to notice that X-15-3 had begun to climb, until Joe Walker, acting as NASA-1, warned Armstrong, "We show you ballooning, not turning. Hard left turn Neil." Try as he might, Armstrong found himself unable to make the turn; the X-15 had bounced off the atmosphere and was now too high for aerodynamic controls to alter its flight path. The aircraft, still moving at Mach 3, sailed on over Edwards still banked hard right at an altitude of 100,000 ft. before descending to a point where Armstrong was finally able to get turned around. Unfortunately, he was now forty-five miles southwest of the base, with no practical emergency landing options. Given the X-15's limited gliding ability, the NASA pilot was initially uncertain whether he could coax the aircraft back to base but quickly calculated that he would just reach Runway 35 at the southern edge of Rogers Dry Lake using a straight-in approach. As chase aircraft raced to join the wayward X-15, Armstrong brought the rocket plane down with little room to spare. During the postflight briefing, one of the chase pilots was reputedly asked how much clearance Armstrong had as he passed over the Joshua trees on the lake bed's boundary. "Oh, at least one hundred feet . . . on either side!" came the reply. The flight had lasted twelve minutes, twenty-eight seconds, making it the longest of the program so far and a record no other pilot wished to break; in Edwards folklore, it became known simply as Neil's cross-country flight.

Ten days later, Joe Walker was back in the pilot's seat for an altitude-buildup flight in x-15-1. Launching over Mud Lake, Walker lit the XLR99 and set up a steep 38° climb. As the rocket plane thundered upward, growing ever lighter as its propellants were expended, the steadily increasing g-forces pinned Walker back into his seat, giving him the uncomfortable impression that he was climbing vertically. Shutting the engine down after eighty-one seconds, he continued to coast upward using the x-15's ballistic controls to alter the aircraft's orientation before peaking at an indicated altitude of 250,000 ft. Looking up from his instruments for a few precious seconds, Walker marveled at the gentle curve of the Earth, easily identifying Monterey Bay to his right; Baja California to his left; and, as the x-15's nose began to fall, the reassuring pale shape of Rogers Dry Lake waiting below. Plunging back toward the desert at an 18° angle of attack, Walker was subjected to -5.5 g's—the "eyeballs out" force of deceleration. With parts of the x-15 now glowing red as they reached a maximum temperature of 1,000°F, his nitrogen-cooled suit maintained a comfortable temperature as it squeezed his extremities, preserving the blood supply to his head and upper body. At 65,000 ft. Walker pulled the x-15 out of its descent, arriving parallel to the runway at 50,000 ft.—the initial high-key point of the landing pattern—before making a steep 360° descent to touchdown, just nine minutes, forty-six seconds after leaving the NB-52's wing. Postflight analysis revealed that the NASA pilot had fallen just short of his planned altitude, reaching 246,700 ft. Such discrepancies were not uncommon as the aircraft's stable platform drifted during flight, but the Fédération Aéronautique Internationale (FAI)—the international governing body responsible for ratifying aviation records—still accepted this as a new official altitude record, crediting Walker with an altitude gain of 201,749 ft. from the point of launch.

Amazingly, Bob White would match Walker's altitude exactly during an altitude-buildup flight to test x-15-3 and its MH-96 control system on

21 June. Although both pilots had fallen just short of the x-15's 250,000 ft. design altitude, there was now little doubt that the research aircraft could comfortably exceed this mark, and in July 1962 Bob White did just that. For a while, x-15-3 looked reluctant to meet the challenge, as technical glitches led to three aborted flight attempts, but finally, on 17 July, everything went right for the air force pilot. After launching on his fifteenth x-15 flight, above Delamar Lake (around seventy-five miles north of Las Vegas), White raced upward in pursuit of a 282,000 ft. target altitude—fifty miles above the earth. As it happened, the rocket plane exceeded expectations as it climbed toward space, building up speed until White reached Mach 5.45. At shutdown, the x-15 had so much energy that it overshot the planned altitude by more than 32,000 ft., offering White a view of the earth that had previously been reserved for NASA's Mercury astronauts and their Soviet cosmonaut counterparts. He later enthused, "You could just see as far as you looked. I turned my head in both directions, and you could see nothing but the earth. It's just tremendous. . . . You have seen pictures from high up in rockets, or these orbital pictures of what the guy sees out there. That's exactly what it looked like. The same thing."

White successfully guided the x-15 back down through what he called "the most dramatically impressive reentry," using the MH-96 control system, but with so much excess energy, he arrived at his high-key point while still moving at Mach 3.5 at an altitude of 80,000 ft. Mindful of Armstrong's recent bounce out of the atmosphere, White made sure to keep the x-15's nose down as he made a long sweeping approach out over Rosamond and back toward Edwards.

Bob White had reached a peak altitude of 314,750 ft. during this flight, setting an official FAI record altitude gain of 269,652 ft. (as with Walker's earlier record, the FAI only recognized altitude gained by the x-15 following its launch from the NB-52, rather than the total altitude from takeoff). The following day, White, Scott Crossfield, Joe Walker, and Forrest Petersen— representing the air force, North American, NASA, and navy, respectively— traveled to the White House to accept the Robert J. Collier trophy from President Kennedy on behalf of the entire x-15 program. During the ceremony, the president remarked that the four pilots had demonstrated courage and skill during their x-15 flights and, through their efforts, had "extended the horizons of knowledge and human endeavor," making them "the kind of

Americans we most appreciate and want the country to be identified with." The ever-modest White accepted the trophy "on behalf of all those people who are not here today, who enabled us to carry out this program," adding, "It was a team effort." Bob White had one more important engagement to keep during his brief trip to the capitol. In June 1959 he had attended a meeting at Andrews AFB, where crew selection for the upcoming Dyna-Soar program had been discussed. During the course of this meeting, it was agreed that air force pilots exceeding the expected maximum altitude of the x-15 (approximately 250,000 ft.) would be eligible for classification as astronauts. The air force had subsequently formalized its qualifying altitude as fifty statute miles, and White now made the short journey from the White House to the Pentagon to receive his astronaut wings, recognizing his achievements as the fifth American to enter space and the first person to fly a winged vehicle into space and back.

The x-15 program soon made what would turn out to be another significant contribution to the space program. On 26 July, Neil Armstrong made his seventh and final flight in the x-15. A month earlier Armstrong had applied for NASA's second astronaut selection. With Project Apollo now forging ahead to meet Kennedy's deadline, the agency needed a new group of astronauts to fly its Gemini spacecraft, which would be used to develop and perfect the techniques required for a lunar voyage. Ironically, Armstrong's application would have been rejected had it not been for a former FRC colleague. Dick Day had left the High Desert in February 1962, moving to NASA's new Manned Spacecraft Center in Houston, as assistant director of the Flight Crew Operations Division. When former HSFS director Walt Williams made Day secretary to the astronaut selection committee, the flight-simulation pioneer found himself responsible for incoming applications. Armstrong's form arrived in Houston around a week late, but Day simply placed it with the valid applications. Speaking many years later, he explained that Armstrong's experience and qualifications placed him way ahead of many other applicants. Both Day and Williams wanted Armstrong in Houston, and they weren't about to let a late application form prevent that from happening. Following a grueling selection process, Armstrong received a phone call from Deke Slayton in early September 1962; the NASA research pilot would now become an astronaut, albeit in a capsule rather than a rocket plane.

Following Armstrong and Petersen's departures, it fell to Jack McKay and Bob Rushworth to continue the aerodynamic heating and hypersonic stability research program in the x-15. Although these flights rarely pushed the aircraft to its performance limits, they were far from routine, and the events of 9 November 1962 provided a terrible reminder of the risks involved. That day's flight, the seventy-fourth of the program, called for Jack McKay to reach Mach 5.55 at 125,000 ft. He was to undertake investigations into boundary-layer heating, while also testing x-15-2's stability and handling at higher speeds and altitudes without the lower ventral fin. Unfortunately, things quickly went awry for McKay as he launched over Mud Lake for his seventh x-15 flight. After lighting the XLR99 and advancing its throttle to 100 percent, the NASA pilot received a call from Bob Rushworth in the control room, asking him to confirm the throttle setting; data was indicating that the engine was only producing 35 percent of its expected output. When McKay confirmed the reading, the primary flight plan was quickly abandoned. Although it was possible that the x-15 could have limped back to Edwards at lower speed, the risk of total engine failure at a point where the McKay had no emergency landing options was simply too high; the flight rules were clear—shut the engine down, jettison the remaining propellants, and head for an emergency landing at the launch lake.

Coolly turning back toward Mud Lake, McKay began emptying x-15-2's tanks to reduce the aircraft's landing weight. Even from a distance, he could easily pick out the black tar lines marking the emergency runway, and using the experience of countless low L/D F-104 approaches, he set up his landing pattern. As the lake bed rushed toward him, the experienced test pilot attempted to lower the x-15's wing flaps to reduce his landing speed, but a second mechanical failure now prevented the flaps from deploying, meaning the aircraft would touch down well above its normal landing speed. Still overweight due to unjettisoned propellants, x-15-2's skids hit the lake bed at a speed of 257 knots, and as McKay pulled back on the stick in an attempt to reduce the nose gear's impact, the left rear gear strut failed. A rapid chain of events now unfolded as the left elevon and wingtip dug into the dirt, yawing the x-15 in that direction. As the ventral stub tore away, the nose gear strut also failed, causing the aircraft to roll to the right, and sensing that the x-15 was about to flip and not wishing to become trapped in the nitrogen-filled cockpit, McKay jettisoned his canopy. The wrecked rocket plane finally came to a stop inverted on the lake bed, and although his helmet had hit the hard-packed clay as the aircraft had rolled, Jack McKay remained conscious but badly injured beneath the battered fuselage. As the pre-positioned emergency vehicles rushed to his aid, fuel seeped from the x-15's fractured tanks, causing clouds of choking ammonia fumes to envelop the scene. Fortunately, the air force rescue helicopter's quick-thinking pilot used his rotor downwash to disperse the toxic gas, allowing help to reach McKay. The ground crew had to dig a hole in the hard clay in order to ease the crushed pilot out from under the fuselage, before the waiting c-130 transport rushed him back to the base hospital at Edwards for medical attention.

Jack McKay escaped the crash with his life, but the lake bed impact had crushed three of his neck vertebrae, leaving him an inch shorter and in chronic pain. Physically, McKay was a bear of a man, and his solid build arguably prevented more serious injury. But it was beyond doubt that the thorough emergency preparations in place for every flight had saved him from a potentially fatal situation. As McKay lay in the hospital, crews scoured Mud Lake to recover the remains of x-15-2, loading the wreckage onto a truck for another trip to Inglewood, where a full investigation and damage assessment could be carried out. Amazingly, Jack McKay was back working within weeks of the incident and flew the x-15 again less than six months later, although the injuries sustained that day never truly healed, tormenting the pilot for the rest of his days.

In the aftermath of McKay's accident at Mud Lake, the program's very next flight saw the departure of the x-15's second government pilot, Bob White. Unlike their NASA counterparts, air force pilots could only serve at Edwards for a limited time before moving on to progress their careers. By late 1962 new challenges awaited White, and on 14 December the original air force x-15 pilot climbed into the rocket plane for the final time.

Launching over Mud Lake, White took x-15-3 out to Mach 5.65 at an altitude of 141,400 ft. to perform high-speed, ventral-off stability tests. By late 1962 many of the x-15's original research goals had been largely accomplished, and flights were now beginning to carry piggyback experiments to high speeds and altitudes—a role that would become increasingly important in the years to come. During this flight, x-15-3's tail carried a new research package; the ultraviolet earth background experiment used a spectrometer and radiometer to measure radiation across the horizon—an important early step toward a planned space-based missile early warning system.

After nine minutes, thirty-six seconds, White brought the rocket plane back to a safe landing on Rogers Dry Lake, completing his sixteenth x-15 flight and ending an illustrious two-and-a-half-year period, which had seen him become the first pilot to exceed Machs 4, 5, and 6, as well as the 200,000 ft. and 300,000 ft. altitude marks in a winged vehicle. Although the modest White was keen to point out that he had achieved these milestones thanks to the combined efforts of a larger team, his coolness and professionalism had helped bring the x-15 through its envelope-expansion phase without major incident. As Bob White departed for West Germany to become operations officer for the Thirty-Sixth Tactical Fighter Wing, his NASA counterpart Joe Walker was also looking toward future assignments.

As the FRC's chief pilot, Walker was soon due to fly the lunar landing research vehicle (LLRV), an ungainly turbofan- and rocket-propelled machine developed by the FRC and Bell Aircraft to research the piloting challenges Apollo astronauts would face as they attempted to land on the moon. The veteran test pilot was also due to fly NASA's research program in North American's massive XB-70 Mach 3 bomber prototype, and so as 1963 dawned, Joe Walker began his final series of x-15 flights, with the aim of taking the research aircraft well beyond its 250,000 ft. design altitude. Walker used x-15-3 with its MH-96 adaptive control system for these altitude flights, the first of which took place on 17 January, when the NASA pilot launched from Delamar Dry Lake toward a planned maximum altitude of 250,000 ft. Following a trouble-free drop and ignition, Walker brought the rocket plane's nose back and headed upward, passing through 100,000 ft. in just over a minute. Less than twenty seconds later, he reached forward with his left hand to shut down the thundering XLR99, but the g-forces pinning him back into his seat made this no easy task. In his postflight comments, Walker explained, "I managed to hook the throttle once with a finger and got enough room to get it shut down finally."

With an excess of energy, the x-15 sped beyond its planned altitude, eventually reaching 271,700 ft. and giving Walker his best views of the earth yet. Coasting beyond the atmosphere, he noted "a right smart amount of miscellaneous material floating inside the cockpit" in the zero-g conditions.

Unlike its Cape Canaveral cousins, there were no sterile white-room preparations for x-15 flights, meaning cockpits were far from spotless. Walker also noticed, "Some kind of search light was shining in the right-hand side of the cockpit instead of the left, and I discovered it was the moon shining." However, the flight was less leisurely than these comments suggest, with one of the aircraft's APUs failing after four minutes and subsequent failures occurring in both the ball nose and rudder servos. Fortunately, these problems did not prevent Walker from bringing x-15-3 back for a safe landing after a flight that lasted nine minutes, forty-three seconds.

Given that he had just flown more than fifty miles above the earth's surface, it seemed reasonable to expect that Joe Walker might be granted the title of astronaut, as Bob White had following his fifty-mile flight. At the time, NASA had no official definition for the lower boundary of space. To become a NASA astronaut, it seemed, you had to fly in a capsule from Cape Canaveral, and the agency viewed its x-15 pilots as just that—pilots. The lack of a lucrative magazine contract or complementary Corvette did not appear to trouble Joe Walker; he still had a job to do (albeit on standard government pay). Together with his fellow pilots, he expressed considerable professional pride in the fact that he actually flew his spacecraft from launch to landing, keeping his feet dry in the process.

In some respects, Walker was fortunate—unlike his air force counterparts, he was at least allowed to accept payments for interviews and appearances. When he wrote about his experiences in the x-15 for the September 1962 issue of *National Geographic* magazine, he received the princely sum of $1,000. Although Bob White was offered the same amount for his contributions to the article, air force rules precluded him from accepting the payment—a far cry from arrangements enjoyed by his USAF colleagues who now flew as NASA astronauts while still retaining their military ranks.

On 19 July, Walker made his second altitude-buildup flight, targeting a maximum altitude of 315,000 ft. Although simulations had suggested that the x-15 could reenter from altitudes as high as 400,000 ft. without its lower ventral, this would require a perfect performance from both man and machine, and given some of the problems that had occurred up to this point, FRC engineers chose 360,000 ft. as the maximum safe altitude. During his twenty-fourth flight, Walker would test how the x-15 handled at the higher angle of attack required to reenter from this altitude. While beyond the atmosphere, he would also perform an experiment similar to one flown during Scott Carpenter's Mercury flight, in which a balloon would be released to determine what effects the rocket plane's wake might have on it.

In order to achieve target altitude, the flight was launched from Smith Ranch, Nevada—the most distant of the remote launch lakes. After racing away from the NB-52, Walker was soon grateful for that extra distance, as a combination of higher-than-predicted thrust and a longer-than-expected burn from the XLR99, together with a slightly higher-than-planned climb angle conspired to produce a 31,000 ft. altitude overshoot. Ever the perfectionist, Walker would later comment, "I was honestly trying for 315,000," but in an aircraft as powerful as the x-15, small errors could lead to big consequences.

Cresting his ballistic flight path, Walker lowered the aircraft's nose to reveal an incredible view of the Mojave Desert. Even from 347,800 ft.—an altitude in excess of sixty-five miles—Walker noted, "It was no trick at all to locate where our North Rogers Lakebed runway was going to be even though I couldn't see the marks." Although Walker attempted to release the balloon, the experiment failed to deploy (as it would on both attempts made using the x-15). Soon, though, it was time for the long reentry, with the pilot commenting later on the "big squeeze" his pressure suit's g-garment gave as he pulled out of the descent to make his spiral approach to Edwards. During this flight, Walker had not only exceeded the air force's fifty-mile space boundary; he had also topped the FAI's preferred one-hundred-kilometer boundary (approximately 62.14 miles)—a point known as the Kármán line in honor of Theodore von Kármán's work to define the limits of aerodynamic flight. But the NASA pilot would go higher still during his final x-15 flight.

Although he was a popular and well-respected figure at the FRC, Walker had a reputation for being obstinate, a stickler for detail, and prone to a quick temper when pushed, but perversely, this made the temptation to play practical jokes at his expense almost irresistible to some of his colleagues. When Walker requested layout changes to x-15-3's instrument panel while preparing for his high-altitude flights, operations engineer John McTigue saw the ideal opportunity to play a prank, asking technicians to make the requested changes but to also paint the panel pink before reinstalling it in the aircraft. When the modifications were complete, McTigue invited

Walker down to the hangar in order to inspect the changes. Standing at a safe distance, the ops engineer waited as Joe Walker took in the new paint job—at which point, his face took on a similar color to the panel. Knowing he'd been had, Walker left the hangar without comment, and McTigue asked for the panel to be returned to its original light-gray coloring. As they prepared for his maximum-altitude attempt, x-15-3's ground crew surprised Walker with a second unexpected paint job, but this one was far more to the pilot's liking. For his final x-15 flight, the crew emblazoned the rocket plane with a reprise of the "Little Joe" artwork that had adorned the x-1E during Walker's time in that aircraft. The tumbling dice, now bearing the legend "Little Joe The II," would travel with Walker to bring good fortune as he attempted the highest aircraft flight in history.

At 9:09 a.m. on the morning of 22 August, the NB-52 lifted away from Edwards, circled to altitude, then headed north, x-15-3 slung beneath its starboard wing. Less than an hour later, the aircraft turned onto its southerly launch heading, over Smith Ranch. Walker had already endured three attempts to get the flight away, but today everything was in the green as he primed the XLR99. After dropping away cleanly and receiving confirmation from his chase that the engine had lit, he brought the stick back and headed for space. Walker was trialing an experimental altitude predictor on this flight, which would hopefully help prevent an overshoot beyond 360,000 ft., while Jack McKay provided a steady stream of altitude and ground track updates from the FRC control room.

With the XLR99's burn time approaching eighty-five seconds, the NASA pilot reached forward for the throttle, but just as he said the word "shutdown," the XLR99 exhausted its supply of propellants and fell silent of its own accord. Glancing at the altitude predictor, Walker reported that it was indicating a peak of 362,000 ft., but his satisfaction was short-lived, as McKay informed him that x-15-3 was actually a little below its planned profile. In his postflight briefing, Walker was asked for his views on the device, to which he responded, "Apparently it worked pretty good for a first whack at it. One of us was so carried away with people beating on him about not overshooting that he took pains not to."

The x-15 eventually topped out at 354,200 ft., marking Walker's second flight beyond the Kármán line. Besides the slight undershoot, the only other problem was a frozen RCS thruster, which complicated the task of orienting the x-15. All too soon the rocket plane's upward momentum was exhausted, and Walker had his hands full ensuring that the aircraft remained stable at its high angle of attack during reentry. Although he later described the x-15 as "joggling" and "bobbling" on its way down, he was able to maintain control, even as his suit again gave "one big squeeze" during pullout. Joe Walker brought x-15-3 in for a smooth lake bed landing with eleven minutes, eight seconds on the clock, having capped his time on the x-15 program with a literal high of over sixty-seven miles.

During his final two flights, Walker became the seventh American to cross the Kármán line and the first person to enter space twice. As he no longer held military rank, Walker also became the first civilian to enter space, but again NASA Headquarters chose not to recognize his achievements. It would take another forty-two years before the agency posthumously awarded Joe Walker the astronaut status he had earned during the summer of 1963, but as x-15 crew chief Charlie Baker saw it, "[Joe] didn't care. We all knew they went there; they all knew they did. They didn't need a whole lot of smoke blowing up their butts to keep doing what they were doing."

9. The Follow-On

The x-15 would never again fly as high as it had during the summer of 1963. As Walker left the program, with the aircraft's original design limits met (and in some cases exceeded), the later months of the year provided a brief opportunity to reflect on the x-15's achievements thus far, while preparing a new role for the hypersonic research plane.

When discussions between the air force's Aeronautical Systems Division (ASD) and NASA Headquarters in July 1961 led to the idea of an extended experimental program for the rocket plane, an x-15 Joint Program Coordinating Committee was formed to examine what types of experiments might be suitable. These ranged from space science and advanced propulsion concepts to a continuation of earlier hardware and materials testing in support of the Apollo program. Although the x-15 could not fly as high as an orbiting satellite or even many contemporary sounding rockets, it did offer its own unique advantages. With some modification, the rocket plane could carry several relatively large experiments at one time, fly them as often as required, and offer the pilot as an expert observer if needed. Of the forty proposals submitted, twenty-eight were selected and forwarded to the Research Airplane Committee. Official approval came in March 1962, allowing NASA to announce the follow-on program in June. Although the first follow-on experiments were flown by Jack McKay in September 1962, the effective conclusion of the x-15's original research phase by mid-1963 allowed the follow-on program to take precedence on future flights. However, the new initiative was not welcomed by all, with Paul Bikle voicing his concerns that any benefits gained from the extended flight program might not justify the high costs and risks involved.

By now the x-15 had become the FRC's major activity, involving over four hundred employees in various capacities, and although the center was pursuing new directions with the LLRV and lifting body programs, nothing of comparable importance to the X-15 lay on the horizon. Since replacing Walt Williams, Paul Bikle had overseen a transformation at the FRC. Although he always encouraged initiative within his workforce and placed great trust in their abilities, the sheer size and complexity of the X-15 program demanded new ways of working. Where Williams had been free to approve ideas verbally, granting his staff the authority to work without administrative obstacles, Bikle was answerable to an increasingly bureaucratic NASA as well as the X-15's main benefactors, the air force. As his workforce expanded and the piles of paperwork grew ever larger, Bikle strived to maintain the center's independent character. Speaking about the director's management style in 2018, X-15 simulator engineer John Perry recalled, "The fact that [Bikle] had the biggest office in the building—the nicest furniture, the nicest view—didn't matter to him as much as [that] he knew the people he was trying to inspire and knew what would likely inspire them to give their best. He was one of the boys. You could go in his office, most of the time he wouldn't be there 'cause he'd be hanging out down in the machine shop, or in the lab, or hanging out in the hall talking to folks like me." When things went wrong or decisions were questioned, employees might receive an invitation to his wood-paneled office for some straight talk—an experience known as the Bikle barrel. "He would help you get up after he smacked you around and ask you, 'What do you need, to try the next thing?'" Perry added, "He just trusted us to tell him the truth."

Although the close, collaborative atmosphere of the Muroc days was largely gone, the FRC remained a popular destination for those wishing to work at the cutting edge of aeronautical research. Many who started their careers as interns during the X-15 years would remain at the center for the rest of their professional lives, building close professional and personal relationships that made the long hours worthwhile. FRC employee picnics, softball games, and hiking trips became key events in the community life of the Antelope Valley, and under Bikle, the links between the AFFTC and the FRC grew stronger, galvanized by the joint test-force approach adopted for the X-15. The frustrations and disagreements that had marred earlier programs were largely replaced by mutual respect and cooperation, and when NASA administrator James Webb presented a group achievement award to the "X-15 Research Airplane Flight Test Organization" to mark the X-15's hundredth flight, this fictitious body represented a tacit means to acknowledge the contributions of the entire team, regardless of their parent organization.

During the latter half of 1963, two new pilots joined the program. Prior to Bob White's departure, the AFFTC had selected Captain James McDivitt as its new project pilot. McDivitt had been among the first class to graduate from the AFFTC's new Aerospace Research Pilot School (ARPS), an initiative to train air force astronauts in readiness for Dyna-Soar and its expected successors. However, before McDivitt transferred to the X-15, he was selected for NASA's second group of astronauts (the so-called new nine), alongside FRC pilot Neil Armstrong. With McDivitt gone, the air force picked Captain Joe Engle to fly the X-15. As a graduate of the third ARPS class, Engle came to the program with a better understanding of exo-atmospheric flight than his predecessors, but before experiencing the rocket plane's full potential, he needed to make the customary familiarization flight, although with the XLR99, this was no longer the slow and low affair it had previously been.

Launching in X-15-1 over Hidden Hills on 7 October, Engle reached Mach 4.2 at 77,800 ft. As he neared Rogers Dry Lake, the new pilot became concerned that he was still too high. Not wishing to overshoot, Engle decided to shed some altitude and figured that he knew just how to do it. "I had done some roll maneuvers, you know, left and right, and it just felt like a dream, so I rolled it over and let the nose dish out, and dropped down so the nose was pointed down as that was the easiest way to get the nose down." After landing safely, Engle thought no more about the incident, but there was no hiding from the aircraft's research instrumentation. When Bob Rushworth learned of the rookie's impromptu maneuver, he was quick to inform Engle that the X-15 was not an aircraft that should be rolled due to the risks of inertial coupling. Paul Bikle took a simpler approach, asking Engle not to do it again in case everyone wanted to try it.

On 29 October a second new pilot made his X-15 debut. Milt Thompson had served as a naval aviator during World War II, flying sorties in the Pacific before moving back into full-time education following the war's conclusion. After studying engineering at the University of Washington, Thompson joined Boeing as a flight test engineer, where he worked on the B-52 bomber. Seeking new challenges, he moved to the HSFS as a flight test engineer in the spring of 1956 before moving to the pilots' office in 1958.

24. U.S. Air Force and NASA pilots with an x-15A-2 in 1965.
Left to right: Pete Knight, Bob Rushworth, Joe Engle, Milt Thompson,
Bill Dana, and Jack McKay. Courtesy NASA.

Having served as a pilot-consultant on the Dyna-Soar program alongside
FRC colleagues Neil Armstrong and Bill Dana, Thompson was named as
one of the six Dyna-Soar astronaut candidates in 1962, but the busy pilot
still found time to fill NASA's vacant seat in the x-15. For his first flight,
Thompson followed the same basic plan as Engle (albeit minus the roll),
reaching Mach 4.1 at 71,000 ft. having launched over Hidden Hills. Both
pilots would make second flights before the end of the year in preparation
for a busy 1964.

   Alongside the introduction of new follow-on experiments, the new year
would also see the return of the second x-15, rebuilt and modified as the
x-15A-2. Following Jack McKay's emergency landing at Mud Lake, the air
force had accepted North American's proposal to rebuild the aircraft as a
hypersonic air-breathing propulsion research test bed. To fulfill this new
role, it was lengthened by twenty-nine inches to accommodate a liquid

hydrogen tank for the test engine, and additional helium and hydrogen peroxide tanks were also fitted. A new skylight hatch was added above the instrumentation bay, allowing experiments to be deployed once the aircraft was beyond the atmosphere, but perhaps the most noticeable additions were two jettisonable drop tanks flanking the aircraft's lower fuselage. In order to test supersonic combustion ramjet engines (scramjets), the x-15A-2 would need to reach speeds approaching Mach 8, something that engineers calculated would be possible if the XLR99 ran for an additional minute. The twenty-three-and-a-half-foot long drop tanks would supply the necessary propellants, but they also introduced new complications. Although they were the same size and shape, the ammonia-filled right-hand tank was considerably lighter than the left-hand LO2 tank, complicating the pilot's task until they could be jettisoned. The tanks needed to be dropped before the x-15 reached Mach 2.6, and each was fitted with an ejector and small rockets to ensure a clean separation, along with a parachute to enable their recovery and reuse.

Given the higher temperatures that the x-15A-2 would be exposed to as it flew beyond Mach 6, the canopy received new oval windows, finally solving the hot-spot problems encountered earlier in the program, but further adaptations were needed to ensure the aircraft's safety. The x-15's original Inconel-X hot structure would not be able to withstand the additional heat load, but new spray-on ablative coatings developed for lifting reentry vehicles appeared to offer a solution. By applying various thicknesses of these materials, the x-15A-2 could be protected as the charring ablator carried excess heat away from the aircraft. But as promising as the technique appeared, the x-15A-2 would be the largest and most complex vehicle ever to receive such a treatment, and the application and refurbishment of the material was only one part of the problem. Tests had shown that ablator from the rocket plane's nose would be redeposited as an opaque residue on the cockpit windows, seriously impairing the pilot's view during approach and landing. To alleviate this, North American devised a shuttered eyelid for the left cockpit window, which would remain closed until the aircraft had decelerated, at which point the pilot would flip it open to reveal the clear glass. Finally, the aircraft's landing gear was strengthened to bear the additional weight that came with the numerous modifications.

The x-15A-2, minus ablative coating, was accepted by the air force on 17

February 1964, and handed over to the FRC for instrumentation installation and final checks prior to its return to flight. Bob Rushworth would perform the aircraft's evaluation phase, testing the basic configuration before moving on to flights with the external tanks. The aircraft would only receive its ablative coating once all the other modifications had been tested in flight. Rushworth finished 1963 with a flight to Mach 6.06 in ship one, the fastest speed yet attained by an x-15, before making the program's one hundredth flight on 28 January 1964. In spite of Bob White's initial reservations, Rushworth had become a stalwart of the x-15 program, and as he looked toward the new challenge of the x-15A-2, he continued to make regular research flights alongside Engle and Thompson in the other aircraft.

The x-15 demanded a huge investment of time ahead of each short flight. Once a researcher's request to investigate a particular flight condition had been approved, the task of devising a detailed flight plan began. Throughout the program, a small band of flight planners (drawn from both the AFFTC and the FRC) spent many long hours in the x-15 simulator, developing the maneuvers needed to obtain each required data point, before working closely with the assigned pilot to ensure that every event from launch until landing had been covered. Once a flight plan was approved, the pilot would then spend up to twenty hours in the simulator, becoming familiar with not only the primary plan but also the various contingency plans, should an emergency occur. Integrated simulator sessions were carried out with the assigned NASA-1 colleague monitoring progress from another room, providing as accurate a real-time experience as possible. Away from the FRC, the pilot would also make numerous low L/D approaches to every lake bed along the proposed flight path, becoming familiar with the local geography and conditions, to remove as many unknowns as possible before flight day.

As planners grew more confident in the x-15's capabilities, ambitious new profiles were devised to gather aerodynamic heating data. As early as May 1963 Rushworth made the first launch on a heading that took him away from Edwards. Pointing the x-15 toward the town of Victorville (some forty miles to the southeast of Rogers Dry Lake), Rushworth accelerated to Mach 5 before beginning a sweeping turn back toward home, gaining heating data by holding a constant angle of attack at 5 g's for around ten seconds. Rushworth later recalled, "At that time, about all I could stand was 5 g's for ten-seconds!" Paul Bikle was less than enthusiastic about these flights, and they were rarely flown. But as the x-15 began to conduct additional research, new profiles were developed. Milt Thompson's fifth flight, which took place on 21 May 1964, was one such example. Researchers had requested that he keep the x-15 just below Mach 3 in order to obtain data for the supersonic transport program. To do this, Thompson would need to throttle the xlr99 back to 30 percent—a level at which the rocket motor was prone to cut out. To prevent these problems, the frc had adopted a minimum 40 percent throttle setting, leaving Thompson to argue for the reduced setting with the director of flight operations, Joe Vensel. After Vensel eventually gave in to the pilot's repeated requests, Thompson launched over Silver Lake, accelerating to Mach 2.9 before pulling the throttle back to the lower setting—at which point the xlr99 quit. In spite of his best efforts to restart the engine, Thompson was forced to make an emergency landing on Cuddeback Lake, running out of runway and leaving the x-15 stranded in the lakeside scrub. History does not record the exact words that passed between Vensel and Thompson on the pilot's return, but Thompson later wrote, "[Joe] raised hell like a father would if his kid had narrowly escaped death due to some stupid action."

On 25 June, Bob Rushworth brought the x-15 fleet back to full strength when he made the first flight in x-15a-2. Given its extensive modifications, initial flights in the rejuvenated rocket plane would be relatively benign stability and control tests without the external tanks. On this occasion Rushworth launched over Hidden Hills and accelerated to Mach 4.59, testing the aircraft's responses both with and without dampers, at one point rolling the x-15a-2 over 120° before recovering. With its basic airworthiness established, ship two received a further modification ahead of its next flight. The original stable platform carried by all three aircraft had proven troublesome throughout the early years of the program. With the x-15 now due to fly well beyond its original projected lifespan, engineers decided that a more capable and reliable unit was needed. When the Dyna-Soar program had been canceled in December 1963, the space plane's advanced digital flight data system (already developed and tested by Honeywell) became available, and the decision was made to fit the system to ships one and three. In the meantime, frc engineers had modified and improved the original Sperry unit, addressing its major deficiencies. The resulting frc-66 Analog Inertial System was better suited to higher Mach numbers than the Honeywell system, and so this unit was installed in x-15A-2.

The FRC-66 received its first in-flight test on 14 August, with Rushworth launching over Delamar Lake on another stability and control evaluation. Having reached Mach 5.23 at 103,000 ft., the air force pilot put ship two through its paces, making control pulses with the dampers disengaged to see how the aircraft handled at the higher speed and altitude. Everything seemed to be going well until Rushworth heard a loud bang and the aircraft began to oscillate wildly. Quickly reengaging his dampers to bring the rocket plane back under control, the shocked pilot reported back to Jack McKay in the control room, "I've got a . . . got a bang and might have been the nose gear that came out." Recognizing the sound, Rushworth had correctly identified that his nose gear door had deployed, but at this stage he was unsure whether the gear had fully extended into the Mach 4.2 airstream. As smoke began to curl into the cockpit, he worked to keep the aircraft stable against the extra drag of the open gear door. Descending toward Edwards, Rushworth was joined by chase pilot Joe Engle, who quickly confirmed that the nose gear had indeed deployed and that the tires looked in bad shape—hardly surprising after their exposure to temperatures as high as 1,000°F. Sure enough, both tires disintegrated as the x-15 touched down. But the wheels remained intact, and the aircraft slid to a halt, with Rushworth exclaiming, "Thanks a lot, Mr. North American!"

Investigating the incident, FRC engineers discovered that a cable linking the gear release handle to the release mechanism had not been lengthened enough during the x-15A-2's modifications. Consequently, as the aircraft expanded in the heat of its high-speed run, the cable pulled taut and tripped the release system. The cable was duly lengthened before the next flight, on 29 September. This time, Rushworth launched over Mud Lake, accelerating to Mach 5.2 at 97,800 ft., when, to the pilot's disbelief, he heard another telltale bang beneath his feet, exclaiming, "Something just let go again! I think it was the little door this time." The x-15's nose gear door had a smaller scoop door that opened as the pilot pulled the release handle. Air loads on this smaller door would then help deploy the main door, at which point the gear would extend. Again, Bob Rushworth had correctly identified the problem, but although the main door held, the superheated air entering the wheel well made short work of the tires, meaning they disintegrated on landing again.

The next pilot to take the x-15A-2 aloft was Jack McKay, making his return to the aircraft in which he had been badly injured during November 1962. Following his recuperation, McKay made seven more x-15 flights, including one on 15 October 1964, during which he tested new wingtip experiment pods on x-15-1. Having made two earlier captive flights to test out modifications to ship two's landing gear system, McKay made his fifteenth overall x-15 flight on 30 November, taking the x-15A-2 to Mach 4.6 before returning to Edwards without incident. With the landing gear problems seemingly licked, Bob Rushworth was surely expecting a smoother flight as he dropped away from the NB-52 on 17 February 1965, but the senior air force pilot seemed fated to suffer every problem the x-15A-2's gear could offer. Having accelerated beyond Mach 5, Rushworth began making control pulses, only to have the right main gear skid deploy. Struggling to bring the aircraft straight and level, he attempted to make Jack McKay aware of his predicament but found that "Jack was talking away and things were going along real nice," adding, "I couldn't seem to get a word in there to tell him that I had a little problem." In its asymmetrical condition, the aircraft was experiencing considerable sideslip as the drag from the extended gear yawed the x-15 to its right. Once again, Joe Engle was in place to inspect the rocket plane as it descended toward the lake bed, confirming that the right gear was down and appeared to be in the correct locked position. Following a safe landing and slide-out, Rushworth exited the aircraft before giving the troublesome x-15 a solid kick. Although this action earned him a quick trip to the Bikle barrel, the FRC director was sympathetic given Rushworth's run of bad luck. As always, the ground crew seized on the unfortunate series of incidents as an opportunity for humor, and when Bob Rushworth next entered x-15A-2's cockpit, he found a small sign reading, "Do not extend landing gear above Mach 4," waiting for him.

When originally accepted by the air force, the three x-15s had been essentially identical with the exception of the interim XLR11 engine installations in ships one and two. By the dawn of 1965, however, each had become unique following numerous modifications during the research and follow-on programs. x-15-1 remained closest to its original form, but it now sported wingtip pods as well as the Dyna-Soar-derived flight data system. x-15-3, with its

MH-96 adaptive control system, was best suited for higher-altitude work, whereas the second aircraft was heavily modified for higher speed and the propulsion test bed mission. As the aircraft diverged from their original configuration, their instrument panels also evolved to reflect these specialized roles, meaning technicians at the FRC were kept busy ensuring that the single fixed-base simulator was correctly configured for each test run. The x-15 pilots were also encouraged to suggest improvements to the simulator, based on their actual flight experiences. But in spite of the many changes, there was one thing about the x-15 cockpit that could not be changed; in Joe Engle's words, "It was sized to a standard MilSpec Crossfield." Scott Crossfield was of average height and build and had devised the cockpit layout accordingly, but this did not suit everyone, especially the tall Engle. As the ejection seat could not be adjusted, custom seat pads and armrests were produced for each pilot, allowing them to reach the controls even when their suit was inflated, but there were still many occasions when timing marks were missed as pilots struggled to reach the throttle under heavy g-loads.

As the three x-15s began carrying follow-on experiments or new aerodynamic instrumentation, the task facing the operations engineers and crew chiefs for each aircraft became ever more complex. Many of the experiments required new wiring to the instrumentation bay and cockpit, and each change needed to be rigorously tested to ensure that it didn't affect any of the aircraft's vital systems or the overall safety of the flight. With such a high workload, the FRC ran both day and night shifts to prepare each aircraft for its next flight, and the operations engineer was responsible for scheduling every work item, to prevent valuable time and effort being wasted along the way. The task could seem never-ending, and FRC ops engineer Vince Capasso remembers, "The night foreman would call, late at night, and he would always ask unrelated questions until he knew I was really awake, then [he would] ask the important question he called about. I am sure he did this with all of us he called late, to be sure our thought process was awake." Despite the long hours, morale remained high among the x-15 team, with the dedicated and highly skilled cadre of technicians working right up until the last minute to ensure that the x-15 would fly. Consultants from North American and the Reaction Motors Division (RMD, as RMI had become known following its acquisition by Thiokol) also remained on hand throughout the program to address any

structural or propulsion issues. Meanwhile, at the AFFTC, air force maintenance crews toiled to ensure that the NB-52s and other support aircraft
were always ready to go on flight day.

Following Walker's maximum-altitude flights in 1963, the X-15s generally stayed at lower altitudes while their improved stable platforms were
evaluated. As a result, 1964 had been a low year, by the rocket plane's standards, with Joe Engle's 195,800 ft. flight in May marking the year's high
point. In the spring of 1965, a new series of high-altitude-buildup flights
began, with Engle pushing X-15-3 back above 200,000 ft. for the first time
in over twenty months on 28 May. With the aircraft's ability to reach and
return from these altitudes long since proven, Engle's flights carried experiments to observe the earth's upper atmosphere and examine how the X-15's
exhaust interacted with conditions at these altitudes. In order to capture
the required data, Engle often used his ballistic controls to angle the aircraft away from its flight path before bringing it back into the correct orientation for reentry; simulator sessions had shown that any failure to align
the aircraft perfectly could lead to roll coupling during reentry—a situation that could rapidly lead to the loss of the aircraft and pilot.

Having qualified for his air force astronaut wings on 29 June during
his third altitude flight, Engle was heading beyond fifty miles again on
10 August, with a target altitude of 266,000 ft. in his sights. Within seconds of lighting the XLR99 following his launch over Delamar Lake, Engle
reported that his yaw damper had failed, before quickly informing NASA-I,
Bob Rushworth, that the damper was back on line. With the glitch apparently fixed, Engle continued his flight, overshooting his altitude slightly
to achieve 271,000 ft. The descent appeared normal, if slightly high on
energy, and the air force pilot soon had X-15-3 safely back on the lake bed
after a seemingly routine flight. However, during his postflight debriefing,
Engle revealed that far from failing just once, the yaw damper had actually
failed another twenty times during the sub-ten-minute flight. Each time
the damper had failed, Engle had simply reached forward to reset it—all
while piloting the rocket plane to an altitude of fifty-one miles and making
a perfect reentry; under these circumstances, a 5,000 ft. overshoot on peak
altitude seemed entirely forgivable. Engle was fortunate to have been flying
the MH-96-equipped X-15-3, allowing him to fly using the right-hand stick
while using his left hand to hit the damper reset button, but the fact that

he had not felt the need to report the recurrent failures said much about Engle's nonchalant style and exceptional piloting abilities.

Although much of an x-15 pilot's time was devoted to the extensive training and planning that preceded each flight, they all continued to participate in other test programs for their respective centers. The air force pilots remained busy wringing out new prototypes or improved models of the Century Series jets for the AFFTC's Fighter Ops Division, while over at the FRC, testing had begun on an entirely new class of aircraft—the lifting bodies. These unconventional, wingless vehicles appeared to offer a practical alternative to the ballistic capsules used by NASA for its Mercury and Gemini programs, and when FRC engineer Dale Reed first raised the possibility of building a piloted lightweight lifting body to validate the basic airworthiness of the concept, he had been quick to enlist the help of Milt Thompson. Following initial flight tests, the lifting body program began to generate more interest within NASA, and Thompson chose to step down from the x-15 in order to dedicate his energies to the new vehicles. On 25 August 1965 he signed off with his fourteenth and final flight, reaching his highest altitude yet as he carried two follow-on experiments to 214,100 ft. in x-15-1.

Thompson's departure left Jack McKay as NASA's sole program pilot, with Bob Rushworth and Joe Engle representing the air force. Both Rushworth and McKay made flights up beyond forty-five miles during the late summer of 1965, with McKay taking the x-15A-2 to 239,800 ft. on 2 September (the modified aircraft's highest altitude to date). During the flight, McKay opened the x-15A-2's large skylight compartment, exposing the ultraviolet stellar photography experiment, which used a star tracker to measure star brightness, free from the distorting effects of the atmosphere. Later that month, McKay went higher still in x-15-3, reaching an altitude of 295,600 ft. during the program's 150th flight. Although he had exceeded fifty miles, McKay, like Walker before him, would not be recognized as an astronaut by NASA until 2005, when he was posthumously awarded astronaut wings. Two days after McKay's trip to space, a new pilot flew the x-15 for the first time. Air force test pilot Captain William "Pete" Knight had followed a similar career path to his blue-suited colleagues, having graduated from the USAF TPS at Edwards before being assigned to the Dyna-Soar program. Following the air force space plane's cancellation, Knight attended the ARPS before joining the x-15 program as replacement for the soon-to-depart Joe Engle. As was customary, Knight's familiarization flight in x-15-1 was a (relatively) slow and low affair, reaching Mach 4.06 at 76,600 ft.

After a series of aborts caused by technical problems and poor weather farther up the High Range, Joe Engle made his sixteenth and final x-15 flight on 14 October, taking x-15-3 beyond fifty miles for the third time. Although he had originally been scheduled to take over the x-15A-2 from Bob Rushworth as the senior pilot's tour at Edwards drew to a close, Engle had decided to pursue a different career path. On 10 September 1965 NASA had announced that it was seeking a new group of astronauts for Apollo and its planned follow-on, the Apollo Applications Program. As an ARPS graduate and x-15 pilot, Engle was well qualified for the role, and after passing NASA's thorough medical, physical, and psychological tests, he was officially named as a member of astronaut group 5 on 4 April 1966. Although the nineteen new recruits were regarded as rookies in NASA's astronaut office, Joe Engle had the unique distinction of being the first NASA astronaut to arrive in Houston having already been awarded astronaut wings, thanks to his flights in the black rocket plane.

Back in the High Desert, the x-15 program—now in its sixth year—was entering its final planned phases. As the follow-on program continued, the x-15A-2 moved a step closer to its ultimate high-speed configuration when Bob Rushworth made the first flight with external tanks fitted. Prior to the flight, the FRC had carried out ground tests to ensure the tanks would separate cleanly without striking the x-15. In a characteristic show of improvisation, a 10 ft. deep pit had been excavated alongside the FRC ramp and filled with wood chips, covered with plastic sheeting. The x-15A-2 was then suspended above the pit with two steel girders of similar length and mass to the tanks attached. Tensioned cables on the girders provided simulated air loads and drag at separation speed, and high-speed cameras were used to capture every detail of separation. Following two successful tests, the x-15A-2 moved to the Rocket Engine Test Facility where the real tanks were used for full-duration test runs with both external and internal propellants.

Rushworth's 3 November test was the shortest and slowest nonemergency flight of the program, lasting only five minutes from launch over Cuddeback Lake to landing at Edwards. With its empty external tanks, the x-15 accelerated to Mach 2.5 before Rushworth hit the Empty Eject button on his left-hand console (a separate Full Eject button was also provided, should the pilot need to jettison full tanks). The tanks tumbled away as planned before their drogue chutes deployed to stabilize them. Although the oxygen tank was destroyed after its main chute failed to open, the test was still regarded as a success, clearing the way for flights with full tanks.

Milt Thompson's replacement, Bill Dana, ended 1965 with his debut flight. Hailing from nearby Pasadena, Dana had served in the air force before studying aeronautical engineering at the University of Southern California. During a visit to Edwards for employment interviews with some of the companies located on Contractors' Row, Dana decided to kill time ahead of an afternoon appointment by taking a drive around the base. Happening upon the HSFS, Dana was impressed by the range of aircraft in the NACA hangars and was hired on the spot as a stability and control engineer. He subsequently reported for work on 1 October 1958—the very day that the NACA ceased operations to be replaced by NASA. Graduating to the role of research pilot, he soon joined Armstrong and Thompson as a consultant on the Dyna-Soar program and flew supersonic transport studies as well as the M2-F1 lightweight lifting body, before moving to the X-15 as NASA's second project pilot alongside Jack McKay. Following an abort on 2 November, Dana made his first rocket plane flight on 4 November, just one day after Rushworth's tank test.

As winter rains flooded the lakes of the High Range, six months would pass before an X-15 flew again. Alongside routine maintenance, this downtime allowed crews to make modifications to all three aircraft, enabling them to carry additional follow-on experiments. By late April the lake beds were dry, but a series of six aborts affecting ships one and two meant it was 6 May before Jack McKay finally made the first flight of 1966. On this occasion, X-15-1 was equipped with four high-altitude experiments, including a micrometeorite collector located in the right wingtip pod, but a turbopump failure within seconds of ignition forced McKay to make another emergency landing. Although Delamar Lake offered a 2.9-mile runway, McKay found that he was not decelerating as quickly as expected after touching down. As he pulled back on the stick to place higher loads on the rear skids, the NASA pilot realized that he would not stop before reaching the shore and, fearing a repeat of his Mud Lake calamity, jettisoned the canopy before ploughing on through one hundred feet of lakeside scrub. Fortunately, the x-15 remained upright, and McKay was able to walk away uninjured.

On 18 May, Bob Rushworth was back in the x-15A-2 for a flight to evaluate ablative materials ahead of the aircraft's forthcoming high-speed flights. With ablator samples applied to its lower ventral, main gear struts, and leading edges, the usually sleek rocket plane sported a more motley appearance as it hung beneath the NB-52's wing on the flight out to Mud Lake. Thanks to their many modifications, all three x-15s had steadily gained weight throughout their flight careers, with the elongated x-15A-2 leading the way (especially if fitted with external tanks). To cope with this extra load, the launch aircraft's underwing pylon and shackles had been strengthened during the winter layoff, and today's flight marked their first true test. Following launch, Rushworth pushed the rocket plane to Mach 5.43 at 99,000 ft., exposing the ablative samples to high temperatures and dynamic loads. While the dark-pink Martin MA-25s ablator performed reasonably well, the Avcoat II material used on the wings' leading edges burned away completely, eliminating it from further consideration.

For Bob Rushworth, it was almost time to move on from Edwards and the x-15 program. With the situation in Vietnam worsening, the air force pilot was due to attend National War College before deploying to Southeast Asia, but Rushworth had one final flight to make before departing. Following his successful test with empty drop tanks the previous November, it was now time to see how the x-15A-2 performed with its maximum propellant load. Unfortunately, before this flight could take place, the entire Edwards community was rocked by one of the darkest days in its history.

On the morning of 6 June 1966, a group of six aircraft rendezvoused in the skies east of Edwards, assuming a V formation with North American's XB-70 bomber at its head. Joining the Valkyrie today were an F-4, T-38, F-5, and F-104, all powered by General Electric engines. The unusual formation had been organized at the company's behest so that the sixth aircraft, a Gates Learjet, could capture striking promotional images. Although North American had originally rejected GE's request for the XB-70 to be involved, the company relented after the air force XB-70 test director, Colonel Joe Cotton, and the deputy for systems test at Edwards, Colonel Albert Cate, both approved the photoshoot. Bad weather on the day meant the original plan to use the majestic Mount Whitney as a backdrop was dropped in favor of an alternate site near Barstow, and as the formation continued to trace a racetrack pattern above the Mojave Desert, the original thirty-minute window came and went as the photographers sought more time to capture the perfect shot. Joe Walker was piloting an FRC F-104N for the photoshoot, keeping station just off the XB-70's partially lowered starboard wingtip, but as the formation finally prepared to break after forty-five minutes, the other pilots were shocked to see Walker's aircraft abruptly roll up over the bomber's delta wing, removing both of the XB-70's vertical tails, before exploding into a fireball. Pilots Al White (North American's project pilot and Crossfield's former X-15 backup) and air force major Carl Cross fought to maintain control of the bomber, but without its tails, the bomber began to roll before snapping into a violent pitch-up, subjecting both men to tremendous forces as they sought to escape from their stricken ship. After much difficulty, White managed to close his escape capsule and eject, but Cross was less fortunate, remaining with the bomber as it smashed into the desert floor. Walker was killed instantly in the collision, the remains of his F-104 falling to earth some distance from the main crash site. As emergency crews rushed toward the rising columns of black smoke, they quickly located the injured White before attending to the grimmer task of recovering Walker's and Cross's remains from their respective aircraft.

Following an exhaustive investigation, an air force accident board concluded that Walker had inadvertently allowed the high-T tail of his F-104 to drift too close to the XB-70's drooped wingtip, neither of which had been in his field of view at the time. As its tail had come within feet of the bomber, the F-104 had been caught by the intense vortices that trailed from the XB-70's angular wingtips. The resulting forces had been too great for Walker to counter, with the F-104's roll into the huge bomber occurring so rapidly that he stood no chance of saving himself. The air force personnel who had approved and facilitated the photoshoot were all reprimanded for their roles in the incident, while the air force revised its regulations to prevent similar tragedies in the future.

Joe Walker had been an immensely popular member of the FRC family and an active member of the Antelope Valley community. Given the risks he had faced during his test-piloting career, it seemed all the more senseless that he should have perished on what had appeared such a routine flight. In the wake of the collision, some of Walker's colleagues petitioned unsuccessfully to have the FRC named in his honor, and testing activities at the center paused as managers reassessed the risks involved in both the x-15 and lifting body programs. But soon it was time to get back in the air again, and on 27 June, Bob Rushworth made a captive flight in the x-15A-2, complete with full external tanks, in preparation for his final flight of the x-15 program. The flight plan for 1 July called for Rushworth to climb away after launch over Mud Lake, accelerating to Mach 2 before pushing over to jettison the now-expended external tanks. From here he would accelerate for a Mach 6 speed run at 100,000 ft., subjecting samples of MA-25s ablator to their sternest test yet. As he fell away from the NB-52 for the final time, Rushworth got a clean light on the XLR99 and pulled the rocket plane into a 13° climb, but just as things appeared to be on track, Jack McKay called with the unwelcome news that the external ammonia tank did not appear to be feeding into the aircraft. If this was really the case, the x-15A-2 would soon become so unbalanced that Rushworth would be unable to control it—there was no option but to jettison the tanks and head for an emergency landing. As the pilot hit the Full Eject button, the tanks fired off at slightly different times, causing the x-15 to pitch and roll before its dampers caught the motions and steadied the aircraft. Although the XLR99 had burned for around thirty seconds, Rushworth elected to return to Mud Lake, where he brought the aircraft in for a safe landing. Postflight analysis revealed that hastily improvised flow instrumentation on the tank had given a false reading—the ammonia had actually been feeding into the x-15 as planned. After thirty-four flights spanning a period of six years, Bob Rushworth's time in the x-15 was over. Although he may not have made any of the early, headline-grabbing envelope-expansion flights, Rushworth had performed much of the program's grunt work, the demanding research flights where his skill and accuracy had proven so valuable.

Following an emergency landing, pilots could expect a reasonably rapid return to Edwards courtesy of the C-130 support aircraft, but the x-15 itself faced a far slower trip home. FRC employees Clyde Bailey and Ralph Sparks were among the team charged with retrieving and returning the research aircraft following remote lake bed landings. After removing all propellants, the partially dismantled x-15 would be lifted by crane onto a forty-foot flatbed trailer. The rocket plane's short twenty-two-foot wingspan enabled it to be transported along the highways of Nevada and California, with the unusual convoy slowly making its way back to Edwards. The 335-mile route back from Mud Lake took the x-15 through Death Valley and the gold-mining country around Johannesburg, and with the California Highway Patrol ruling that the convoy could not travel after 4:00 p.m., these trips became multiday adventures full of desert gas stations and cheap motels. Following Rushworth's final landing at Mud Lake, two motorists had an uncomfortably close encounter with the x-15A-2 as they sped toward Tonopah, camper in tow. Driving at the head of the convoy, Bailey tried to draw the couple's attention to the trailer-bound rocket plane, but when they failed to heed his warnings, the flimsy walls of their camper proved no match for the x-15's Inconel-X wing. Apparently oblivious to the damage, the couple continued without stopping as Bailey and Sparks did their best to explain to a highway patrolman what had happened. Some days after arriving back at the FRC, Clyde Bailey received a call from the irate motorists. Apparently the x-15 had taken the roof clean off their rented camper, and now they were seeking compensation for the damage. However, all threats of further action were hastily withdrawn as Bailey informed them that they had driven into a $70 million government airplane.

Following Rushworth's departure, the program entered a routine of sorts with Jack McKay continuing research flights in x-15-1, Bill Dana embarking on a series of high-altitude flights carrying multiple experiments in x-15-3, and Pete Knight taking over the x-15A-2. Among them, the three pilots logged a dozen flights over a ten-week period, with Jack McKay's final flight in the x-15 occurring on 8 September 1966.

Although the flight plan had called for the veteran NASA pilot to go out on a high with a flight to 243,000 ft., a fuel-pressure problem forced the unlucky McKay to make his third emergency landing at a remote lake bed—a record for any x-15 pilot. After twenty-nine x-15 flights, the larger-than-life McKay bowed out of the program. Since joining the HSFRS in 1952, he had made more rocket research aircraft flights than anyone except Scott Crossfield, earning a reputation as a solid and dependable research pilot. As one NASA engineer told journalist Richard Tregaskis during the early days of the x-15 program, "[McKay's] the kind of test pilot that'll fly for you all day long. He doesn't like to criticize the airplane. If it'll fly, he'll fly it—and maybe enjoy it. Then he goes home, and maybe has a few on the way." Sadly, the injuries sustained during his 1962 x-15 crash continued to take their toll on Jack McKay, ultimately contributing to his death in 1975, aged just fifty-three.

October 1966 saw a twelfth pilot fly the x-15. Major Michael Adams had joined the air force in 1950, flying forty-nine fighter-bomber missions over Korea during the later stages of that conflict. Having earned a degree in aeronautical engineering from Oklahoma University in 1958, he went on to study astronautics at MIT before attending the USAF TPS at Edwards in 1962. Graduating at the top of his class, Adams attended the ARPS before being assigned to the base's Manned Spacecraft Operations Division, where he was selected as an astronaut for the Manned Orbiting Laboratory (MOL), the air force's post-Dyna-Soar attempt to orbit military space station. By mid-1966 the MOL program's uncertain prospects led Adams to transfer across to take up the second air force x-15 seat.

Adams endured five aborts caused by poor weather and technical issues before finally making his first familiarization flight in x-15-1 on 6 October, but this turned out to be anything but routine. Having launched over Hidden Hills, he was eighty-nine seconds into his flight when a sudden loud bang was followed by a premature engine shutdown. Unbeknownst to Adams, a bulkhead in the x-15's ammonia tank had ruptured, and a combination of this and his push-over maneuver had disrupted the propellant feed, causing the XLR99 to quit. Unable to restart the rocket engine, Adams felt that he could still make it back to Edwards, but the experienced Pete Knight advised, "Let's make it Cuddeback, Mike." Adams took the emergency in his stride, landing without incident. It may not have been the ideal way to open his x-15 career, but he had shown that he possessed a cool head in a crisis. With his first flight out of the way, Adams began preparations for a research flight in x-15-3 the following month.

In the meantime, the other pilots remained busy, with Bill Dana taking x-15-3 to 306,900 ft. on 1 November—an overshoot of 39,900 ft. This would be the program's final flight beyond 300,000 ft., but like Joe Walker and Jack McKay, Dana would face a long wait before receiving his NASA astronaut wings. Pete Knight, meanwhile, continued to push the x-15A-2 toward its new target velocity, reaching Mach 6.33 on 18 November, the fastest speed yet recorded by an x-15. For this flight, Knight had used full external tanks to run the XLR99 for 136 seconds, a significant increase over the 80 to 90 seconds available using only internal tanks.

By now the three research aircraft were beginning to show their age. When North American had first delivered the x-15s, few would have imagined that they would still be flying eight years later, but the follow-on program and the modifications to x-15-2 had prolonged the rocket plane's lifespan well beyond original estimates. Milt Thompson would later observe that although their total flight time remained low, the constant cycle of maintenance and inspections simply wore the aircraft out. The high costs associated with x-15 operations had also come under renewed scrutiny, and by the end of 1966 the hypersonic research aircraft's days looked to be numbered.

With winter rains again softening the desert lake beds, it was late March before the 1967 flight campaign got underway. For Pete Knight it was time to test two more essential elements for his upcoming high-speed flights. Before applying the ablative coating, engineers wanted to test the eyelid device developed to protect the left windscreen glass from ablator residue. As it would only be opened during the landing pattern, Knight needed to verify that the reduced view still allowed him to judge his approach accurately. When Bob White had suffered from a shattered window earlier in the program, he had proven that landing using only one window was possible, if far from ideal. The second new element was a dummy scramjet attached to the aircraft's heavily modified ventral stub. By 1967 much of the earlier optimism surrounding air-breathing hypersonic propulsion had dissipated as the technical complexities and high costs involved became more widely appreciated. Although it looked increasingly unlikely that the hypersonic research engine (HRE) sponsored by Langley and the FRC would reach maturity in time to fly on the x-15, the FRC had created a dummy engine shape in order to investigate the aerodynamic effects of carrying the experimental payload.

On 8 May, Knight launched over Hidden Hills in the x-15A-2 (minus external tanks) to test the aircraft's stability with the barrel-like dummy scramjet attached. As he put the aircraft through its paces, reaching a maximum speed of Mach 4.75, Knight found that the x-15A-2 handled reasonably well until he opened the eyelid during his approach to Edwards. As the doors flipped open, they created a slight canard effect, causing the aircraft to pitch and roll slightly, movements that Knight now needed to counteract in addition to his other tasks. The approach was also marred when the left windscreen glass fogged and the dummy scramjet's parachute failed to operate correctly after it was jettisoned. Postflight inspections also revealed that ablative material attached to the leading edge of the ventral stub had been badly eroded by shockwaves from the dummy engine—a finding that should have raised concerns about the installation. Instead, x-15a-2 was withdrawn to receive its ablative coat while its siblings continued to fly a variety of follow-on experiments to high altitude.

One payload that was regularly flown during this period was the Western Test Range monitoring experiment, a classified attempt to use x-15-mounted sensors to track a Minuteman II missile fired from Vandenberg AFB, on the California coast. Unsurprisingly, getting the x-15 to altitude at the exact moment a missile was launched proved extremely challenging, and although the experiment was attempted on ten separate occasions, success proved elusive. During one such attempt on 29 July 1967, Pete Knight found himself facing one of the most serious in-flight emergencies of the program. Just over a minute into the flight, Knight had raced through 80,000 ft. when, in his words, "all of a sudden the engine went 'blurp' and quit." Within seconds, his instrument panel lit up with warning lights before going dark—both of the x-15-1's APUs had failed, taking their associated generators with them. Without electrical or hydraulic power, Knight no longer had dampers or aerodynamic controls, leaving him only the ballistic control jets—largely ineffective at his current altitude—with which to control the still-climbing x-15. Although an emergency battery allowed him to attempt an APU restart, he needed to wait for the turbines to spin down first, a process that took around thirty seconds. Without dampers, the wallowing x-15 topped out at 173,000 ft., but the coolheaded Knight had a plan: "When it rolled over to the left again, I looked down, and I was almost directly over Mud Lake. I thought, 'Well, if we get anything running, we will go back to Mud,' and I had already made that decision."

Down below, neither Mike Adams as NASA-1 nor the five chase pilots had any idea what had happened. When the x-15's generators failed, all telemetry had ceased—even the radar transponder had disappeared, leaving some to wonder if the aircraft had exploded during ascent. As the chase planes raced for the emergency lakes in the hope of spotting the missing x-15, Knight was battling to hold his reentry attitude while attempting to restart the apus and generators. Falling earthward, he heard the left apu spin back into life, but the generator remained stubbornly off-line, denying him electrical power. The apu did, however, restore the hydraulic controls, and even without dampers the air force pilot was still willing to take his chances with an emergency landing at Mud Lake rather than risk a high-speed ejection. Although still out of radio contact, Knight was finally spotted by the eagle-eyed Bill Dana as he approached Mud Lake. After sliding to a halt on the lake bed, Knight decided to free himself from the aircraft, but after opening the canopy and removing his helmet, he found himself unable to disconnect his suit's hoses. Knight pulled the emergency release in frustration, and the seat's headrest fired up into the canopy before falling back and gashing the test pilot's head, the only injury he sustained during the incident.

Although the exact cause of the power failure proved elusive, the most likely culprit appeared to be electrical arcing from the Western Test Range monitoring experiment. Pete Knight was awarded the Distinguished Flying Cross for bringing the unpowered x-15 home from an altitude beyond thirty-two miles, but the busy pilot was soon back in action as x-15a-2 emerged ready for its final push beyond Mach 6.

Following its previous flight, the aircraft had been thoroughly cleaned to remove oil and paint from its Inconel-X surface, before the gaps between its panels were painstakingly taped over in readiness for the ablative material. Although most of the x-15 would be coated with the sprayable ma-25s, premolded sections of a second ablator, esa-3560-11a, were used on areas such as the wing leading edges where temperatures would be highest. Once these had been applied, the rest of the rocket plane received repeated applications of ma-25s, building up layers of varying thickness depending on the expected heat load. After being left to cure, the entire aircraft was sanded to remove any irregularities, while ensuring that every surface had been covered to the correct thickness. When cured, the ma-25s was dark pink—possibly not a color that nasa or the air force would have chosen for their would-be record breaker. Fortunately, any embarrassment was spared as x-15a-2 received a final white protective coating. The now-gleaming rocket plane's transformation had required a huge amount of labor, calling into question whether the technique would be suitable for larger, more complex vehicles, while the additional weight and drag that came with the ablator meant that x-15a-2 was now unlikely to fly beyond Mach 7.

25. The heavily modified x-15a-2 in its ultimate form, complete with external propellant tanks and ablative coating. Courtesy NASA.

Pete Knight made the first flight of the newly protected aircraft on 21 August, but although the coating generally fared well during its first full-scale test, shockwaves from the dummy scramjet once again impinged on the ventral stub, charring and badly eroding the ablator. After its white coat had been refurbished, the aircraft was finally ready to push on into uncharted territory using its full range of modifications. The flight plan for 3 October called for Knight to climb to 62,000 ft., pushing over after sixty-five seconds to jettison the now-expended tanks at a speed of Mach 2. From here he would continue a gentle climb to 100,000 ft., accelerating steadily as the aircraft burned off its remaining propellants. Hitting maximum velocity, Knight would then perform a series of high-speed stability maneuvers before decelerating during the approach to Edwards.

On the day of the flight, things got off to a slightly uncertain start as Knight hit the launch switch and reached for the throttle, only to realize that x-15a-2 was still attached to its pylon. After hitting the button a second time, the rocket plane fell from its shackles, and Knight got a good light on the XLR99, rotating into a 15° climb as he rapidly pulled away from the NB-52. Around one minute later, he reduced X-15A-2's climb angle before sending the now-spent tanks tumbling toward the desert. With its weight and drag significantly reduced, the rocket plane continued to surge forward, finally coming level at 102,000 ft. Shutting the engine down after 139 seconds of powered flight, Pete Knight was now traveling at Mach 6.72—the fastest speed ever achieved by an aircraft. Streaking toward Rogers Dry Lake, he began working through the maneuvers on his test card, unaware that his aircraft was sustaining serious damage. Again, shockwaves from the dummy scramjet were impinging on the X-15's ventral stub, but at the blistering speeds the rocket plane was now traveling, the effects were far more severe than they had been on previous flights. With its protective coating applied, engineers had estimated that the aircraft's skin would reach maximum temperatures in the region of 600°F—well within Inconel-X's structural limits, but the severe shockwaves now pounding the ventral were actually generating temperatures in excess of 2,700°F. Under such extreme conditions, the ablative material was quickly eroded, exposing the underlying skin to the blowtorch-like heat. Tough as it was, the Inconel-X stood little chance, and it soon began to melt and roll back, creating a number of large holes and exposing the aircraft's internal structure to the superheated air.

Knight's first indication of a problem came when the H2O2 HOT warning illuminated on his instrument panel. To a rocket plane pilot, hot peroxide meant only one thing—the risk of explosion. Distracted by the warning, Knight was now approaching Edwards higher and faster than he had planned. Reaching the high-key point, he attempted to jettison the aircraft's remaining propellants, yet the telltale contrails of unused ammonia and LO2 failed to appear behind the X-15. Opening the eyelid, Knight now had a clearer view of the lake bed, and as he rolled in on approach, he received a reminder from Bill Dana to arm and jettison the dummy scramjet. Hitting the jettison button, Knight was surprised not to feel the jolt that usually accompanied separation, but observers confirmed that the scramjet was no longer attached to the aircraft. Bringing X-15A-2 to a halt on the lake bed, Knight found it curious that the recovery crews seemed more interested in the rear of his aircraft than in helping him from the cockpit, but when he finally saw the battle-scarred ventral for himself, he began to appreciate just how close he had come to disaster.

As engineers examined the charred rocket plane, the fate of the dummy scramjet remained a mystery, until some fine detective work by flight planner Johnny Armstrong pointed to the Edwards bombing range. The influx of hot gases had destroyed the engine's explosive bolts, causing the scramjet to separate from the ventral as Knight made his 180° turn in the landing pattern. After getting permission to enter the bombing range, Armstrong found the battered and broken scramjet within yards of his predicted impact point.

The indications of shockwave impingement on the ventral stub during earlier flights had been ignored, or at least not fully appreciated, and as a result, x-15A-2 had come perilously close to a major structural failure. Had maximum velocity been sustained for a few more seconds, the damage could have resulted in the loss of the aircraft and possibly its pilot, but for a second time Pete Knight's luck had held. Sadly, although x-15A-2 was repaired, other events would soon mean that its fifty-third flight would also be its last.

Two weeks after his record-breaking speed run, Pete Knight was back in an x-15, earning his astronaut wings during a research flight to 280,500 ft. in x-15-3, but nine days later, tragedy struck on what would be that aircraft's final flight. On 15 November 1967 Mike Adams launched over Delamar Lake for his seventh x-15 flight. With the xLR99 firing, he began the climb to 250,000 ft., where he would activate three different experiments located in the aircraft's tail and wingtip pods. Around sixty seconds later, the mh-96 control system's dampers failed, but Adams quickly reset these and continued to climb. x-15-3, with its unique flight control system, had been fitted with new cockpit displays and additional instruments as part of a wider evaluation of energy-management techniques. These included a computerized attitude display developed by NASA Ames, allowing the pilot to manage the aircraft's energy more accurately during the flight's boost and reentry phases. With x-15-3 now performing mainly as a high-altitude experiment carrier, this display had been further modified to help the pilot maneuver the aircraft into the precise attitudes required by certain experiments once beyond the atmosphere. Adams now switched the display into this second mode, but as he neared maximum altitude, he reported that the x-15 did not appear to be responding to its ballistic controls as expected. Endeavoring to follow his flight plan while dealing with the control issues, Adams began reading the Ames display as if it were still in its primary boost guidance mode, rather than the secondary experiment-pointing mode. As a result, the x-15 soon began to yaw to the right as the pilot attempted to correct his perceived heading error, his every control input making the situation worse rather than better.

Back in the control room, those monitoring the flight seemed unaware of the growing problem, with Pete Knight continuing to reassure Adams that things were looking good. As the aircraft continued to yaw, at one point turning a full 180° from its intended heading, it began to reenter the atmosphere, where the increasing dynamic pressure forced the x-15 into a hypersonic spin. Desperately, Adams called out, "I'm in a spin," yet although other witnesses, including Milt Thompson, clearly heard the call, Pete Knight appeared not to register or understand the message, asking Adams to repeat himself twice.

Fighting for control in the rapidly thickening atmosphere, Adams managed to arrest the spin, leaving the inverted x-15 diving toward the earth at Mach 4.7, but just as it looked like the pilot might recover the aircraft, the MH-96 forced the x-15 into a severe pitching oscillation. Within seconds, the rocket plane succumbed to the ever-increasing aerodynamic loads, breaking up at around 65,000 ft. before falling to the ground near the town of Johannesburg. Less than five minutes after leaving the NB-52, x-15-3 lay broken among the California scrub. When emergency teams reached the scene, they found Adams's body in the crushed cockpit. After 190 flights, the x-15 program had suffered its first fatality.

A joint accident board was immediately convened by the air force and NASA, but one vital piece of evidence was missing. The x-15 always flew with a camera attached to the inner rear of the canopy, from where it could record the aircraft's instrument panel. Unfortunately, the canopy had separated from the forward fuselage during the midair breakup. Analyzing the aircraft's trajectory, a search team managed to locate it, but to their dismay the camera had been torn loose and was nowhere to be found. After a week's delay due to bad weather, the search resumed, and the camera was found. But its film cartridge was missing. FRC engineer Vic Horton refused to concede defeat, organizing a further search of the area that finally yielded the missing film. Although damaged by exposure to the elements, the recording provided investigators with vital clues as to what had happened that November day.

Once all the evidence had been examined, the complex series of events could finally be pieced together. As Adams had climbed away following launch, a non-flight-qualified component used in the right wingtip pod experiment began to emit an electrical disturbance. This interfered with the MH-96 flight control system, causing the initial damper failure and the poor control response Adams noted later in the flight. Although ground controllers subsequently testified that they were aware Adams was experiencing problems and that they requested the experiment be terminated, this information was never passed on to the pilot by NASA-1. Seemingly distracted by his high workload and ongoing control problems, Adams used the Ames boost guidance display in the incorrect mode, making what he believed to be corrective control inputs, while failing to notice that the x-15 was yawing away from its intended heading. With the aircraft incorrectly aligned for reentry, the increasing dynamic pressure caused the rocket plane to spin before the MH-96 became stuck in the pitch oscillation—a condition traced back to a design oversight in an earlier modification.

Adams had previously mentioned experiencing disorientation during high-g climbs on earlier flights, a phenomenon that many x-15 pilots reported during the program. This led the accident board to suggest that vertigo may have impaired Adams's performance during the flight, although a 2014 analysis carried out by NASA's Engineering and Safety Center found little evidence to support this hypothesis.

During his final x-15 flight, Adams's altitude exceeded fifty miles and therefore qualified him for air force astronaut wings. These were presented to his widow without fanfare at a small ceremony held during January 1968. In 1991, following a long campaign by aviation historian Dennis R. Jenkins and others, Mike Adams's name was included on the Space Mirror Memorial, erected at Kennedy Space Center in Florida to honor those who have paid the ultimate price during America's exploration of space.

During a meeting held on 5 July 1966, the Aeronautics and Astronautics Coordinating Board recommended that air force funding for the x-15 program should cease at the end of 1967. If NASA wished to continue flying the x-15 beyond this point, it could still expect the cooperation of the AFFTC, but the civilian agency would have to fund the program itself. As this deadline drew closer, however, the air force had still not completed two of the follow-on experiments (the Western Test Range monitoring experiment and a high-altitude infrared background experiment), and so a compromise was reached, allowing the program to continue into 1968.

FRC director Paul Bikle had long been opposed to extending the x-15's operational life, but the prospect of flight-testing the HRE and the possible conversion of x-15-3 into a delta-winged hypersonic cruise test bed had garnered enough support to outweigh the center director's objections. However, Mike Adams's death and the loss of x-15-3 in November 1967 put a tragic stop to these plans. With only one airworthy x-15 remaining and NASA's budget in decline, it seemed highly unlikely that the agency would approve additional funding for the program. When the HRE flight-testing program was officially abandoned in December 1967, both NASA and the air force agreed that x-15 flight operations would cease at the end of 1968, although the aircraft should be maintained in a flyable condition into 1969.

With the lake beds once again dry following winter rains, flights resumed on 1 March 1968, with Bill Dana making another unsuccessful attempt to perform the Western Test Range monitoring experiment. The x-15 flew seven more times during 1968, with piloting duties alternating between Dana and Pete Knight. FRC research pilot John Manke had been due to join the program, but with only a single aircraft remaining, there was no longer a need for additional pilots. On 24 October, Bill Dana launched on his sixteenth x-15 flight, climbing to 255,000 ft. in another attempt to capture an elusive Minuteman II rising from Vandenberg AFB. Things seemed to be going to plan until the experiment lost power and retracted. As he brought the x-15 down onto Rogers Dry Lake to end the program's 199th flight, Dana could not have known that the x-15 would never fly again. With time running out before the end-of-year deadline, repeated attempts to launch the two hundredth flight were thwarted as technical problems and bad weather along the High Range conspired against Pete Knight. A final attempt on 20 December 1968 was canceled before the NB-52 even left the ground, as snow began to fall at Edwards AFB.

After more than nine years, time had finally caught up with the hypersonic research aircraft. The x-15 had come a long way since John Becker's team had circulated their design study in 1954. Having achieved a top speed of Mach 6.7 and a maximum altitude of 354,200 ft., the x-15 had significantly exceeded its original design goals, while broadly validating existing laboratory models regarding hypersonic flight. During the follow-on phase, the three aircraft had assumed a new role, routinely carrying experiments to high speeds and altitudes and returning them safely—a capability no other system of the time could easily match. Although at the program's outset it had seemed likely that the x-15 might carry the first human into space, the post-Sputnik space race drastically altered America's attitude toward the high frontier. As the aircraft's original parent agency, the NACA, had been replaced by NASA, the names of Crossfield, Walker, and White faded in the nation's psyche to be replaced by the heroes of Mercury. Away from the spotlight, the dedicated men and women of the AFFTC and FRC continued to write new chapters in the annals of aerospace knowledge, leaving a legacy of research and experience that would prove vital when NASA turned again to the idea of a reusable winged spacecraft.

As the teams at Edwards marked the end of the x-15, raising drinks in the bars of Rosamond to departed friends and future challenges, few beyond the High Desert noticed the rocket plane's passing. One day after x-15-1's final attempt at flight two hundred, the world was transfixed as astronauts Borman, Anders, and Lovell embarked on a Christmas voyage to the moon. Their *Apollo 8* flight marked humanity's first tentative step beyond our earthly environs, a timely reminder of just how quickly the technology of spaceflight had progressed in the years since Scott Crossfield had first flown the x-15.

10. Failure to Launch

Across its nine-year career, the x-15 surpassed all expectations, to become arguably the most successful research aircraft in history. At the program's outset, it was tasked with verifying predicted aerodynamic and heating models; examining how a vehicle's hot structure would tolerate high temperatures and loads during hypersonic flight; investigating high-speed stability and control as it exited and reentered the atmosphere; and finally, demonstrating that a pilot could perform in both high-g and zero-g conditions. Once government flights with the XLR99 engine began in early 1961, Joe Walker and Bob White were rapidly able to expand the x-15's flight envelope, achieving its design speed and altitude before the end of 1962. Although Walker later took x-15-3 to 354,200 ft. and Pete Knight reached Mach 6.72 in the much-modified x-15A-2, breaking records had never been the aircraft's raison d'être.

With its original research goals largely fulfilled by 1963, the x-15 was able to fly for five more years, enabling a wealth of additional research to be carried out under the extended follow-on program. The next phase of research—demonstrating that a winged vehicle could make a controlled lifting reentry from near-orbital velocities—would fall to a future research aircraft, and initial planning for a round three successor was already underway by the time the first x-15 emerged from Inglewood in September 1958.

By the mid-1950s, the air force had begun to consider its future role in space, viewing operations beyond the atmosphere as a natural extension of the service's existing remit. Around the same time, engineers at Ames and Langley were embarking on preliminary research into possible configurations for a round three research aircraft. Although the air force and the NACA had a cultural affinity with winged vehicles, both organizations recognized that development of an orbital space plane would present significant challenges, with flight-testing unlikely to occur before the early 1960s. Following discussions held in February 1956, the air force initiated Project 7976 to examine the potential for a winged spacecraft, alongside the more expedient ballistic capsule approach to orbital flight. By December 1957, with America still reeling from the shock of Sputnik, eleven companies had submitted their concepts for a manned satellite.

Having anticipated these requirements, Harrison Storms had already proposed what he believed to be the perfect solution. In the chief engineer's mind, North American was already building the nation's first spacecraft—the x-15. Although the rocket plane was limited to brief arcs beyond the atmosphere in its present form, the configuration had already undergone huge amounts of wind tunnel testing and was well understood. Following discussions with Charlie Feltz and Scott Crossfield, Storms proposed launching a stripped-down x-15 atop a cluster of now-surplus boosters from North American's recently canceled Navajo cruise missile program. The modified x-15 would employ a variety of exotic heat-resistant materials in place of the standard Inconel-X structure, allowing it to withstand the intense temperatures of reentry. Flights would launch from Cape Canaveral, with three Navajo boosters acting as a first stage, before a fourth booster at the center of the cluster took over as the second stage. The x-15's own XLR99 would then provide the final push to an apogee of 400,000 ft. After circumnavigating the globe, the x-15 would reenter at the orbit's perigee of around 250,000 ft. while passing above the southern United States. The pilot would then eject over land before the rocket plane flew on to a watery demise in the Gulf of Mexico.

The plan failed to impress the air force, being rightly viewed as little more than a risky stunt without significant research or operational value. However, as the conquest of space became a matter of national importance in October 1957, Storms submitted a more detailed orbital x-15 proposal as North American's official response to Project 7976. Now christened the x-15B, the rocket plane would carry two crewmembers on a three-orbit mission. Substantially larger than the original x-15, the x-15B would use two Navajo boosters as its first stage, with a third booster acting as the second stage. In place of the XLR99, the aircraft would be equipped with an XLR105 rocket motor, developed by North American's Rocketdyne Division for the Atlas ICBM. Producing 75,000 lbf. of thrust, the XLR105 was capable of placing the X-15B into a low but stable orbit. Toward the end of the third orbit, a pair of 5,000 lbf. solid rocket motors at the rear of the aircraft would fire to deorbit the rocket plane, before the X-15B reentered to make a controlled landing on Rogers Dry Lake.

By March 1958, Project 7976 had expanded to cover the air force's near and midterm spaceflight plans under the title Man in Space (MIS). A development plan published on 2 May detailed its intention to pursue the simpler ballistic capsule route for the initial Man in Space Soonest (MISS) phase. North American's proposal had been rejected for a variety of reasons, not least of which were serious doubts over the X-15B's ability to perform a stable reentry from orbital velocities. Although Storms may not have cared to admit it, the X-15 configuration had never been designed for this regime, and the costs and development time required to create a feasible orbital X-15 could not be justified. A second major reason for the rejection was the lack of a suitable launch vehicle. While North American still favored the Navajo booster, its record during earlier test flights had been less than encouraging, and the dynamics of clustering multiple boosters together had not been thoroughly investigated.

Undaunted, Storms returned for a third attempt in late 1959. This time, he proposed launching a slightly improved version of the X-15B atop a Saturn booster. Although this new plan allowed the X-15B to carry out limited orbital operations such as launching small satellites or inspecting targets of interest, many of the concerns raised regarding the previous proposal remained unaddressed. In continuing to push for an orbital X-15, Storms was also ignoring one inescapable fact; the air force was already working on a far more capable winged vehicle designed from the outset to withstand the challenges of lifting reentry from speeds above Mach 20. Dyna-Soar was the space plane the air force was banking on to become the foundation of its operational future in space.

The Dyna-Soar program ran between October 1957 and December 1963 and sought to develop a vertically launched space glider capable of performing a variety of military missions before making a controlled lifting reentry and horizontal landing. Dyna-Soar had its roots in the wartime work of Eugen Sänger and Irene Bredt, but the real impetus behind the various programs that eventually coalesced to become System 464L (as the project was officially known) lay in the postwar work of Walter Dornberger at Bell Aircraft.

As early as 1952 Bell had submitted its first proposal for a suborbital hypersonic bomber missile, better known as BoMi, to the air force. Although the official reaction to Dornberger's revolutionary weapons system was discouraging, with the WADC questioning its technical feasibility, additional studies and revised proposals would continue for many years, using a combination of company funds and air force development contracts. BoMi went through numerous configuration changes as Bell's understanding of the hypersonic flight regime improved during the 1950s, with the early three-stage piggyback concepts giving way to a sharply swept delta-winged dart mounted atop a conventional booster by 1956. As various air force systems requirements continued to shape Bell's boost-glide vehicle, other companies were also encouraged to submit concepts for piloted high-speed exoatmospheric bombers and reconnaissance vehicles under the program names Brass Bell and RoBo (for Rocket Bomber).

Recognizing that more research would be required before such systems would be truly feasible, the air force also began to work closely with the NACA, which was already considering what form its round three successor to the X-15 should take. By now, two main schools of thought had emerged within the civilian agency regarding the optimum configuration to tackle speeds beyond Mach 10. At Ames, a team led by Alfred Eggars and H. Julian Allen had devised a high-winged, flat-topped delta shape, with downturned wingtips, whereas on the other side of the country, John Becker's Langley team favored a flat-bottomed, delta-winged glider that would be less dependent on heavy heat shielding. After announcing a new Hypersonic Weapons Research and Development Supporting System (HYWARDS) program in November 1956, the air force and selected contractors began to work in partnership with the NACA to develop a new boost-glide research vehicle based on the Langley configuration, capable of achieving Mach 15 or above, as well as hypersonic lifting reentry.

By late 1957 the air force decided to consolidate all its hypersonic boost-glide efforts, announcing the new Dyna-Soar program just days after Sputnik took to the heavens. Dyna-Soar (a contraction of Dynamic Soaring—an alternative term to describe the boost-glide profile) would consist of three stages, starting with the initial research vehicle previously covered under HYWARDS, which would lead to a more advanced technology demonstrator that could perform limited operational missions. The final stage of the program would be a fully operational weapons system capable of undertaking a range of orbital reconnaissance and bombing missions by 1974. In January 1958 the air force invited the cream of America's aerospace industry to bid for the Dyna-Soar phase one contract, covering the initial research vehicle. With Becker's group at Langley continuing to offer guidance to the various bidders, the NACA acknowledged that this vehicle would become its de facto round three research aircraft.

But even as the air force Dyna-Soar project office began narrowing down the field to select its preferred contractors for the hypersonic glider and its launch vehicle, the program began running into political and financial barriers, exacerbated by confusion over the vehicle's true purpose. In November 1959, Boeing was selected to build the Dyna-Soar glider, with Martin providing a booster based on its Titan I ICBM. The system was scheduled to make its first piloted suborbital flights by May 1964, although these would be preceded by drop tests out at Edwards as well as unpiloted suborbital flights. Unfortunately, as the 1960s dawned and the nation looked toward NASA's Project Mercury to launch the first American into space, Dyna-Soar came under repeated pressure to justify its existence as either a research vehicle or a weapons system. Squeezed by limited finances, the project office issued a dizzying flurry of program plans and revisions in an attempt to keep up with the Department of Defense's perceived priorities. Plans continued to change, even as Boeing unveiled its final design for the sleek black space glider in 1961, with Dyna-Soar now switching from the Titan I to the more powerful Titan II and the shorter suborbital phase one flights being dropped in favor of a streamlined program that would jump straight to single-orbit test flights.

By now Dyna-Soar was not only at risk due to the Department of Defense's apparent lack of conviction in the need for such a vehicle; it was also coming under threat from within the air force itself as new programs were proposed to perform many of the same roles as the delta-winged glider. As the first NASA astronauts began to fly in their Mercury capsules and President Kennedy set America on a course for the moon before the end of the decade, the air force found itself struggling to justify why it needed its

26. A mock-up of Boeing's x-20 Dyna-Soar vehicle. Courtesy USAF.

own human spaceflight program, especially as the bombing and reconnaissance roles could now be filled by ICBM's and satellites, respectively. More changes to the proposed launch vehicle, which eventually saw Martin's proposed Titan IIIC selected, led to increased costs and sliding schedules, and although the program was able to name its first batch of astronauts in 1962 (including FRC pilots Milt Thompson, Neil Armstrong, and Bill Dana, alongside future AFFTC x-15 pilot Pete Knight), Dyna-Soar's prospects were becoming increasingly uncertain. Drawing the focus of Kennedy's secretary for defense, Robert McNamara, the vehicle now received a new designation, becoming the x-20, to reinforce its role as a research aircraft and increase NASA involvement. Out at Edwards, engineers from the AFFTC and FRC began to plan Dyna-Soar flight profiles and research objectives, while the pilots stepped up their training for the demanding launch and reentry phases. But when McNamara requested a study comparing Dyna-Soar's suitability for the air force's specified missions to that of NASA's new two-person Gemini capsule in January 1963, the writing was on the wall for Boeing's space plane.

Rumors of Dyna-Soar's cancellation had been swirling around Washington for some time, but the confirmation finally came on 10 December 1963, when McNamara announced that Dyna-Soar would be canceled and its funds redirected toward a new Manned Orbiting Laboratory program. In the end, Dyna-Soar succumbed to a combination of factors. At the program's outset in 1957, the air force had reasonably assumed that it would play a leading role in America's manned space program, but when President Eisenhower handed this responsibility to the newly created NASA in 1958, a parallel military effort became increasingly difficult to justify. Had Dyna-Soar been conceived as a pure research program like the X-15 or its predecessors, it might have found a more receptive, if limited, audience within Air Force Headquarters and the Department of Defense, but that had unfortunately not been the case. Dyna-Soar's cancellation effectively ended NASA's hopes for a round three vehicle to research lifting reentry from orbit; there would be no immediate follow-on to the X-15.

Speaking in February 1965, Walter Dornberger expressed his opinions on McNamara's decision, predicting, "The cancellation of the Dyna-Soar space glider program will prove one day to have been a serious setback. We now have no experimental manned spacecraft capable of accomplishing tasks required in the predictable future." Even though Bell Aircraft had not been selected to construct the vehicle, the wall of Dornberger's office bore a large painting of Boeing's Dyna-Soar, complete with a brass plaque that read simply, "Dyna-Soar, born 1952, Walter Dornberger, died 1963, Mac the Knife."

Although the Dyna-Soar never flew, the program did advance our understanding of the challenges a hypersonic lifting vehicle would face as it returned from orbit. Many of the exotic materials such as zirconium, carbon, molybdenum, and columbium that Boeing intended to use on the vehicle, were tested during the Aerothermodynamic Elastic Structural Systems Environmental Tests (ASSET) program initiated by the Air Force Flight Dynamics Lab (AFFDL) in January 1961. ASSET used small unpiloted delta-winged glide vehicles launched atop Thor missiles to expose materials and structures to the stresses of hypersonic reentry, with the vehicle's aerodynamic design based of the forward section of Boeing's Dyna-Soar, complete with blunt nose and rounded leading edges.

The first ASSET flight took place on 18 September 1963—just months before Dyna-Soar's cancellation; although the two programs were not officially linked, McNamara proposed an expansion of ASSET during his 10 December announcement, seeing it as a more cost-effective means to satisfy many of Dyna-Soar's research goals. Five more ASSET flights were made, and although one failed due to booster problems, the program was hailed as a success, having fulfilled all its major goals by the time of the final flight in February 1965.

For the AFFTC and the FRC, Dyna-Soar's demise brought an end to the higher-and-faster era that had begun with the x-1. In the absence of a round three vehicle, the x-15 flew on, retaining its place as the highest and fastest flying of the piloted research aircraft. The x-15 directly benefitted from the Dyna-Soar cancellation, with the now surplus MH-96 adaptive flight control system and stable platform finding their way onto the round two rocket plane.

Unfortunately, the lack of a round three program and the air force's move away from higher-speed operational aircraft meant there were fewer direct beneficiaries of the x-15 research than one might have imagined at the program's outset. Many of the rocket plane's most valuable contributions came in the form of the operational lessons learned by the AFFTC and the FRC during its development and operation. With Walt Williams, Gerald Truszynski, and Dick Day all transferring from the High Desert across to NASA's orbital spaceflight program during its formative years, the influence of the x-15 program on Project Mercury and its successors was clear to see. The establishment of the fully integrated High Range, with its central control room, formed the blueprint for the control centers and tracking networks that played such a major part in the success of NASA's space program, with the role of NASA-1 reprised as the capsule communicator, or CAPCOM. The extensive use of simulators for both flight planning and pilot training was quickly adopted by the Mercury, Gemini, and Apollo programs, with integrated simulations involving crew and controllers becoming a central element of mission preparations.

When NASA's Mercury astronauts entered the Johnsville centrifuge during their training, they were following in the footsteps of the early x-15 pilots. The work that Scott Crossfield and his colleagues had done in assessing a pilot's ability to perform in a high-g environment helped define the aeromedical standards used during the space program. But beyond the transfer of training and support experience, the x-15 also made two major contributions to NASA's plans for its post-Apollo spacecraft—the Space Transportation System, or space shuttle.

The new spray-on ablative materials that emerged during the 1960s, bringing with them the promise of easy application and refurbishment, appeared to offer great potential for reusable space vehicles. However, the experience gained through using such materials on the x-15A-2, a complex vehicle with moving control surfaces and numerous access panels, proved disappointing. Weeks of spraying and sanding were needed to ensure that every part of the airframe had been coated to the correct depth, a time-consuming process that would need to be repeated after every flight. Although the ablator's performance was generally regarded as satisfactory during the two flights on which it was used, NASA realized that the approach would prove impractical for a larger operational vehicle with high predicted flight rates.

One major debate during the shuttle's development centered on a need for air-breathing engines for use during approach and landing. The FRC's pilots and engineers would argue that an unpowered shuttle could make an accurate landing simply by managing its energy during reentry and descent, just as the x-15 had done on so many occasions, but in making this case, they were also able to draw from their experiences with a new type of vehicle—the lifting body.

Part 3

The Lifting Bodies

11. Look Ma, No Wings!

On the morning of 15 March 1958 throngs of scientists and engineers began to gather at the Ames Aeronautical Laboratory in Sunnyvale, California, for the latest in a series of NACA conferences on high-speed aerodynamics. Since 1946 these conferences had provided the NACA a means to share its latest research on high-speed flight with the aviation industry, military, and staff from its own centers, but on this occasion, many of the forty-six papers that were presented reflected a bold new direction for the civilian organization. Whereas Soulé and Becker's talk of space leaps had met with some resistance just four years earlier, the NACA now stood ready to embrace the post-Sputnik realities of the space age, and much of the conference's first session focused on the different schools of thought regarding human spaceflight that had emerged from Langley and Ames during 1957.

First to speak was Max Faget, the diminutive bow-tied Langley engineer who had become the leading advocate for nonlifting ballistic reentry vehicles. Building on earlier research into the hypersonic characteristics of blunt bodies, Faget believed that a simple conical capsule not only could withstand the rigors of reentry but would also prove small and light enough to fly on the nation's largest available booster at the time (the Atlas ICBM) without the need for additional stages. With deceleration forces predicted to reach 8.5 g's during reentry, the pilot would need to remain supine in a form-fitting couch in order to remain conscious, while the simple ballistic capsule would offer no means of aerodynamic control as it plummeted back to the earth, meaning recovery forces would have to monitor a large landing footprint. But even taking these drawbacks into account, Faget remained confident that his approach offered the best prospect for placing an American into orbit in the short term, concluding, "It appears that, as far as reentry and recovery is concerned, the state of the art is sufficiently advanced so that it is possible to proceed confidently with a manned satellite project based upon the ballistic reentry type of vehicle."

Faget's Langley colleague John Becker presented a very different vision for orbital spaceflight. Pointing to the work his team had done in support of the air force HYWARDS program, along with the NACA's independent round three studies, Becker favored a flat-bottomed, delta-winged vehicle capable of making a lower-drag lifting reentry. Although more technically challenging than Faget's ballistic capsule, a delta-winged glider (as proposed for round one of the Dyna-Soar program) would only subject its pilot to a maximum g-load of around 2.5, allowing upright positioning in the cockpit. A vehicle of this type would also be capable of maneuvering to either side of its original ground track—an ability known as cross range—opening up the prospect of runway landings across a large proportion of the United States. Although Becker conceded that these gliders would be heavier than Faget's capsules and require more sophisticated heat shielding to offer protection during their extended reentry periods, these problems appeared surmountable, and the resulting vehicle might well open the way to more flexible and economical space access in the future.

Somewhere between these two approaches, however, lay a third option—a wingless lift-generating vehicle that promised the low-g and cross range of Becker's hypersonic gliders while also taking advantage of the heat-deflecting blunt-body principle behind Faget's conical capsules. This hybrid approach, presented by Ames engineer Thomas Wong and colleagues, drew on earlier research into nonsymmetrical reentry shapes—slender half cones that generated lift without the need for protruding horizontal surfaces such as wings or conventional tails. These unusual configurations became known as lifting bodies, and they would go on to form the basis of a major new flight research initiative out in the High Desert.

The lifting body story began with the pioneering work of H. Julian Allen during the early 1950s. As chief of the High-Speed Research Division at Ames, Allen and his colleague Alfred Eggars had demonstrated that a large detached shockwave would form ahead of a blunt body at hypersonic speeds. This shockwave would dissipate much of the heat generated during reentry before it came into contact with the vehicle itself, with the remainder being carried away by ablative materials. Following its publication in 1953, the first major application for Allen's blunt-body theory was the development of blunted conical warheads for ballistic missiles, but the researchers had also noted that these high-drag shapes generated small amounts of lift. Picking up on this, Eggars and Ames colleague Clarence Syvertson went on to show that if one side of the cone was removed to create a flattened upper surface, the resulting shape could attain a high enough L/D to perform a lifting reentry. By adding small control surfaces to the rear of the vehicle, the length and direction of the glide could also be controlled, and this led to the initial Ames M1 lifting body.

Understanding that these half-cone configurations would endure far lower g-forces during reentry than ballistic vehicles would, Ames researchers responded to the growing NACA and military interest in manned satellites by developing a piloted lifting body configuration using the M1 shape during 1957. However, one major problem remained; although lifting bodies performed well at hypersonic velocities where high drag was an advantage, their ability to remain stable during the transition from supersonic to subsonic speeds seemed far from certain. When wind tunnel testing revealed that the M1 configuration had a worrying tendency to tumble at lower speeds, additional research yielded a slender, boat-tailed half-cone shape that offered improved stability and lower subsonic drag. This configuration, known as the M2, was subsequently fitted with vertical fins to aid directional stability, earning it the nickname the Cadillac Model, and the unconventional vehicle soon attracted the attention of the FRC's Dale Reed.

Having earned a degree in mechanical engineering in his native Idaho, Reed journeyed to the HSFRS in 1953, recognizing that the NACA facility was the place to be for a young engineer wishing to work at the forefront of aeronautical research. He began his career calculating structural loads on many of the station's round one research aircraft, including the D-558-II and the X-IE, before moving into hypersonic-heating research when the X-15 arrived, but it was Reed's continuing interest in stability and control that led to his fascination with lifting bodies. By 1962 various groups within both NASA and the aerospace industry had begun to examine lifting body configurations, but Reed noted that a great deal of skepticism still surrounded the ability of these vehicles to remain stable at low speeds. Although some engineers were proposing deployable wings or jet engines for use during the approach and landing phase, practical research into the low-speed characteristics of lifting body shapes remained relatively scarce. As a keen model

27. NASA Flight Research Center engineer Dale Reed holds his
M2 model in front of the M2-F1. Courtesy NASA.

aircraft builder, Reed decided to tackle the problem head-on by building
a two-foot-long version of the M2 Cadillac Model out of balsa wood, in
order to test the configuration for himself.

The hallways of Building 4800 at the FRC soon echoed to Reed's footsteps, as colleagues were treated to the incongruous sight of the tall, bald-

ing engineer towing his M2 model at the end of a length of string in order to demonstrate the shape's inherent stability. Free flights from the center's rooftops followed, but although many of his fellow engineers were impressed by the model's performance, director Paul Bikle and chief of research Thomas Toll remained unconvinced that these encouraging results warranted a full-scale test program. Undeterred, Reed set about recruiting a handful of like-minded colleagues to help him put together a more detailed proposal, and his first volunteer soon arrived in the shape of junior engineer Dick Eldredge. Between them, Reed and Eldredge were able to work out the finer structural details of their proposed vehicle and devise a convincing test plan, but they recognized that Bikle was more likely to view their proposal favorably if it had the support of one of the center's research pilots. Fortunately, Dale Reed knew just the right man to approach.

Milt Thompson had already shown a willingness to work with unconventional flying machines, having been one of the instigators of the FRC's Parasev research program. Built as a proof-of-concept vehicle for the flexible Rogallo wing (then intended for NASA's Gemini program as a replacement for parachute recovery), the Parasev had made its initial flights while being towed across the lake bed by one of the center's carry-all trucks, before progressing to higher-altitude tests using a tow aircraft. Reed was considering a similar two-stage testing process for his lightweight lifting body, and after hearing Reed and Eldredge's pitch, Thompson was quick to lend the project his full support. As the trio continued to work on their proposal, Thompson suggested that they should also enlist the support of one of the lifting body concept's originators, Alfred Eggars, as he not only would bring added gravitas to their case but was also in a position to arrange wind tunnel time at Ames. When Eggars responded positively to Reed's approach, the group asked him to attend their upcoming presentation to Paul Bikle, where they hoped to gain the director's support. When Reed, Eldredge, and Thompson presented their proposal using movies that Reed's wife, Donna, had made of the M2 model in flight to demonstrate the vehicle's potential, both Bikle and Eggars agreed to support a modest research program, with Eggars pledging the required tunnel time and Bikle providing funding for the vehicle's construction and support during flight tests.

Under Paul Bikle's leadership, FRC employees were always aware that NASA Headquarters was located more than two thousand miles away, and the M2-F1 program (manned configuration 2, flight version 1) provided a perfect illustration of this maxim. Given the limited, low-cost nature of the program, Bikle chose to finance the M2-F1's construction using his discretionary funding. By defining the lifting body as a full-scale wind tunnel model, the director saw no need to involve Headquarters at this stage, although, as Milt Thompson remembered, "If the aircraft just happened to be capable of flight, that was something that was out of management's control." Bikle favored a Skunk Works approach (as pioneered by Kelly Johnson's Advanced Development Projects Division at Lockheed) for the M2-F1's construction, with the vehicle's tricycle inner structure being constructed in-house by the FRC's talented fabrication team under the supervision of operations engineer Vic Horton. This structure would then be fitted with a series of plywood outer shells, starting with the M2 but possibly extending to other configurations. As the FRC lacked specialist woodworkers, Bikle recommended that work on the shell should be contracted out to accomplished local sailplane maker Gus Briegleb.

As work on the inner structure got underway, a sign reading Wright Bicycle Shop appeared on the large canvas curtain separating the M2-F1 team from the rest of the fabrication shop. Behind this curtain, engineers and technicians worked together, rapidly solving any problems that arose during the tricycle framework's construction, while using off-the-shelf components, such as a Cessna landing gear system, to save time and reduce costs. Meanwhile, Briegleb's small team were making good progress on the M2 shell at his El Mirage workshop. The sailplane maker had originally quoted Bikle a price of $5,000 to construct the shell within four months, but the center director believed that the task would prove more complex than Briegleb was anticipating so promptly doubled this budget. As work on the M2-F1's plywood shell continued, the FRC began tow testing the now-complete tricycle framework to check its brakes and steering.

Just four months after Bikle gave his approval, and with less than $30,000 spent, the M2-F1 was ready for testing to begin. Just 20 ft. long, with a width of 14.4 ft. (including the outboard elevons), the M2-F1 lived up to its lightweight tag, tipping the scales at a mere one thousand pounds. Along with its Perspex canopy, the lifting body also featured a glazed nose and side panels to improve the pilot's forward vision when the vehicle was at higher angles of attack during the crucial landing flare. Roll control came from

the elephant-ear elevons, with rudders at the rear of each vertical fin handling yaw. A large body flap fitted to the upper rear of the M2-F1's fuselage
had originally been intended to provide both pitch control (if both halves
were moved symmetrically) and additional roll control (if the halves were
moved asymmetrically), but engineers decided to bolt the two halves of the
flap together to provide pitch control only. In order to manage this unusual
array of the control surfaces, Eldredge devised a central swash plate system
to which all the control rods were linked, meaning rapid control configuration changes could be made if required. Although a center fin was also
fabricated in order to provide additional directional stability, this was later
discarded after tests suggested that it was unnecessary.

With its white finish and red lateral stripe, the M2-F1 stood in stark contrast to the dart-like X-15s and F-104s that sat alongside it in Hangar 4802.
Although the lightweight lifting body soon gained the unflattering but
descriptive nickname the Flying Bathtub, it held the potential to be every
bit as revolutionary as the famed research aircraft that had gone before it.
Now all that was needed was a means to tow the M2-F1 on its initial lake
bed test runs, and the FRC's solution proved almost as unconventional as
the lifting body itself.

The ultimate aim of the M2-F1 program was to demonstrate that a lifting
body, with its inherently low L/D and corresponding high sink rate, could
be controlled during approach, flare, and landing. In order to achieve this,
the wingless glider would be towed to a release point above the lake bed
by the FRC's trusty C-47 support aircraft, but before committing to these
altitude flights, the team needed to learn more about the M2-F1's handling
by making ground tows across the lake bed.

By now, Reed's team had been supplemented by engineers Ken Iliff,
Bertha Ryan, and Harriet Smith, and between them the new volunteers
began to analyze the lifting body's likely flight characteristics. When Iliff
calculated that the M2-F1 would need to reach 85 mph before it would leave
the ground, Reed realized that none of the FRC's existing support vehicles
offered the performance needed to get the glider aloft. Taking the problem to Bikle, the pair soon agreed that a new, more powerful tow vehicle
would be required, but fortunately, the director knew just whom to turn to
for advice on the subject. Walt "Whitey" Whiteside was a longtime associ-

ate of Bikle's, with the two men's careers having run in parallel since they had both joined the USAAF's Flight Test Division at Wright Field in the 1940s. During Bikle's time as technical director of the AFFTC, Whiteside had served as his assistant director of maintenance, and after becoming director of the HSFS, Bikle wasted no time in bringing the tough Californian across to NASA. There had always been a lively automotive culture at Edwards, with car clubs and impromptu drag races playing an important role in the social fabric of the Antelope Valley. With his love of hot rods and off-road motorbikes, Whiteside played an active part in this scene and so was more than happy to help when Bikle asked him to source a high-performance tow vehicle for the M2-F1. Speaking in 1997, Whiteside recalled, "I had followed the races and knew where all the shops were . . . so I broke loose and just took off and went all over Southern California looking for what can put the most horsepower on the ground so that it can tow something out on the lake bed."

Whitey's final selection was a 1963 Pontiac Catalina convertible fitted with the largest triple-carburetor engine GM could offer. With engine and transmission modified and tuned for maximum performance, Whiteside also sourced the right wheels and tires for maximum acceleration across the lake bed's surface. Roll bars and radios were then fitted, with the front passenger seat being reversed to allow the "copilot" a good view of the M2-F1 (a second, sideways-facing observer's seat was also fitted in the rear of the car). As accurate data on tow speeds would be vital during test runs, the Pontiac was equipped with air data instruments to supplement its standard speedometer. Finally, to ensure that the white convertible maintained the government issue look of the FRC's other support vehicles, its hood and trunk were painted a high-visibility yellow, while the doors were labeled "National Aeronautics and Space Administration."

Like any good research vehicle, the Pontiac needed a few calibration "flights" to ensure that its instrumentation was in good order, and Whiteside would venture across into neighboring Nevada, where speed limits did not yet apply on the long desert roads. As possibly the only government-purchased muscle car in existence at the time, Whitey's creation would be able to tow the thousand-pound M2-F1 up to a speed of 110 mph in a mere thirty seconds—it was time to see if the Flying Bathtub would really fly after all.

There had been some debate among the project team as to how the M2-FI's control system should be rigged for lateral (roll) control, with two options being considered. The first of these used left or right stick movements to deflect the outboard elevons, while the rudder pedals would control the vehicle's two vertical rudders. The second method reversed this arrangement, using the stick to move the rudders, while the rudder pedals operated the elevons. Although the first method seemed more intuitive to many of the team, Milt Thompson's experiences in the simulator led him to select the less conventional second option. Despite warnings from Iliff that this arrangement might lead to serious control issues, the team agreed to support their pilot's decision.

Tow testing got underway on 1 March 1963, with Whiteside upping the Pontiac's speed with each successive run. At 60 mph, the M2-FI's nose lifted away from the lake bed, giving Thompson a chance to get a feel for the aircraft's controls while checking his forward visibility in a nose-high attitude. Once satisfied, he gave Whitey the go-ahead to make a run at takeoff speed. As Whiteside reached 86 mph, the M2-FI strained to break contact with the lake bed at the end of its one-thousand-foot towline, but rather than lifting cleanly away from the ground, the glider bounced unsteadily from left gear to right, stubbornly refusing to remain level until Thompson lowered the nose. After making more runs with the same results, the team were forced to abandon the tests, with the pilot conceding that he felt the M2-FI would have rolled had it become airborne. Although Thompson believed that the vehicle's landing gear was at fault, Reed had little doubt that the unconventional roll control arrangement was the real culprit. Before any more tow attempts were made, the M2-FI was loaded onto a trailer and driven north to Ames for a series of tests in the huge forty-by-eighty-foot tunnel.

By comparing the aerodynamic characteristics of the full-size vehicle with results from earlier subscale tests, Reed was sure the team could isolate the lifting body's control problem while also gaining a more accurate estimate of its actual L/D. The M2-FI was mounted in the center of the tunnel's cavernous test section on three twenty-foot-high instrumented pylons, but there was one small snag. With no means to operate the control surfaces remotely, someone would need to be in the cockpit during the test runs. Having not enjoyed his first experience of sitting in the plywood lifting body as the 135 mph airstream roared past, Dale Reed was happy to leave

STEP ON IT WHITEY, HE'S GAINING ON US ! !

28. A humorous depiction of an M2-F1's ground-tow test from the
Flight Research Center's X-Press newsletter. Courtesy NASA.

this task to Thompson and Eldredge, and between them the duo would
endure long hours in the isolated cockpit before their colleagues eventually released them at the end of each test session. Analysis of the resulting
data revealed a few unexpected anomalies that were quickly fixed by the
engineers. The tests appeared to support Reed's intuition regarding the roll
controls, and the swash plate system was duly rerigged to the more conventional arrangement before the aircraft was retested with more promising
results. The Ames tests also showed that the full-size vehicle, with its permanently extended landing gear and other protrusions, had a somewhat
lower-than-expected L/D of three, meaning it would descend one foot for
every three it moved forward.

When ground tows resumed on 5 April, Thompson found that he was
able to lift the M2-F1 away from the lake bed and attain level flight without any of the earlier stability problems, maintaining an altitude of 10 ft.
for the whole four-mile run. As tow speeds increased, Thompson was soon
reaching 200 ft. before releasing the tow rope to practice his landings,
but with such a high predicted descent rate, the team recognized that the
approach and flare were still likely to prove challenging during the upcoming air-tow flights. In order to provide some margin of safety, a small 250

lbf. solid rocket motor was fitted in the M2-F1's lower fuselage to provide instant L/D by lowering the glider's sink rate and extending the flare maneuver. As an additional precaution, a lightweight ejection seat was also fitted, allowing the pilot to leave the lifting body if a heavy landing or loss of control looked imminent.

On 16 August 1963 Milt Thompson stood ready to make his first air-tow flight, but the morning did not start well. After the Pontiac had towed the M2-F1 to its takeoff point at the southern end of the lake bed, the team were left waiting as the C-47 tow plane failed to arrive. When Jack McKay finally showed up in the dual-engine transport some forty-five minutes later, the one-thousand-foot towline was hooked up to the M2-F1, and following final checks, the first takeoff was attempted. As the M2-F1 reached 80 mph, Thompson pulled back on the stick to raise the glider's nose, but within seconds, the towline fell away from the C-47, thwarting the attempt. If the rope had come loose just seconds later, with the lifting body between 100 and 200 ft. but still carrying insufficient speed to make a landing flare, Thompson would have been left with little option but to eject. Once the C-47's tow hook had been modified and tested, the unshaken test pilot made a second attempt, and this time the towline held. As the M2-F1 rose into the air, taking up position above the C-47's wake, Thompson gave McKay the go-ahead to lift the transport from the lake bed, and the two aircraft gently climbed away toward the lake's northern shoreline. Reaching the north end of the lake, McKay circled up to the agreed 5,000 ft. release altitude before turning south to line up with Runway 18, Thompson's intended destination. After ensuring that his instrumentation and cockpit camera were operating, Thompson instructed the C-47 to increase its speed to 120 mph before he released the towline and began the steep plunge toward the lake bed. The M2-F1 was flying free at last, but with a descent rate touching 4,000 ft. per minute, this first flight would be a brief affair. As the FRC workforce crowded onto the center's rooftops and parking lots, craning for a glimpse of the strange wingless aircraft, Thompson flared out of his dive and brought the lifting body in for a smooth landing on the hard clay runway.

The flight had taken less than two minutes from release to touchdown, but the significance of Thompson's achievement was clear. The M2-F1's debut had provided the first proof that a lifting body shape could make an approach and landing without the additional complexities of deployable wings or jet engines.

Milt Thompson made sixteen more glides in the M2-F1 before the end of 1963, gradually probing the aircraft's stability and handling across a range of flight conditions. By December he felt confident enough in the unconventional glider's performance to begin checking out other pilots, starting with fellow FRC pilot Bruce Peterson and air force flight test legend Colonel Chuck Yeager. Peterson had joined the FRC as an engineer in August 1960, but like Thompson, he soon made the switch to the pilots' office, where he was initially assigned to the Parasev program. With Peterson having mastered the art of flying under tow during his time in the kitelike glider, Thompson felt that the young research pilot would be well suited to the lightweight lifting body program, and after a series of buildup runs behind the Pontiac during November 1963, Peterson was soon ready to make his first air tow.

Yeager had returned to Edwards AFB in July 1962 as commandant of the USAF TPS. As this role now encompassed training air force astronauts via the recently formed ARPS, the veteran test pilot was interested in evaluating the lifting body as a potential training vehicle for his students. Peterson and Yeager both began their air-towed flights on 3 December, with Yeager declaring, "She handles just great!" after winning a wager with Thompson by making a pinpoint landing on his first flight. Unfortunately, later in the day, Peterson landed heavily, and when the landing gear shock struts, with their now-chilled oil, failed to absorb the impact, both rear wheels sheared off the vehicle, putting an end to the day's activities. "It really was a hilarious scene," Thompson later recalled. "Those wheels seemed to roll on for miles."

The repaired M2-F1 was back in the air the following month, and as Peterson and Yeager completed their checkout flights, the FRC's Don Mallick also made two familiarization flights in the lifting body. Throughout 1964 the M2-F1 became a regular sight in the skies above Edwards, with Thompson and Peterson totaling twenty-nine flights between them as they sought to complete the vehicle's aerodynamic analysis. Although the M2-F1 was now being towed to around 12,000 ft. before being released, its flights remained brief affairs with the pilots barely able to perform four test maneuvers before they had to set up for landing. As the data obtained during these flights was compared to earlier wind tunnel results, confidence in the lifting body concept grew, but the tests also revealed some worrying characteristics. As the M2-F1 lacked conventional horizontal surfaces such as wings or a tail to help damp lateral movements, it had always been expected to roll easily. Although the outboard elephant ears did provide limited lateral stability, their use tended to yaw the vehicle into a sideslip that actually increased the lifting body's tendency to roll if large rudder movements were applied in an attempt to straighten the vehicle up—a condition known as effective dihedral. Pilots were advised to avoid excessive rudder use, but problems could still occur if larger adjustments were made to control the lifting body's position while it was being towed, as would be dramatically demonstrated on later flights.

Following Thompson's initial free flights, Paul Bikle had approached NASA Headquarters to discuss a more advanced research vehicle. By this time, Dale Reed had begun working on plans for an all-metal, rocket-propelled version of the M2, which became known as the M2-F2. This heavyweight vehicle would be carried to higher altitudes by one of the X-15 program's NB-52 launch aircraft, before using a trusty XLR11 rocket engine to propel it on flights into the transonic and supersonic regimes. To Reed and Bikle, the Ames M2 shape remained the obvious candidate for this next step given the work that had already been done with the M2-F1, but a Langley team working under Eugene Love had developed their own, highly promising lifting body configuration—the Horizontal Lander 10 (HL-10).

With the FRC, Ames, and Langley all now involved in lifting body research, NASA Headquarters assumed a more active role in coordinating these related programs, with the agency's Office of Advanced Research and Technology (OART) naming Fred DeMerritte as its lifting body program manager in 1963. One of DeMerritte's first actions was to invite representatives from each of the interested centers to join him in a small planning group, laying the groundwork for the heavyweight lifting body program. Having already submitted the FRC's proposal for an advanced lifting body, Bikle recommended that in keeping with previous research aircraft programs, two aircraft should be procured. As well as doubling the potential for research, this would also ensure that the program could continue if one of the vehicles was lost. Although DeMerritte agreed with Bikle's logic, he suggested that rather than procuring two identical vehicles, the program would be better served by flying two different configurations—the M2-F2 and HL-10.

Once this approach had been agreed, the FRC drew up detailed specifications for each vehicle, and by February 1964 the center had circulated a request for proposals to twenty-six prospective contractors. Following a short assessment period, Northrop's Norair Division was awarded the contract for the M2-F2 and HL-10, due in no small part to the company's willingness to keep costs and complexity down by employing the same Skunk Works approach that the FRC had used for the M2-F1. Wherever possible, both vehicles would share common systems and components, with many of these being off-the-shelf items from other aircraft, including Northrop's own T-38 jet trainer. The company would assign a small team of engineers and technicians to work closely with the FRC's own people throughout the design, construction, and testing phases, a relationship that would be simplified by the Northrop plant's nonunion status. Whereas some industry representatives had suggested a cost per vehicle in the region of $15 million, Norair's fixed-price contract would see the new lifting bodies delivered for a mere $1.2 million apiece. The M2-F2 would be constructed first, with delivery to the FRC due to take place during spring 1965. The HL-10 would follow six months later.

As the new lifting body program got underway, Paul Bikle met with his AFFTC counterpart, Major General Irving "Twig" Branch, to discuss a joint test operation along the same general principles as those used for the X-15. Between the M2-F1's encouraging flight tests and the recent cancellation of Dyna-Soar, air force interest in lifting bodies was on the rise, so the joint arrangements suited both parties. The two men signed a memorandum of understanding on 19 April 1965, and this led to the formation of a joint FRC-AFFTC Lifting Body Flight Test Committee, with Bikle as its chairman, to manage the program and its research objectives. The joint program also meant air force pilots would fly the NASA vehicles, with Captain Jerauld "Jerry" Gentry selected as the AFFTC's project pilot. In order to gain some lifting body experience ahead of his flights in the M2-F2 and HL-10, Gentry was to check out in the M2-F1 once the vehicle had been assessed by the Fighter Test Division's chief pilot, Major Don Sorlie.

By the summer of 1965 Gentry had completed his familiarization ground tows behind Whiteside's Pontiac and was due to make his first air-tow flights along with FRC pilot Bill Dana on 16 July 1965. Conditions that morning did not look good, with rain turning the lake bed's usually hard surface layer into sticky mud as a solid overcast hung above the desert. Assessing the situation, Milt Thompson decided that flights could still proceed as long as conditions did not get any worse, and after making a quick shakedown flight himself to check the vehicle, he turned the M2-F1 over to Bill Dana. After an uneventful takeoff and climb to altitude, Dana released the towline and pushed the M2-F1's nose over to pick up speed, but as he did so, water rapidly began to pool in the aircraft's nose window. Somehow, rain had seeped into the lifting body's hull, and as Dana recalled, "The water in the nose window formed a giant lens and the lens demagnified what I saw through it. So I was looking through this big lens at the lake bed and the lake bed runway [and it] looked like it was four to five miles away."

Fortunately, the water drained away as quickly as it had arrived, allowing him to use the nose window to judge his flare as he raced toward the lake bed. After this close call, lesser pilots might have called it a day. But Gentry was keen to press on with his first flight, and the aircraft were soon ready for their next takeoff. Having set off across the lake bed behind the C-47, Gentry soon reached 80 mph, rising steadily to take up his position above the tow plane's wake, but as soon as both aircraft were airborne, Thompson noticed the lifting body was beginning to rock from side to side. In his attempts to keep the lifting body in the perfect position behind the tow aircraft, Gentry was making constant elevon and rudder inputs, but in doing so, he was fighting the effective dihedral, causing the aircraft to roll to ever-greater angles. As Thompson and Dana watched with growing alarm, the pilot-induced oscillation (PIO) reached a point where the lifting body became almost inverted. Having seen enough, Thompson urged the air force pilot to eject, but Gentry elected to stay with his aircraft and, having released the towline while upside down, coolly completed his barrel roll to bring the M2-F1 down onto the lake bed just as the aircraft came level again.

Having landed safely from a seemingly unrecoverable position, Gentry felt sure he knew what had gone wrong, and Thompson, too stunned to put up much of an argument, agreed that he could make another attempt. Fortunately, before the preparations for a second flight were completed, an angry voice boomed over the radio, "Get that thing back in the hangar, now!" Having watched Gentry's hair-raising first flight from Joe Vensel's office, Joe Walker was in no mood for a repeat performance. The chief pilot's intervention proved all the more fortuitous when a postflight examination revealed that Gentry's landing had damaged the M2-F1's plywood shell at the attachment points for the main gear, meaning a complete failure could well have occurred had the second flight been attempted.

It would be April 1966 before any other pilots checked out in the M2-F1. That month, air force X-15 pilot Joe Engle and FRC pilot Fred Haise made ground tows behind the Pontiac, although both were soon due to leave Edwards for the Apollo program. In the meantime, having checked Gentry out in a regular sailplane to build his experience of flying on a tow rope, Thompson was back on the lake bed the following August to watch the air force pilot's belated second attempt to fly the M2-F1. As the lifting body rose into the air, Thompson and Dana again stared on incredulously as the telltale rocking motion began again. As Gentry rolled inverted, Thompson yelled at him to eject, but the test pilot coolly released the towline and rolled the glider level to make another improbable landing. Although the M2-F1 was undamaged, Paul Bikle wisely decided that it was time to retire the lightweight lifting body before someone got hurt.

Dale Reed had been faced with a difficult choice as he began planning for the heavyweight lifting body program during the summer months of 1963. Whereas the M2-F1 had been a relatively small FRC in-house project, the involvement of NASA Headquarters and external contractors in the heavyweight program would inevitably bring whole new layers of administrative and operational complexity. Recognizing these upcoming challenges, Paul Bikle met with Reed to discuss the engineer's options. Although Reed had been instrumental in promoting the value of a lifting body flight test program, before making it a reality with the M2-F1, Bikle recognized that his friend's real passion lay in innovation rather than fighting the day-to-day bureaucratic battles that would await him as project manager. The director suggested that Reed should remain with engineering, while John McTigue transferred across from his current role as an X-15 operations engineer to manage the lifting body program. Under this arrangement, Reed would remain a project engineer, but he would also retain the freedom to pursue other ideas.

When Reed concurred with Bikle's suggestion, McTigue finished up his X-15 duties before picking up the reins to drive the heavyweight program forward. Nicknamed Tiger John by his FRC colleagues, McTigue had earned a reputation for ensuring that his aircraft were always delivered on time and in peak condition for every flight. Impressed by the ops engineer's attention to detail and obvious flair for management, Bikle felt that McTigue was ready to take on a bigger challenge, and having already worked with Reed on some aspects of the heavyweight M2 proposal, Tiger John was well prepared as the FRC finalized its specifications for the M2-F2 and HL-10.

Although outwardly similar to its predecessor, the M2-F2 was different in many respects. When wind tunnel testing at Ames suggested that the elephant-ear elevons would not survive the heat of reentry, the decision was made to omit the outboard control surfaces from the M2-F2 in order to more accurately represent a possible follow-on orbital vehicle. In place of these surfaces, the engineers at Ames added a new lower body flap at the rear of the aircraft to provide pitch control, allowing the split upper body flaps to be used independently as ailerons. To ensure adequate control, a complex interconnect system would link the aircraft's rudders and ailerons, while both upper and lower body flaps could also be extended symmetrically, acting like a shuttlecock to provide additional stability during transonic flight. Although both Reed and Iliff voiced concerns that the novel control arrangement might introduce new stability problems, the system was implemented as the M2-F2 took shape in Northrop's Hawthorne plant in southern Los Angeles.

While sharing the M2-F1's body width and being only slightly longer than its plywood predecessor, the M2-F2 needed to accommodate an XLR11 motor along with the associated propellant tanks and turbopump system. Two smaller hydrogen peroxide–fueled landing rockets were also included to provide a controllable source of instant L/D. Although the M2's half-cone shape offered plenty of internal volume, the propulsion system dictated that the M2-F2's cockpit be moved farther forward than that of the M2-F1, changing the pilot's position with regard to the aircraft's center of gravity. This change did, however, bring one major advantage; with the addition of a special adapter fitted to the NB-52's X-15 pylon, the lifting body's cockpit now sat forward of the bomber's wing, meaning that the M2-F2's ejection seat could still be used should catastrophe strike while the two aircraft were still mated.

Throughout development, the FRC lifting body team made the daily commute to Hawthorne, taking advantage of regular shuttle flights between the plant and Northrop's facility at Edwards. McTigue and his Northrop counterpart, Ralph Hakes, quickly developed an excellent working relationship, allowing decisions to be made quickly with the minimum of paperwork. When problems did occur, the Northrop and NASA teams worked together to find practical solutions without incurring long delays. At one point when the M2-F2 became tail heavy due to the multitude of systems housed in its broad rear fuselage, Northrop proposed that rather than simply adding ballast, the entire cockpit structure should be strengthened to act as a counterbalance, which would also offer the pilot more protection in the event of a crash—a decision that would later prove extremely fortuitous for one pilot. However, as development of the M2-F2 forged ahead during the early months of 1965, the Langley-designed HL-10 had run into problems.

Having worked through nine previous iterations of their lifting body configuration since 1962, Eugene Love's team felt sure that their HL-10 design represented a major improvement over both the ballistic capsules favored by Max Faget and the Ames M2. The broad delta-shaped HL-10, with its upturned tip fins, large vertical tail, and flush glazed cockpit, was arguably the most elegant of the early lifting body designs. Unlike the M2, with its half-cone lower fuselage and flat, deck-like upper surface, the HL-10 featured a flattened base and upper surfaces that gently curved away from its straight back—a negative camber design that its creators believed would improve hypersonic handling, while offering a subsonic L/D of around four. The FRC's Ken Iliff had asked Love's team to ensure that their design included large ailerons to improve pilot control and prevent the adverse aileron yaw seen on the M2, and as the HL-10 moved toward production during 1963, the Langley group were so confident in their configuration that they were already proposing a twelve-person variant of the HL-10 as an orbital taxi for future space stations. During its low-g return from orbit, the lifting body would offer enough cross range to enable landings across a huge swathe of the United States, while the use of refurbishable ablative coatings opened up the prospect of a truly reusable spacecraft rather than the single use capsules then in use.

With so many years of research having gone into the configuration's development, it came as something of a surprise to DeMerritte, Reed, and

Northrop (who were already nearly halfway through construction) when Love's team revealed that the HL-10 would need to be modified. Additional wind tunnel tests had shown that the lifting body's L/D and directional stability fell below original expectations, but the Langley engineers believed that these shortfalls could be rectified through changes to the control surfaces and by increasing the size of the tip fins. Although far from enthusiastic about the lateness of the request, the FRC and Northrop accepted the modifications, and development continued. Unfortunately, these would not be the last fixes that the Langley lifting body would require before it lived up to its early promise.

On 15 June 1965 the completed M2-F2 was rolled out from the Hawthorne plant before making the winding journey up through the San Gabriel Mountains to reach the FRC the following day. Like the M2-F1 before it, the heavyweight lifting body was soon heading to Ames for tests in the forty-by-eighty-foot tunnel, with the resulting data being used by Ryan and Smith to program the FRC's simulator. On its return to Edwards, the M2-F2 entered a lengthy checkout process during which much of the aircraft's plumbing was brought up to FRC standards by technician Dave Stoddard, before all its systems were checked individually prior to being tested as an integrated system. As the M2-F2's initial flights would be short glides to test its control responses and stability, the rocket engine was not installed at this stage, buying John McTigue time to rectify an unexpected problem.

By the mid-1960s, very few of the old XLR11 or LR8 engines remained in storage at Edwards, with many now residing in museums or on static display in their original aircraft. After investigating possible alternatives and finding the costs prohibitive, McTigue came up with a plan: "I went out and got parts for the rocket engine out of an airplane out at the college in Lancaster [the D-558-II displayed at Antelope Valley College]. I got parts out of the airplane in front of our building [the Bell X-1E on display outside the FRC]. I went back to Wright Field and got engines out of their museum. I went down to San Diego and got engines out of their museum. . . . I brought all those engines back. There was like eight of them, and parts."

After having the engines rebuilt by Thiokol, the FRC's own rocket lab (still under the direction of X-1 veteran Jack Russell) subjected them to a thorough series of pressure tests before declaring the reclaimed rocket engines fit for use. "They weren't the best engines in the world . . . and every once in a while, we'd lose a chamber," McTigue explained later, but Tiger John now had his XLR11s for the relative bargain price of $50,000.

As promised, Northrop delivered the HL-10 just over six-months after the M2-F2—no mean feat considering the midprogram modifications. The second heavyweight lifting body arrived at the FRC on 18 January 1966 and immediately went into an eleven-month period of wind tunnel testing and systems checkouts. With all three x-15's still flying, the arrival of the two new lifting bodies looked likely to stretch the FRC's engineering resources to breaking point. In order to alleviate this pressure, the AFFTC agreed to share simulation and flight planning for the new vehicles, with an experienced air force team including x-15 planners Bob Hoey and Johnny Armstrong taking on responsibility for the M2-F2, while a team at the FRC would do the same for the HL-10.

During March 1966 the M2-F2 began a series of captive flights beneath the wing of an NB-52, ensuring that its systems would still function after being cold soaked during the climb to altitude. The fifth of these flights was completed on 7 June, and although the aircraft's dampers were still occasionally failing, the project team felt ready to move on to the first glide test. Those plans were soon put on hold, however, as the tragic events that unfolded in the skies above Barstow the following day caused everyone to pause and take stock. Joe Walker's death in the XB-70 collision stunned the FRC, leading to a brief suspension of flight-testing. The loss of the center's hugely respected chief pilot on what had appeared to be a routine flight caused a reexamination of the risks represented by the ongoing flight research programs.

With time to reflect on the results of the captive flights, John McTigue decided that the M2-F2's intermittent stability augmentation problems needed to be solved before he would sanction a glide flight, and so the lifting body's systems were torn down, modified, and retested. After an additional captive flight on 6 July to validate the fixes, McTigue passed the M2-F2 as ready—it was time to fly.

12. The Heavyweights

As the M2-F2's first flight drew closer, Milt Thompson's long hours in the simulator lab allowed him to identify some worrying flight characteristics that would need to be explored with extra caution. With such low L/D, the lifting body had to make a very steep approach in order to generate the speed required to flare and land safely. This meant flying at low, often negative angles of attack, but under these conditions, the M2-F2 became susceptible to a roll instability as the vehicle's effective dihedral produced an adverse yawing movement. Any attempt to fight this instability could easily make the situation worse, triggering a PIO, which would see the lifting body rolling wildly as it raced toward the lake bed.

While in flight, the pilot could adjust the interconnect ratio between the lifting body's control surfaces via a selector in the cockpit, increasing or decreasing the sensitivity of the aircraft's responses. Although decreasing the ratio appeared to lower the risk of a PIO in the simulator, Thompson knew that he would have to "separate the real from the imagined" the hard way, by testing the aircraft's control responses in flight. Together with AFFTC planners Bob Hoey and Johnny Armstrong, the NASA test pilot devised a flight plan that would allow him to perform a practice flare while still at altitude. If the aircraft remained controllable during this maneuver, Thompson would continue with the flight. If, on the other hand, the M2-F2 became uncontrollable, he would eject, allowing the vehicle to tumble to the lake bed alone. "It's nice to find that out early in the flight, since I only had three and a half minutes to learn how to fly the vehicle before I had to land it," Thompson commented later.

As dawn broke over the High Desert on the morning of 12 July 1966, Milt Thompson climbed into the M2-F2's cockpit knowing that today's short glide might determine not only the fate of the M2-F2 but also that of

29. The M2-FI (left) and M2-F2 (right) pose together, showing the clear family similarities between the lightweight and heavyweight vehicles. Courtesy NASA.

the entire lifting body program. The wingless vehicles had already gained many supporters, but some within NASA (including Thomas Toll, the FRC's former chief of research who had now returned to Langley) believed that piloted flight tests represented an unnecessary risk. If the M2-F2 proved itself to be unflyable, these critics were likely to call for the cancellation of the low-speed lifting body program. Fortunately, Thompson had plenty to occupy his mind once strapped in to the aircraft, as he ran through the lengthy preflight checklist with crew chief Bill LePage.

The day's flight plan called for the NB-52 to climb to 45,000 ft. and head north along the eastern edge of the lake. The M2-F2 would launch close to the air force's Rocket Propulsion Laboratory up on Leuhman Ridge, and from there Thompson would continue north before making a 90° left turn along the northern shore of the dry lake, where he would push the M2-F2's nose over, building up speed for his practice flare. Coming level at approximately 18,000 ft., he would then use the vehicle's hydrogen per-oxide landing rockets to regain speed before a second 90° turn lined him up with Runway 18. Pushing over once more, Thompson would then dive toward the lake bed at a 30° descent angle, building his speed up to 345 mph before flaring at 3,000 ft. and touching down gently at an estimated 230 mph. At least that was the plan.

As the NB-52 approached the launch point, Thompson switched across to the lifting body's internal systems, taking care to ensure that cockpit heat was turned to maximum—the M2-F2 was not an aircraft in which he wanted to battle against windshield icing. With all systems ready, the NASA pilot counted down from five before falling away from the pylon with a far less violent jolt than he was used to from his x-15 flights. Less than thirty seconds later he made the first of his planned turns, bringing the lifting body parallel to the lake's northern shore before pushing the nose over to build up speed. After successfully performing the practice flare and testing the landing rockets, Thompson rolled into the final leg of what had up to now been a satisfactory first flight.

Pushing the M2-F2's stick forward to build up his approach speed, Thompson felt the aircraft becoming sensitive to roll inputs—nibbling at the edges of the PIO zone. In an attempt to stabilize the situation, he reduced the control interconnect ratio, but as he did this, the aircraft began to sway from side to side. Reducing the ratio further, the test pilot was soon being thrown around the cockpit as the lifting body rolled to increasingly severe bank angles. Realizing that he was in a PIO and making the condition worse by instinctively fighting the movements, Milt Thompson figured there was only one surefire way to stop the problem—he let go of the controls. Plummeting toward the lake bed in a 30° dive, this may have appeared an almost suicidal act, but Thompson knew that if he stopped feeding the PIO, the rolling movements would dissipate. Taking a split second to scan his controls, the pilot realized to his horror that rather than reducing the interconnect ratio, he had actually been increasing it, thus explaining the rise in control sensitivity that had preceded the PIO. Switching back to his intended ratio, Thompson regained control just in time to begin his flare, reducing the M2-F2's rate of descent until, 50 ft. above the ground, he deployed the landing gear and sank down onto the lake bed. Relieved to have made it down in one piece, the NASA pilot let the M2-F2 roll out across the dry lake for a mile and a half, joking with the chase pilots that he was trying to reach the bars of nearby Rosamond.

Milt Thompson would get his drink soon enough, but the postflight party would have to wait until the debriefing had taken place. Apologizing to the team for the nerve-racking approach, the pilot explained the cause of the problem; the interconnect ratio controller in the aircraft worked in the opposite direction to the one in the simulator. This simple oversight could well have cost the life of a less skilled pilot, but Thompson had followed his training, taking time to reassess the situation rather than continuing to fight the worsening PIO. As the team watched cockpit camera footage from the flight, Northrop engineers were shocked to see Thompson release the stick as the lake bed raced toward him. The pilot himself was more surprised to see his helmet smashing against the aircraft's canopy during the most severe oscillations—he had simply been too busy to notice at the time.

With the M2-F2's second glide flight just a week away, the team wasted no time in updating the simulator using data and pilot observations from the dramatic debut, but there were still opportunities for some trademark FRC humor to lighten the mood. During the first launch, Vic Horton had been in the NB-52 acting as launch panel observer. With the M2-F2 still engineless, Horton's workload had been light that day, but he had been responsible for activating a small downward-looking camera in the launch pylon adapter, to capture valuable views of the separation. Unfortunately, the flight engineer had omitted to do this, much to the frustration of launch dynamics engineers Berwin Kock and Wen Painter. Determined to prevent the same thing from happening on the second flight, Kock and Painter presented Horton with a box labeled LAUNCH PANEL SIMULATOR during a preflight briefing. The box featured a large switch handle fashioned from a pencil, with OFF and ON labels above and below it. When the unimpressed Horton pulled the handle down to its ON position, a banana rolled out of the box. As laughter rang around the briefing room, Horton launched the banana in Kock and Painter's direction. But the engineers had made their point, and Horton activated the camera during the second flight on 19 July.

This time, Thompson followed a simpler ground track involving a single 90° turn, allowing him more time to feel out the M2-F2's stability through a series of test maneuvers. After launching near the town of Boron, the first leg was taken up with stick pulses using different interconnect ratios (following the first flight, the simulator had thankfully been modified to reflect the aircraft's actual ratio controller), before Thompson made his left turn to line up with the lake bed runway. Decreasing the vehicle's angle of attack as he steepened his descent, he continued making small roll inputs at ever-lower ratio values, feeling for the control boundaries. Then, as Thompson pulsed the stick with the ratio set to the lowest planned value, the M2-F2 rolled in the opposite direction. Although this reaction had been predicted in the simulator, it still took all the pilot's self-control to stop from making a second input in the same direction in an attempt to bring the aircraft level. With the second flight safely concluded, Thompson made two more glides during August before his final lifting body flight on 2 September 1966.

Before the M2-F2 flights had even begun, Thompson informed Paul Bikle that he was considering a move away from the cockpit and back into engineering, the role that had originally brought him to the NACA back in 1956. Having left the x-15 program during 1965 to concentrate on the lifting bodies, the pilot had looked to the future and concluded, "There was no planned follow-on to the x-15 or lifting bodies. I wasn't interested in boring holes in the sky, so it was time to quit." With Thompson's decision made, Bikle was keen for him to make the transition as soon as possible, and with new pilots already training to fly the M2-F2 and HL-10, the research pilot made a clean break after his fifth glide flight, quitting flying entirely, lest he be tempted to return to the program. With his big personality and unquestioned skill in the cockpit, Milt Thompson's departure was a huge loss to the pilots' office, but he would continue to play a leading role at the FRC, becoming the chief of the Research Projects Office in January 1967.

Having already flown the M2-F1, Bruce Peterson now became NASA's project pilot for the M2-F2, making his debut in the heavyweight lifting body just two weeks after Thompson's swan song, with a glide flight on 16 September. Following his troubles in the lightweight M2-F1, the AFFTC still harbored doubts about Gentry's suitability to fly the heavyweight vehicle, but after Bikle, Thompson, and Don Sorlie discussed the matter, it was agreed that Gentry could fly the M2-F2 once Sorlie had assessed the lifting body for himself. Following his third flight on 5 October, Sorlie gave the vehicle and Gentry his blessings, and the air force test pilot made his first flight in the metal M2 a week later. With both project pilots now checked out in the M2-F2 and the HL-10 due to make its first flight before the end of 1966, it appeared as though the lifting body program at Edwards was set fair for a busy 1967.

Following his discharge from the navy in 1955, Robert W. Kempel (or simply Bob, as he was most commonly known by) had enrolled as an architecture student at Compton College in Southern California. While frequenting the college library, wondering if architecture truly was his vocation, he was regularly distracted by the latest copies of *Aviation Week and Space Technology* magazine. Enthralled by the latest aeronautical advances, Kempel became especially interested in the groundbreaking flight test activity taking place at Edwards AFB. As he pored over these reports, Bob made a life-changing decision; rather than pursuing a career in architecture, he changed his major field of study to engineering, with the hope that he, too, could someday test these exotic aircraft. With a natural aptitude for mathematics and physics, Kempel thrived in his new discipline, and after applying to California State Polytechnic University, on California's central coast, he was accepted into their aeronautical engineering program, where he also undertook light-aircraft flight instruction before graduating with his bachelor of science degree in 1960.

Keeping his promise to focus on flight-testing following graduation, he joined the FRC and was assigned to the X-15 program in the stability and control group. This group included Roxanah Yancey, the pioneering human computer who had arrived at Muroc in 1948, and soon Bob was soaking up every detail he could while plotting and cataloging wind tunnel and flight data. As he developed a better understanding of flight test jargon, especially regarding the X-15's hypersonic stability and control challenges, the neatness and accuracy of Kempel's data plots suggested that his architectural training had not been a total loss, and by 1963 his supervisor was relying more and more on the young engineer. Unfortunately, with nobody else in training to help shoulder this workload, Bob was becoming frustrated and bored by the limited opportunities that his situation offered him to delve deeper into the many facets and mysteries of flight vehicles' dynamics and control.

The AFFTC had a very small flight-dynamics research group, working on both the X-15 and Dyna-Soar, that had an opening for a civilian engineer. Initially, the civilian interviewer was very suspicious about Bob's wanting to quit NASA to join their small group, but after a detailed security and personal investigation, Bob was offered the position and tendered his resignation at the FRC. The working atmosphere at the AFFTC was much more dynamic, fast paced, and gung-ho, and that satisfied Bob's thirst for knowledge: "I learned hand-over-fist—just as fast as I could; knowing most of the jargon helped too. This is where I really learned my trade," he remembers now. "I learned how to develop the six-degrees-of-freedom equations of motion for an airplane, how to simplify these equations for individual flight condition analysis, how to mechanize these equations on an analog computer, how to solve these equations on a digital computer, using punch cards, et cetera. All this was really a worthwhile education. The air force working environment was much more aggressive and stimulating than NASA."

Initially assigned to the Dyna-Soar program, Kempel was aware that the space plane's future was in doubt, but the professional challenge of investigating the X-20's stability and control during reentry more than made up for any worries about the program's long-term prospects. When McNamara's axe fell on the hypersonic glider at the end of the year, Kempel hoped to transfer to one of the other lifting reentry programs underway at the center, but he was soon redirected to the AFFTC's vertical takeoff and landing projects, including the X-19—an ungainly marriage of helicopter and airplane then being developed for the air force by the Curtiss-Wright Corporation. After traveling to New Jersey for meetings with Curtiss-Wright engineers, Kempel quickly lost faith in the X-19, bluntly telling his program manager, "I hope our pilot isn't in it when it crashes." When his requests to move back into lifting reentry came to naught, Kempel reluctantly concluded that it was time to move on again. Fortunately, a conversation with a former FRC colleague had revealed that NASA needed a stability and control engineer for its HL-10 group, and so after serving his thirty-day notice at the AFFTC, Kempel headed back to his original employer and on to the challenges of wingless flight.

When compared to the air force's M2-F2 team, which included highly respected X-15 veterans Bob Hoey and Johnny Armstrong, NASA's HL-10 group seemed new to the game. As he began to analyze the Langley configuration, Kempel would also "fly" the FRC's rarely used M2-F2 simulator to gain a better understanding of that vehicle's dynamics. As a licensed pilot, the engineer quickly grew concerned with the M2-F2's apparent lack of stability, noting, "When I'd get into the approach condition, I'd crash, and I didn't know what was wrong. I thought it was me, my being hamfisted." Raising the subject with Bertha Ryan, his counterpart on the M2-

F2 project, Bob was surprised to hear that she did not use the simulation, regarding it as "too artificial." After having his concerns dismissed, Kempel pledged to do everything he could to ensure that the HL-10 would be a far more stable and controllable aircraft. Armed with data from earlier wind tunnel tests, he set to work creating a simulation of the HL-10's flight characteristics, using the techniques he learned during his years at the AFFTC.

By isolating snapshots of the lifting body's behavior under different flight conditions, Kempel and his colleagues gradually developed their simulation, and once complete, this indicated that the HL-10 would indeed be a far more benign flying machine than the M2-F2. However, after presenting their work to the wider lifting body team, they began to notice an unusual reaction. Pilots and engineers alike dismissed the favorable simulation, often commenting, "It can't be that good," the inference being that the HL-10 team had been overly optimistic in their analysis of the data. At one point, Ryan simply told Kempel that he had been "doing it all wrong" by following the training he had gained during his time at the AFFTC, rather than following standard NASA techniques. "I had a hard time," Kempel recalls, "management didn't believe my results. They didn't believe that the HL-10 could be that superior over the M2-F2." After going back and rechecking his calculations, Bob remained resolute in his conclusions, but the skepticism remained. "The manager of the program [McTigue], I'd pass him in the corridor, and he'd walk by . . . he wouldn't say anything; he'd just look at me and shake his head 'no.' What could I do?" By December 1966 the HL-10 was fully checked out and ready to settle the debate in the air.

For the first flight, Bruce Peterson would follow a similar flight plan to the one Milt Thompson had flown during the M2-F2's debut. Launching on a northerly heading 45,000 ft. above Leuhman Ridge, the HL-10 would make two 90° turns before landing on Runway 18 just over three minutes later. Although an attempt on 21 December had been aborted due to technical glitches, Peterson was aloft the very next morning, dropping away from the NB-52's starboard wing at 10:30 a.m. to give the HL-10 its first true test. Within moments of launch, he reported a high-frequency buffeting from the lifting body's control surfaces, which grew progressively worse as the vehicle's speed increased. Making the first turn, Peterson noted that the HL-10's controls appeared to be excessively sensitive to both pitch and roll inputs, with the vehicle continually moving around—

"acting squirrelly" in test pilot parlance. Things got worse when the NASA pilot made his practice flare during the second leg. As he increased the vehicle's angle of attack on pull-up, Peterson found that all roll control disappeared, allowing him to move the stick hard left with no response. Pushing the nose over to pick up speed, he set the aircraft's stability augmentation system to its lowest rate in an attempt to retain enough control to get the lifting body back on the ground. Landing safely after a shorter-than-planned flight, Peterson advised that the lifting body needed significant improvement before it flew again. To the watching engineers and managers, it seemed their suspicions about the simulation had been warranted after all.

With the Christmas holidays upon them, the disappointed HL-10 team grabbed a few days' respite before reconvening to work out what had gone wrong. According to their calculations, the lifting body should have remained stable and controllable throughout the flight, but clearly that had not been the case. A study of the flight data did not appear to reveal any fundamental problems, and consequently, their focus fell on mechanical improvements to the control system in order to reduce the sensitivity that Peterson had experienced during his flight. Once the center's technicians had made the recommended modifications, preparations began for a second flight, but one member of the group was not so sure that the HL-10's problems had been solved so easily. Wen Painter remained convinced that they were missing an underlying issue and that the answer was waiting for them in the data—it was just a matter of finding it. As stability and control lead for the project, Bob Kempel began a painstaking comparison of the simulated and actual flight data, looking for any anomalies that might point to the true cause of the HL-10's problems.

By analyzing twelve snapshots of data, Kempel uncovered significant differences between the expected and actual performance, all of which pointed to a deeper aerodynamic problem in the lifting body's design. When Peterson had raised the HL-10's nose, increasing the aircraft's angle of attack, the control surfaces rapidly lost their ability to control the vehicle. When the angle of attack was reduced, control returned. Following Kempel's analysis, the FRC team reached the inescapable conclusion that the airflow over these control surfaces must have been disrupted. Whereas wind tunnel tests had suggested that airflow over the HL-10's tip fins would remain smooth, or attached, under these conditions, it now appeared that the flow had actually become separated as the angle of attack had increased, rendering the aircraft's control surfaces ineffective until the airflow reattached at lower angles of attack.

Presenting their conclusions to their Langley colleagues, the FRC engineers were surprised by the reaction of project aerodynamicist Bob Taylor. Kempel recalls, "He had one of those nice mechanical pencils. Well, he jumped up from the conference table, and he threw his pencil on the floor and just came apart in pieces. He was very upset with himself because he said, 'I knew that was going to be a problem.'" Although Kempel had been unable to detect a flow separation in the data he had used to build the simulation, it appeared that Langley engineers had seen enough during their wind tunnel tests to indicate that the problem might exist. With the issue identified, the two teams now set about devising a solution that would get the HL-10 flying as originally intended. It would be another fifteen months before the Langley-designed lifting body made it back into the air, but in the meantime, the FRC hoped to begin powered flights in the M2-F2 to keep the overall program moving along.

Following Jerry Gentry's fourth flight in the M2-F2, which had taken place during November 1966, the lifting body was grounded to have its XLR11 engine installed. In order to accommodate the rocket, small modifications to the vehicle's lower body flap and upper split flaps were needed, and once these were complete, an additional glide flight was scheduled to assess any aerodynamic impact. Falling away from the NB-52 on 2 May 1967, Gentry was immediately aware that the reconfigured flaps had increased the M2-F2's drag, eating into the vehicle's already limited performance. After a short but uncomfortable flight, during which the air force pilot felt close to losing control of the lifting body, Gentry informed engineers that he would not fly the M2-F2 again until its performance was improved. Before committing to more modifications, the project team decided to make a second glide flight, giving Bruce Peterson a chance to assess the situation. The flight took place on 10 May, and although things began well, Peterson soon encountered trouble, feeling as if the M2-F2 was "nibbling" at the edge of a PIO. In an attempt to alleviate the problem, the NASA pilot changed the control interconnect several times, but he could not seem to find a satisfactory setting, commenting, "Boy, there's some glitches."

Having made his second turn to line up with the lake bed runway, Peterson pushed over into a 30° dive at a very low angle of attack, but in doing so, he entered a PIO, with the M2-F2 making two severe rolls. As chase pilot John Manke urged him to eject, Peterson managed to regain control of the lifting body, but he was now off course, with no runway markings to help him judge his height above the lake bed. To make matters worse, the air force rescue helicopter now appeared directly in his path. Although Manke assured his colleague that he would clear the other aircraft, the already overworked Peterson was clearly distracted, exclaiming, "That chopper's going to get me, I'm afraid." As the lake bed rushed toward him, Bruce Peterson fired his landing rockets, extending his flare as he attempted to drop the landing gear, but time had run out. Before the gear had a chance to fully extend, the lifting body hit the hard clay, shearing off its ventral antenna. With the flow of telemetry suddenly halted, the engineers in the FRC control room instinctively turned to a monitor showing a live feed from the lake bed just in time to see the M2-F2 skid a short distance before flipping up into a sickening series of rolls. When the lifting body finally came to a halt, having left a trail of wreckage in its wake, the shocked engineers were left fearing the worst.

Amazingly, rescue crews found Bruce Peterson alive as they reached the inverted wreckage. Northrop's decision to strengthen the M2-F2's cockpit as a means to balance out the vehicle's center of gravity had protected the NASA pilot from the worst of the impact, but Peterson had sustained severe facial injuries, including the loss of his right eyelid. After being rushed to the base hospital, he was transferred to March AFB before finally being taken to the University of California in Los Angeles, where specialist eye care was available. Sadly, during his long recovery, Peterson contracted a secondary infection that cost him the sight in his right eye, bringing a premature end to his career as a research pilot (although he did later resume some limited flight-support duties after obtaining an FAA waiver). Following his return to the FRC eighteen months later, visitors to Peterson's office noticed that his picture of the M2-F2 now hung upside down. "Well," the eye-patched Peterson would say, with a dry test pilot humor, "that's how she looked last time I saw her." The footage of Bruce Peterson's horrific M2-F2 crash would later become famous as part of the title sequence for hit television show *The Six-Million-Dollar Man*.

The lifting body program was now left with no flightworthy vehicles and two strikes against it in the eyes of its critics. But as NASA efforts faltered, awaiting the return of the modified HL-10 and a decision regarding the fate of the wrecked M2-F2, a third wingless configuration was already taking shape at the Martin Marietta plant in Baltimore, and unlike the M2-F2 or the HL-10, this lifting body could already boast a spacefaring pedigree.

With NASA's Mercury program well underway by mid-1960, the American aerospace industry began to consider a new generation of advanced spacecraft to meet the anticipated future demand for regular spaceflight operations. Having lost out to Boeing in the 1959 Dyna-Soar selection, the Martin Company was still eager to play a major part in the air force's spacefaring plans, and the Baltimore-based company was soon awarded a modest research contract by the U.S. Air Force's Space and Missile Systems Organization (SAMSO) to investigate the suitability of lifting body shapes for various military applications. Although ballistic film-recovery capsules had already been developed for the secret Corona reconnaissance satellite program, the air force was interested in a more advanced capsule, capable of performing a controlled lifting reentry. Martin's initial response to this requirement was a design based on the Ames M1 shape, but when SAMSO subsequently requested that the capsules be recoverable during every orbit regardless of ground track, the company was forced to consider higher cross-range configurations. Based on existing research, the most promising candidates appeared to be a modified version of the Ames M2 shape or a flat-bottomed configuration known as the A3, then under development by the Aerospace Corporation. However, Martin was soon able to consider a third option emanating from its own chief scientist, Hans Multhopp.

Having studied aeronautical engineering under some of Germany's most gifted engineers during the 1930s, Multhopp spent his wartime years working for forward-thinking aircraft manufacturer Focke-Wulf, where he became chief of the company's advanced design bureau prior to Germany's surrender in 1945. Following the war, Multhopp was taken to Great Britain, where he continued his work into advanced swept-wing configurations at the Royal Aircraft Establishment in Farnborough before an offer from the Martin Company brought him to the United States in 1949. After playing a key role in the development of Martin's unusual T-tailed XB-51 jet bomber, the senior aerodynamicist turned his attention to the challenges of hypersonic lifting reentry as part of the company's unsuccessful Dyna-Soar team during the late 1950s. Now, using the Aerospace Corporation's A3 as his starting point, Multhopp devised a more rounded and refined configuration that he believed would offer both superior cross range and better stability than its competitors. Although initially known as the A3-4, Martin soon reclassified the new configuration SV-5 (SV standing for space vehicle) to avoid confusion.

Multhopp's work on the SV-5 was rewarded in March 1962, when the air force asked Martin to develop a small research vehicle capable of demonstrating a six-hundred-nautical-mile cross range during reentry from orbital speeds. With the U.S. Air Force's Space Systems Division (SSD) already favoring a lifting body configuration for their proposed SAINT II satellite-inspection vehicle, interest in wingless shapes was strong within certain parts of the service. As Martin continued to refine the SV-5 through an exhaustive series of wind tunnel tests during 1963, the newly named Precision Recovery Including Maneuvering Entry (PRIME) project became part of a larger SSD initiative, the Spacecraft Technology and Reentry Tests (START) program alongside the existing ASSET program. Where ASSET would test the structures and materials originally developed for Dyna-Soar, PRIME would demonstrate that low L/D shapes could achieve the levels of reentry maneuverability that aerodynamicists were predicting. The four planned flights would see the small SV-5D vehicles launched from Vandenberg AFB across the Western Test Range atop Atlas boosters. After reaching an altitude of approximately 500,000 ft., the SV-5D would perform a series of pre-programmed cross-range maneuvers as it raced across the Pacific before descending under parachute near Kwajalein Atoll in the Marshall Islands.

Just six and a half feet long, the SV-5D's main structure was fashioned from aluminum, titanium, and beryllium, with three different types of Martin-developed ablative material being applied to its various surfaces depending on predicted heat loads. Control came from a split underbody flap, which offered pitch control when used symmetrically or roll control if the flaps were used asymmetrically. The rudders and upper flaps were fixed in what Martin felt was the optimum position for reentry. When the vehicle decelerated to Mach 2, a drogue ballute would deploy to stabilize the vehicle before the main parachute slowly lowered it toward a waiting C-130 aircraft. If all went well, the specially equipped cargo plane would snatch the lifting body from the air, using the same techniques devised for the Corona program. Should the midair recovery fail, a flotation system was also included to allow for water recovery.

The first of the four planned PRIME flights took place on 21 December 1966, with the lifting body only required to make pitch maneuvers on this occasion. Although the flight itself was a success, the vehicle was lost when the main parachute failed to deploy correctly. During the second flight on 5 March 1967, the SV-5D executed an impressive 654-mile cross-range maneuver, but more parachute problems, together with a flotation-bag failure, prevented recovery. However, the third flight on 19 April was more fortunate. Having performed a 710-mile cross-range maneuver as it decelerated from Mach 25, the lifting body was snagged by the waiting C-130 crew within five miles of its predicted recovery point. After inspecting the SV-5D's ablative heat shield, Martin felt confident that the lifting body could be refurbished to fly again, but that turned out to be unnecessary. Following the success of the first three flights and with little budget remaining, the air force canceled plans for a fourth flight; the PRIME program had proven that the SV-5 shape could maneuver during a controlled reentry from near-orbital velocities down to Mach 2. But how well would Multhopp's design handle the transition from supersonic to transonic and subsonic flight?

Encouraged by both the M2-F1 flight tests and NASA's subsequent commitment to the heavyweight lifting body program, the air force decided to move forward with a third and final element of the START program—the Piloted Low-Speed Tests (PILOT) project. PILOT would see a larger, piloted SV-5P vehicle perform a series of research flights from supersonic velocities down through the approach and landing phase (essentially mirroring the M2-F2 and HL-10 programs). Measuring just over twenty-four feet in length, with a width of thirteen feet, the new SV-5P lifting body would be fitted with a refurbished XLR11 engine and be launched from an NB-52, but unlike its NASA contemporaries, it would not need a glazed nose, as its bubble canopy and sloping forward section offered pilots far better forward vision even at high angles of attack. Having solicited proposals from both Northrop and Martin, the air force elected to stay with Martin due to the company's superior understanding of the SV-5 shape, and the contractor duly received approval to begin construction of a single SV-5P on

2 March 1966, with delivery set for August 1967. In addition to the sv-5P, Martin also decided to build two additional sv-5J jet-powered demonstrators at its own expense, aware of Yeager's interest in obtaining a lifting body trainer for the ARPS.

Unlike the sv-5D PRIME vehicle, with its fixed upper flaps and rudders, the piloted sv-5P featured eight movable surfaces, providing control across a wide range of flight conditions. In addition to the split lower flaps, the aft of the vehicle was also equipped with split upper flaps that could again be used either symmetrically or asymmetrically. The two tip fins carried split rudders, while a third vertical fin had been added to improve airflow and provide directional stability. Although the new lifting body also used an interconnect system to govern the movement of its various control surfaces, Martin had developed a largely automated system with preset positions for supersonic, transonic, and subsonic flight. As construction of the sv-5P continued in Baltimore, the FRC and the AFFTC agreed that the new vehicle would be managed by the existing Lifting Body Flight Test Committee, meaning that the PILOT flight tests would be conducted on the same collaborative basis as the M2-F2 and HL-10 tests. The AFFTC's Johnny Armstrong was named as project manager, while the FRC's Norm DeMar became operations engineer. On 11 July 1967 the air force officially designated the sv-5P as the X-24A and transferred direction of the program to the Research Technology Division at Wright Field. The X-24A was rolled out from the Martin plant on 3 August before being partially disassembled and shipped to Edwards AFB on 24 August.

Following its arrival in the High Desert, the new lifting body underwent initial ground tests before heading to Ames in February 1968, to take its turn in the forty-by-eighty-foot tunnel. Unlike the M2-F2 and the HL-10, which had been tested with just their bare aluminum finish, the air force and NASA were interested in evaluating what effect a scorched and roughened ablative surface might have on the vehicle's L/D during a return from orbit. In order to achieve the desired texture, fiber netting was taped to the lifting body's skin and then sprayed over with a sand-and-glue mixture before the netting was removed, leaving a reasonable approximation of what engineers had seen on the recovered PRIME sv-5D. When the treated X-24A arrived at Ames, the tunnel technicians were not thrilled by what they saw. FRC engineer Jon Pyle remembers, "They had an absolute fit, and you can understand that. Here was a vehicle covered in sand, and you're going to blow air at it. You're literally blowing some of the sand off, and it runs around and wipes out the propellers on your wind tunnel!" But the tests did provide some interesting results: "We got a tremendous amount of data out of that. Although [the simulated ablative] had a significant effect, it wasn't that detrimental." However, the reduction in the x-24a's l/d that Pyle and his colleagues observed was enough to discourage the air force from attempting to fly the x-24a with the finish still applied.

Before the x-24 could fly, frc technicians faced a lengthy process of installing research instrumentation and upgrading some of the vehicle's systems. Although frc engineers and technicians had worked alongside their Northrop counterparts during the construction of the m2-f2 and the hl-10, the situation had been somewhat different with Martin. While the air force had encouraged frc input throughout the manufacturing process, Martin's Baltimore plant was a union shop, and this prevented the nasa team from carrying out any specialist installation work. Consequently, when the x-24a arrived at Edwards, many of its systems (including most of the cockpit instrumentation) had to be replaced or reconfigured in order to bring it in line with its stablemates. Checkout testing also revealed that, in their eagerness to avoid the center-of-gravity problems that Northrop had encountered with the m2-f2, Martin engineers had actually made the x-24a slightly nose heavy. This problem was largely eliminated as frc technicians strengthened servos and support structures for the lifting body's control surfaces, but some ballast was still added to balance the vehicle. It was not until the early months of 1969 that the x-24a flight review board finally declared the new research aircraft fit to fly, but by that time, the heavyweight lifting body roster at Edwards was edging back toward full strength as the hl-10 and m2-f2 teams worked through their stability and control problems.

Following the frc team's discussions with their Langley counterparts regarding the hl-10's flow-separation problems, the long process of devising a suitable engineering fix began. Eugene Love and his team were eager to make good for their design's aerodynamic deficiencies, and after months of wind tunnel testing, the Langley group presented two possible solutions to keep the airflow attached to the vehicle's tip fins and afterbody. The first of these involved reshaping the inner surfaces of the fins to give them a more rounded profile like the upper surface of a wing, while the second approach involved fitting outward-cambered extensions to the leading edges of the tip fins. Although the Langley engineers were able to provide comprehensive test data suggesting that either approach could solve the problem, they were unwilling to recommend which of them should actually be adopted; that task fell to the FRC's Bob Kempel, who said, "I slaved over the data—I plotted it all out by hand. I just took the set that looked most linear, with no unexpected bumps or curves, and chose that." Kempel's analysis led him to favor the leading-edge extensions, and once his recommendation had been approved by the program's managers, Northrop fitted outward-canted fiberglass extensions to the HL-10's tip fins.

By this time, however, Kempel was also deeply involved in investigations into the conditions that lay behind Bruce Peterson's horrific crash in the M2-F2 on 10 May 1967. "Following the gear-up landing, I, by default, became the stability and control engineer on the program," he remembers. "Using my digital program and my archived aerodynamic model of the M2-F2 in its approach flight condition, I discovered that the normally benign roll and spiral modes had coupled at the low angles of attack (between +2° and -2°) needed to maintain the 345 mph speed required during approach. The resulting roll-spiral mode meant that the pilot's normal control techniques now drove the airplane-pilot combination unstable—a very undesirable situation." Upon making this discovery, Bob felt incredulous that the program team had knowingly allowed pilots to fly into such a precarious situation on approach. Distressed that the condition had not been flagged as a major concern, Kempel developed a bias against the M2 configuration, and when word reached him in late 1967 that John McTigue was hoping to rebuild the wrecked M2-F2, the normally mild-mannered engineer became furious: "I marched down to [McTigue's] executive office and barged in unannounced. I told him I wouldn't recommend letting that airplane get any higher than when it was attached to the right wing of the B-52! And if we did fly the thing (the M2-F2), we would need additional wind tunnel tests. John said very little to me and just let me rant and rave about the M2-F2 until I left his office."

Meanwhile, preparations were underway for the HL-10's long-awaited return to flight. Having sat as a "hangar queen" for more than a year, the aircraft was finally cleared to fly again in March 1968, with Jerry Gentry selected to perform the first test of the modified lifting body. During this vital flight, the HL-10's control surfaces would carry rows of woolen tufts, providing a visual indication of airflow under different flight conditions. If the tufts lay flattened in a uniform direction, that would indicate a smooth attached flow, meaning the problem had been solved, but if they appeared to dance about wildly, suggesting turbulent air, then the engineers would have to return to the drawing board.

On 15 March 1968 Gentry dropped from the NB-52 and immediately began to put the HL-10 through its paces. As the engineers looked on nervously, the air force pilot made a series of pitch and roll maneuvers along with a practice flare, testing the vehicle at different angles of attack before touching down safely on Runway 18 after a four-and-a-half-minute flight. During the postflight debriefing, Gentry praised the HL-10's performance, even going so far as to say that it had handled better than an F-104 during approach and landing. After fifteen months of investigation, analysis, and modification, the HL-10 was back. Soon pilots were queuing up to fly the simulator, and for Bob Kempel, their enthusiastic responses came as vindication for his earlier efforts: "It made me feel pretty good, because managers couldn't walk down the corridor and shake their heads 'no' at me anymore!" Gentry made five more HL-10 glide flights in the three months that followed before he was joined by the FRC's John Manke, whose debut flight in the aircraft took place on 28 May.

The resurgence in lifting body activity during 1968 came as a welcome boost for the FRC as it looked toward its post-X-15 future. The move toward space activities that had accompanied the formation of NASA and the arrival of the hypersonic rocket plane was now giving way to more prosaic research into lower-speed aerodynamics and flight control systems, but the lifting bodies still offered the center a means to contribute toward NASA's future spaceflight plans. Although the FRC had continued to expand throughout the 1960s, the close camaraderie and humor shared by its workforce was still very much in evidence, especially when it came to the now lighthearted rivalry with their AFFTC counterparts. Since the outset of government flights in 1960, the X-15 had always carried both air force and NASA markings to reflect the joint nature of the program, but although that cooperative arrangement also extended to the lifting bodies, neither the M2-F2

30. Test pilots Jerauld Gentry, Peter Hoag, John Manke, and Bill Dana
with the HL-10 lifting body. Courtesy NASA.

nor the HL-10 carried any air force livery. As a proud blue-suiter, Jerry Gentry decided that this situation needed to be rectified, and so began the legend of Captain Midnight.

The story began when FRC technicians entered Hangar 4802 one Monday morning, only to discover that a mysterious intruder had decorated the HL-10 with the legend "US AIR FORCE" in large stick-on letters. As rumors began to spread about the perpetrator, now known as Captain Midnight, a heavily retouched photo appeared. It showed Gentry—now wearing a mask, cape, and a flight suit bearing the "midnight" insignia—standing with Captain Charles Archie in front of the HL-10 (complete with its USAF markings and insignia). The photo was captioned, "The Midnight Skulker Strikes Again *or* It Really Is a Joint Lifting Body Program!" With the mystery apparently solved, reprisals weren't long in coming. While inspecting the M2-F2 one day, Gentry had casually let slip how much he disliked the color of the zinc chromate paint used as a primer on the aircraft's baremetal interior. Little did the air force pilot know that this seemingly trivial comment had been filed away for future use by the NASA technicians.

When Gentry arrived at the FRC one morning to make a lifting body flight, those same technicians were ready and waiting to take their revenge. As the pilot entered the air force support van to don his pressure suit, his trusty 1954 Ford was spirited away to the FRC paint shop, where it received a coat of the yellow-green primer. Once the paint was dry, the pranksters added large flower-power decals to the Ford's doors and hood before placing it back exactly where Gentry had left it. Upon returning to his vehicle some hours later, the air force man was initially confused by the eye-watering vision that stood in its place, before coming to the awful realization that this was indeed his ride home. After recovering from the initial shock, Gentry took the prank in good humor and continued to drive the unmistakable Ford for the rest of his time at Edwards.

With the modified HL-10 now back in the air, attention turned back to the M2-F2 and the possibility of returning the heavily damaged aircraft to flight. In the aftermath of Bruce Peterson's accident, John McTigue shipped the lifting body's twisted remains back to Northrop's Hawthorne plant for the contractor to assess the extent of the damage. Unfortunately, the news was not encouraging. Not only had much of the M2-F2's external shell been destroyed, but it appeared that some of the key members that made up the aircraft's underlying structure had also been damaged. If the lifting body was going to be repaired, it would need to be done from the ground up. Although some, including Bob Kempel, had made clear their objections to repairing the M2-F2, there was still a groundswell of support for getting the original heavyweight lifting body back in the air. While both Dale Reed and Milt Thompson openly acknowledged that the M2-F2 exhibited poor flying characteristics, in their minds that very quality made it valuable; after all, if they could successfully tame the troublesome M2, then most other lifting bodies would be easy by comparison.

Having kept the full extent of the vehicle's damage largely hidden from NASA Headquarters (with the exception of program manager Fred DeMerritte), McTigue sought the advice of Paul Bikle, and the director came up with a characteristically creative solution to the problem. Northrop should continue with their detailed inspection of the M2-F2 by removing all the damaged parts and repairing them where possible (even if this repair involved the FRC's own technicians fabricating a replacement part). By working in this low-key manner, DeMerritte was able to forward small amounts of funding while still being able to justify the effort as an inspection. Having given the M2-F2 little thought for over a year, Bob Kempel was summoned to McTigue's office during the later months of 1968, with the program manager asking his stability and control engineer to accompany him on a visit to Northrop's Hawthorne plant, where the M2-F2 was being rebuilt. Given his antipathy toward what he felt was an unsatisfactory vehicle, Kempel initially declined the offer, but when Tiger John persisted, Kempel eventually relented to the manager's wishes. The next day, the duo drove down to Hawthorne, where Kempel found himself impressed by the progress the joint Northrop-NASA team had made on the reconstruction. Following lunch, McTigue beckoned Bob Kempel into a back room to see a large-scale, highly polished model of the M2-F2 with a center fin added—a new configuration that McTigue referred to as the M2-F3. "This was the first I had heard of an M2-F3," Bob recalls. "The center fin didn't strike me other than just being interesting," but unknown to Kempel at the time, the addition of the new fin represented the culmination of a major research effort back in the same wind tunnels at NASA Ames where the M2 configuration had first taken shape.

Even as Northrop was embarking on the "inspection" process, a team led by aerodynamicist Jack Bronson had begun to investigate potential modifications that might fix the lifting body's marginal stability and cure the roll-spiral coupling (and consequently the PIO problem) for good. The earlier decision to build the M2-F2 without outboard elevons had left the heavyweight vehicle reliant on its split upper flaps for roll control. Whenever one of these flaps had been raised, a high-pressure zone was created above it, and this placed a force on the adjacent vertical fin, causing adverse yaw and pushing the lifting body into the sideslip condition where PIOs were most likely to occur. By placing a vertical fin between these upper flaps, Bronson's team believed that any high pressure resulting from flap movements would exert an equal force on the outer and center fin, removing the adverse yaw (ironically, the M2-F1 had originally been conceived with a similar arrangement, but this was never actually flown). Although initially unconvinced, Kempel agreed to analyze the results of forthcoming M2-F3 wind tunnel tests, and much to the engineer's surprise, when his simulation was complete, it did appear that the roll-spiral coupling condition had been alleviated. Although the M2-F3 would still have very low L/D and would generate more base drag than the other heavyweights, its general flying characteristics now seemed closer to those of the modified HL-10.

The rebuilding process continued to move along slowly until January 1969, when NASA Headquarters officially announced that the M2-F2 would be rebuilt as the M2-F3. To reduce costs, the vehicle was returned to the FRC so that the center's various workshops could work together using Bikle's favored Skunk Works technique. Along with its new center fin, the M2-F3 was also fitted with its XLR11 engine (mounted 90° from vertical to avoid the flap modifications required on the M2-F2), an improved stability augmentation system, and an experimental reaction control system.

Following the HL-10's return to flight in March 1968 and with first flights for the X-24A and M2-F3 imminent, the AFFTC and FRC teams could finally begin a step-by-step program of lifting body research that would continue into the mid-1970s. As one lifting body began its rocket-powered envelope-expansion flights, so the next would begin its glide flights, until each vehicle had yielded a full set of research data that could be compared to that of its contemporaries. As Bill Dana and Pete Knight brought the X-15 program to an end during the latter half of 1968, Jerry Gentry and John Manke were preparing for the first powered lifting body flights in the HL-10.

Up until now, all lifting body flights had been short unpowered dives toward the lake bed, but the advent of powered flights brought interesting new challenges. As originally conceived, the wingless shapes were designed to return from orbit, remaining stable and maneuverable as they gradually decelerated from Mach 25 to a subsonic approach and landing. But now they were being asked to accelerate under power, passing through the troublesome transonic region as they climbed to higher altitudes before beginning their descents. To face up to these challenges, the HL-10 team planned a gradual buildup to Mach 1, echoing the approach used for the XS-1 over two decades earlier.

The first faltering step in this process took place on 23 October 1968, when Gentry was forced to make an emergency landing on Rosamond Dry Lake following ignition problems with the refurbished XLR11 engine. A second attempt on 13 November proved far more successful, with Manke using two of the engine's chambers to reach Mach 0.84. When Gentry transferred across to the X-24A program at the end of 1968, Manke was joined by Bill

Dana, and the former x-15 pilot made his first powered lifting body flight in the HL-10 on 25 April 1969. One week earlier, Manke had taken the lifting body out to Mach 0.994 and hoped to exceed Mach 1 on his next flight, but before that could happen, both Manke and Bob Kempel were involved in a comedy of errors involving the HL-10 simulator.

Following Gentry's departure, the AFFTC's Major Peter Hoag took over as air force project pilot for the HL-10. As Hoag became familiar with the lifting body's characteristics in the FRC simulator, he became concerned about how well the aircraft would handle as directional stability fell away at very high angles of attack. Knowing that the HL-10 would remain controllable, Kempel asked the simulator engineer to program a second version of the simulation to demonstrate this to Hoag. Although the simulator engineer explained that the best he could do would be to set directional stability to zero across all flight conditions, a second data tape was made and used to set Hoag's mind at rest. Unfortunately, this second tape remained in the simulator lab, and when John Manke called in one lunchtime to make some extra practice runs ahead of his upcoming supersonic flight, a substitute engineer unwittingly loaded this version into the computer rather than the standard HL-10 simulation. As the engineer left the lab to grab some food, Manke began the simulation.

As the pilot hit the start button and began to push the HL-10's nose over, the complete lack of directional stability at the now-reduced angle of attack rapidly sent the aircraft out of control. Having "crashed" the simulator, Manke wanted answers, and with the HL-10 engineers still on their break, he vented his dissatisfaction to the center's management instead. Bad news had a habit of traveling fast, and before the lunch break was out, word of the incident had reached NASA Headquarters. As Kempel and his colleagues returned, they were summoned into one of the wood-paneled management offices to explain why they were trying to kill a highly qualified research pilot. "Fortunately for us," Bob jokes now, "Mr. Bikle was gone that day." Quickly realizing what must have happened, the engineers and their inquisitors headed down to the lab, where the correct simulation was loaded for Milt Thompson to try. "Milt flew it with no problem whatsoever," Kempel recalls. With the misunderstanding ironed out, Manke's preparations resumed, and on 9 May he made the first supersonic lifting body flight, taking the HL-10 to Mach 1.127 at an altitude of 53,300 ft.

The Langley-designed lifting body would eventually reach a speed of Mach 1.86 during a flight by Hoag on 18 February 1970, before Dana hit a maximum altitude of 90,303 ft. nine days later. These would be the highest and fastest marks for any lifting body, and throughout its career, the modified HL-10 was regarded by air force and NASA pilots alike as the best handling of the lifting bodies. After all the early doubts regarding Bob Kempel's simulations, it turned out that it really was that good after all.

After ten years of research, development, and modification, 1970 was the year when the lifting body program truly came of age. As the decade began, the HL-10 was already nearing its maximum speed and altitude marks, but it was soon joined by the X-24A and M2-F3, with all three aircraft logging powered flights before the year was out.

Having transferred across from the HL-10 to become the AFFTC's project pilot for the X-24A, Jerry Gentry was ready to make the first glide flight in the air force lifting body by April 1970. With its squat, bulbous appearance, the X-24A was much maligned as the ugly duckling of the lifting body fleet (often described as a "flying potato"), but although it may have lacked aesthetic charm, Multhopp's SV-5 was the only shape to have performed a hypersonic reentry from Mach 25 down to Mach 2. It now fell to Gentry and his colleagues to demonstrate that it could manage just as well during the final phase of flight. As had been the case for Thompson in the M2-F2 and Peterson in the HL-10, the X-24A's first glide flight would only offer Gentry a few scant minutes to compare the aircraft's actual flight characteristics to those predicted by the simulation. Although the aircraft's automatic system governing the interconnect between its eight control surfaces had the potential to reduce the pilot's workload, Gentry soon had his hands full as he made the lifting body's first flight on 17 April.

After separating smoothly from the NB-52, the test pilot began to feel out the X-24A's control responses, but barely a minute into the flight, he reported the same "nibbling" feeling he had experienced in the M2-F2. Having gingerly guided the lifting body through a series of maneuvers during the descent, he rolled in to make his approach but still felt unhappy with the vehicle's lateral stability. Fearing a loss of control, Gentry increased his angle of attack, reducing the X-24A's speed before firing the hydrogen peroxide–fueled landing rockets to make a successful flare and landing.

Subsequent investigations revealed that the automatic interconnect system had failed soon after launch, but there also seemed to be an underlying problem with the vehicle's lower flaps, which was causing a roll instability. During his second glide flight, on 8 May, the automatic interconnect system functioned as designed, yet Gentry still felt that the x-24a remained only marginally stable as he descended toward the lake bed.

Recognizing that they were dealing with a more fundamental problem, the x-24a team dug into their data and soon came up with an answer. During earlier small-scale wind tunnel tests, the model's mounting sting appeared to have disrupted airflow around the lower flaps, leading to erroneous data. Once again, full-scale flight-testing had proven its worth by revealing the true aerodynamics of the vehicle. With the problem now identified, Johnny Armstrong and his team were able to amend the x-24a's control system before the third glide flight, on 21 August. In an incident reminiscent of George Jansen's launch of the unsuspecting Bill Bridgeman back in January 1951, the B-52 pilot (flying his first launch mission) accidentally hit the launch switch while attempting to arm the system. Fortunately, Gentry took the unexpected separation in his stride, calmly announcing, "I've been inadvertently launched," before proceeding with his planned maneuvers. Although the control system changes appeared to have solved the earlier problems, Gentry still complained of some minor roll instability during his descent to Runway 17 (the premature drop meant the x-24a had insufficient energy to reach its intended landing point, so an alternate runway was used). After examining at the data, Armstrong felt sure that Gentry had actually encountered turbulence rather than a control problem, so for the fourth glide flight, he placed a camera below the flight path to match flight footage with the pilot's reports. Sure enough, when this footage was analyzed, it showed the x-24a passing through a wind shear just as Gentry reported a slight instability. With their lack of horizontal surfaces to damp out lateral movements, all the lifting bodies proved sensitive to gusts and turbulence during their flight tests.

Jerry Gentry was now joined by John Manke following his flights in the HL-10, and once the NASA pilot had completed his checkout flight in October 1969, the x-24a team began preparing for their first powered flights, scheduled for the early months of 1970. Although the x-24a was lighter, with a greater fuel capacity, than its NASA counterparts, its high-speed potential was limited by the extra drag generated as its control surfaces extended to improve transonic and supersonic stability. The Martin lifting body was also more sensitive to high or low angles of attack than the M2-F3 or the HL-10, but with these limitations known ahead of time, Johnny Armstrong and his team were able to devise their flight plans accordingly. As with the HL-10, the first powered flights would use only two of the XLR11's four chambers, before later flights moved on to use the engine's full 8,480 lbf. thrust in order to investigate transonic and supersonic flight conditions. Gentry made the lifting body's first rocket flight on 19 March 1970, reaching a modest Mach 0.86 without major incident.

Gentry and Manke continued to alternate flights, pushing the vehicle higher and faster until on 14 October, twenty-three years to the day after Yeager's historic Mach-busting flight in the Bell X-1, John Manke eased the X-24A to Mach 1.18 at 67,900 ft. Jerry Gentry bettered his colleague's feat, on 20 November, reaching Mach 1.37 on his thirteenth and final flight in the aircraft. The air force pilot was soon due to leave Edwards for Southeast Asia, with his place in the X-24A going to Major Cecil Powell, but before leaving for his combat tour, Gentry would make one final lifting body flight.

After almost two and a half years of reconstruction and checkout, a process that had cost the FRC an estimated $700,000, the M2-F3 was now ready to fly. Bill Dana, one of the center's most experienced research pilots, had already flown both the lightweight M2-F1 and the HL-10, but he was a vocal critic of the center's decision to keep flying the M2-F2 once its inherent lack of stability became apparent. Following Peterson's crash, many pilots remained wary of the rebuilt M2, however good its wind tunnel data looked, and although Dana agreed to become NASA's M2-F3 project pilot, there remained a great deal of apprehension at the FRC as his first flight approached. As he walked past the reconfigured lifting body one day, it occurred to flight planner Jack Kolf that the black NASA lettering on the aircraft's outboard fins could easily be changed to the pilot's name, and following a quick trip to the center's paint shop, where a stencil was made, the M2-F3's center fin soon bore the legend DANA. It was many days before Paul Bikle noticed the personalized markings during one of his regular trips down to the hangars, and although the impromptu paint job was soon removed, Kolf's gesture had helped to relieve tension. The aircraft's first glide flight took place on 2 June 1970, and to the great relief of both Bill Dana and the project engineers, the modified lifting body proved far more stable than its predecessor.

Whereas the M2-F2's pilots had needed to make constant inputs to keep the squirrelly aircraft under control, the M2-F3's center fin and improved stability augmentation system kept it solid and stable throughout the flight, and although Dana acknowledged that the M2-F3 did not fly as well as the HL-10, he now felt confident that it could be flown safely.

Following two more glides, the go-ahead was given for powered flights to begin, and on 25 November, Dana launched from the NB-52, igniting two of the XLR11's chambers to carry the M2-F3 to Mach 0.8 at 51,900 ft. before the pilot carried out a unique experiment. Having been impressed by the aircraft's landing performance during his glide flights, Dana believed that it would be possible to land the vehicle without using its nose window, and although this had little bearing on the M2-F3 flight program itself, it would be important on any orbital successor where a glass nose would not be an option. To test his theory, Dana had asked the ground crew to cover the inside of the M2-F3's nose window with thick paper, leaving a cord attached so he could rip the covering away if needed. Although he made a safe, if heavy, landing with the paper in place, Dana recommended that the experiment should not be repeated.

Gentry, Manke, Dana, and Hoag made a total of twenty-two lifting body flights during 1970, with the M2-F3's powered debut capping an incredibly productive year for the entire program, and away from the flight line, the FRC also made an important contribution to NASA's next major spaceflight initiative—the Space Transportation System, or space shuttle. After stipulating that the proposed spacecraft would land horizontally, the general assumption within both NASA and the industry was that the shuttle would carry air-breathing engines for use during its approach and landing phase. As these engines would need to remain retracted (or at least protected) until the spacecraft had reentered the atmosphere, their inclusion looked likely to add a considerable amount of complexity—not to mention weight—to the vehicle. To the FRC's engineers, it seemed that their colleagues from the Manned Spacecraft Center and Marshall Space Flight Center were discounting lessons learned over the previous decade.

Having routinely demonstrated that the X-15 could perform unpowered low L/D approaches from the edge of space, the FRC could now also apply its recent heavyweight lifting body experience to advocate a similar technique for the shuttle. In order make its case, the center hosted a symposium titled Flight Test Results Pertaining to the Space Shuttlecraft for representatives from the shuttle program and interested contractors on 30 June 1970. During the day-long event, FRC engineers and pilots from both the NASA center and the AFFTC presented ten papers covering relevant aspects of the x-15, lifting body, and other pertinent flight test programs (including a series of low L/D approaches flown in the NB-52 by Fitz Fulton), with the aim of convincing the assembled audience that the shuttle should not rely on air-breathing engines to execute its approach and landing.

In the weeks running up to the symposium, the HL-10 concluded its flight career with two low-level approach and landing tests flown by Pete Hoag. Using three hydrogen peroxide landing rockets in place of the XLR11 engine, the HL-10 was able to make shallow, airplane-like powered approaches at an angle of 6° rather than the regular 18°. For these flights, the vehicle was released at its normal 45,000 ft. launch altitude and flew a normal glide flight path to a predetermined point, where the small rockets were ignited, extending and shallowing final approach. Hoag reported that the lower angle had actually made the landing harder to judge when compared to the usual steep, unpowered technique.

The FRC also contended that, since a future shuttle's air-breathing engines could only be used once the vehicle became subsonic, it would need to glide down to that point anyway. Would it not be simpler and safer to continue the unpowered approach rather than maneuver the space plane into a position where it had to rely on jet engines? Although the center stopped short of recommending that the air-breathing engines should be removed from consideration, the symposium put forward a persuasive case that even if they were retained, the default approach and landing technique should be unpowered, with the engines only being used to correct errors or allow an emergency go-around capability.

In his closing remarks, Milt Thompson offered this note of reassurance: "If all the other NASA centers, in conjunction with the Department of Defense and industry, can get the shuttle off the ground, into orbit and ensure that it survives the entry, we at the Flight Research Center can guarantee that it can be flown to the destination and landed safely."

With the HL-10 now retired and the x-24A midway through its envelope-expansion flights, much of the activity over the next two years centered on

31. U.S. Air Force test pilot Cecil Powell with the x-24a in 1971.
Courtesy NASA.

the M2-F3. Following its powered debut, the rejuvenated rocket plane was back in the air in February 1971, with a familiar face in the cockpit. As Jerry Gentry was the only remaining active pilot to have flown the troublesome M2-F2, the M2-F3 team were keen to have him fly the lifting body before his deployment to Vietnam in order to get a pilot's comparison of the two configurations. On 9 February, Gentry made his sole glide flight in the M2-F3, happily confirming that the earlier aircraft's vices appeared to have been tamed by the modifications. In doing so, the air force pilot became the only person to fly the M2-F1, the M2-F2, the HL-10, the x-24a, and the M2-F3.

As Gentry departed, Bill Dana resumed his buildup flights in the M2-F3, while John Manke and Cecil Powell continued to put the x-24a through its paces. Although Manke took the aircraft to its highest speed of Mach 1.6 on 23 March, the later stages of the x-24a test program were beset by engine problems as McTigue's salvaged xLR11s began to show their age. With chambers failing to ignite and even a small engine fire on one flight, the team decided to avoid any further risk by withdrawing the vehicle following its twenty-eighth flight on 4 June 1971. Although it would never fly again in its current form, the much-maligned "flying potato" would later return to the skies above Edwards following a dramatic metamorphosis.

Bill Dana put the M2-F3, the only lifting body left on flight status, through the same series of envelope-expansion flights previously performed in the HL-10 and the X-24A. Although now far less challenging to fly, the M2-F3 still encountered the occasional incident, the most dramatic of which occurred in June 1970. With preparations well underway for a transonic buildup flight, FRC ground crews began to fill the lifting body's propellant tanks as it hung beneath its launch aircraft's wing. Although similar operations had been undertaken routinely at Edwards since 1946, crews always remained vigilant as the hoses were connected and the propellants began to flow. With the water-alcohol mixture and LO2 flowing into the little rocket plane, an alert technician noticed liquid pouring from the M2-F3's oxidizer vent and, recognizing the liquid as the water-alcohol mixture rather than LO2, immediately raised the alarm. It quickly became apparent that a valve in the propellant system had failed, allowing the water-alcohol fuel to enter the LO2 tank, where it combined with the oxidizer to create a highly unstable and potentially explosive mix. The ground crew now had a potentially catastrophic situation on their hands, as any jolt might trigger a blast that could destroy both the M2-F3 and its NB-52 mothership.

Once the immediate area had been safely evacuated, the Edwards Flight Operations Office quickly sent out word that no supersonic flights were to be made over the base, lest the shock of a sonic boom trigger an explosion. In order to diffuse the situation, the M2-F3's vent valves would need to be opened, allowing the oxidizer to boil off and leaving only the less reactive water-alcohol fuel. At considerable personal risk, crew chief Bill LePage and ops engineer Herb Anderson headed back into the danger zone. Armed with specially padded tools and taking extreme care not to make any sudden movements, the two men gingerly opened the valves before retreating to a safe distance. It took several days for the oxidizer to boil off, but eventually the ground crew were able to demate the M2-F3 and purge its tanks before repairing the faulty valve ahead of the next flight.

Once back in the air, Dana began pushing the lifting body further into the transonic zone, with flights to Mach 0.93 and 0.97, before he took the M2-F3 supersonic for the first time on 25 August 1971. Unfortunately, engine problems also began to take their toll on M2-F3, with Dana experiencing a small fire that necessitated an emergency landing on Rosamond Lake during his next flight. By the end of the year, the NASA pilot had taken the modified lifting body to Mach 1.27 at a maximum altitude of 70,800 ft., demonstrating that the M2 shape was capable of flying the last leg of a return from orbit. Although the M2-F3 still showed a tendency toward pitch problems as it approached Mach 1, the team were generally satisfied with its performance.

During the rebuild, engineers had taken the opportunity to install an experimental electronic flight control system—the Command Augmentation System (CAS). Whereas the pilot would normally fly the aircraft using the center stick, the CAS would be operated from a smaller side stick, assessing the pilot's inputs alongside flight data to determine how far the control surfaces should be moved. With no direct mechanical linkage between the pilot and the control surfaces, the CAS was essentially an early digital fly-by-wire system, although the pilot could regain direct control at any time by switching back to the center stick. The CAS could also be used to operate the M2-F3's reaction control system, allowing the pilot to assess its effectiveness under a variety of flight conditions. Following a six-month hiatus in lifting body operations caused by a combination of winter rains and maintenance on the NB-52 launch aircraft, Bill Dana made the first CAS research flight on 25 July 1972. Although the system never completely lived up to engineers' expectations, it did simplify certain piloting tasks, such as pitch control, by allowing them to hold a chosen angle of attack.

On 5 October, Dana made the one hundredth lifting body flight—a real milestone for what had begun as a low-key, in-house research effort a decade earlier. Before the year was out, two new pilots checked out in the M2-F3, with the AFFTC's Cecil Powell and NASA's John Manke both going on to make supersonic flights in the lifting body before its swan song on 20 December 1972. Although it had met with some opposition at the time, the decision to rebuild and reconfigure the damaged M2-F2 had paid off handsomely, giving the joint NASA and air force test team a third point of comparison alongside the HL-10 and the X-24A. Although the retirement of the M2-F3 effectively marked the end of the heavyweight lifting body research program as originally planned, it wouldn't be long until a very different wingless shape would grace the skies above Rogers Dry Lake.

13. Racehorses and Unrealized Plans

On 13 May 1971 Paul Bikle retired from his role as director of the FRC. Having replaced Walt Williams in 1959, Bikle's nearly twelve-year tenure had spanned the entire x-15 flight program as well as the majority of the lifting body program. Under his direction, the FRC had not only continued its work to advance the state of the art in aviation; it had also become heavily involved in research that directly supported NASA's spaceflight activities, including the planned space shuttle. Bikle's long-serving deputy De Beeler took temporary charge of the center before Lee Scherer transferred across from NASA Headquarters to take the role permanently in October 1971.

Bikle's departure marked the end of an era for the center. During the 1960s, the FRC became so synonymous with the x-15 program in the minds of politicians that Bikle had been called on to defend its continued existence beyond the end of that program. By encouraging a wide variety of research, including supersonic transport–related studies using the XB-70 and preparations for the Apollo moon landings via the LLRV program, the director managed to ensure an ongoing role for the facility. Bikle's support for lifting body flight research had seen Dale Reed's modest proposal for the M2-FI grow into a major experimental program during the second half of the decade, but as the 1970s dawned and the realities of NASA's post-Apollo budget contractions, alongside the wider fiscal pressures affecting the American economy, reached the High Desert, the future looked less certain. Lee Scherer's appointment marked a change in the nature of the center and its role within NASA. Paul Bikle and Walt Williams had both been experienced flight test engineers, willing to pursue programs based on their own practical experience and ready to place their faith in the abilities of the engineers, technicians, and pilots that worked with them. Under their leadership, the FRC had developed a very individual, independent character when compared to its sister centers.

When Scherer took over, he brought a different style of management, which saw the center more closely aligned to NASA Headquarters. As the administrative links between the FRC and Washington were strengthened, more of the decisions that affected the center's activities were being made by senior managers on the other side of the country. Ironically, the X-15's success in validating the accuracy of laboratory research techniques meant that some within NASA began to question the need for actual flight research as the agency looked toward the shuttle. Unsurprisingly, the FRC still maintained that flight-testing could play a valuable role in the development of new systems. During his closing comments to the FRC's influential 1970 Shuttlecraft symposium, Milt Thompson spoke about the importance of verifying new configurations in free flight, pointing to the success of the heavyweight lifting bodies and suggesting that a similar approach might be useful for the shuttle. The FRC had already recommended to OART that subscale vehicles representing likely shuttle orbiter configurations should be flight-tested at the center, allowing different designs to be verified in the air just as the M2 and HL-10 configurations had been. Initially, many at the FRC and Langley had hoped that the proposed shuttle would use a lifting body configuration, with the HL-10 being considered the most promising candidate for further development, but as NASA's Phase A concept definition studies progressed, a combination of cross-range requirements and the need for a longer payload bay began to favor other configurations, eventually leading to the selection of a delta-winged design.

In 1972 the FRC proposed the construction of a single-seat, subscale shuttle orbiter prototype that could be flown to verify the design's performance across different sections of the planned return profile. By using various combinations of XLR11 and XLR99 rocket engines, an incremental flight test program could test the orbiter's transonic, supersonic, and hypersonic characteristics, answering many questions regarding the proposed approach and landing techniques. Although the minishuttle proposal gained some prominent supporters—including Robert Gilruth, then director of the Manned Spacecraft Center, and NASA's chosen orbiter contractor, North American Rockwell (the company formed following Rockwell-Standard's 1967 merger with North American Aviation)—it went no further, following concerns about possible budget increases and the program's willingness to rely on more sophisticated simulation techniques.

The late 1960s and early 1970s also saw FRC engineers involved in some highly speculative proposals to launch one of the existing lifting body configurations into space in order to perform piloted flight tests from orbit down to a runway landing. Dale Reed went so far as to discuss these ideas with Wernher von Braun, with the rocket engineer promising to commit unused Apollo launch vehicles to the cause or possibly to carry a lifting body as an additional element on one of the support missions for NASA's forthcoming *Skylab* space station. However, Paul Bikle refused to support such a scheme, seeing orbital flights as being beyond the FRC's remit.

By the end of 1972 the M2-F3 had made its final flight, and with the HL-10 and the X-24A already retired, this might have spelled the end for the lifting body flight-testing program, had it not been for the Air Force Flight Dynamics Laboratory (AFFDL) and NASA's Langley laboratory. The first generation of lifting bodies had all been blunt-nosed shapes with large high-volume bodies. Such shapes were seen as being well suited to ferrying cargo and crew to and from future orbital outposts. With hypersonic L/D ratios no greater than 1.4, they would still offer a cross-range potential of 350–500 miles to either side of their orbital path, giving more flexibility when it came to reentry opportunities to a wide range of landing locations. Referring to their utilitarian nature but unspectacular performance, Dale Reed would later call these vehicles "plough horses," but by the mid-1960s, new research was pointing to a very different type of lifting body configuration, one capable of achieving performance well beyond that of the first generation of shapes.

The AFFDL came into being in 1959 as one of four laboratories working under the Wright Air Development Division (WADD), successor to the earlier WADC. Later reorganizations brought the laboratory under the control of the Air Force Systems Command (AFSC), following that organization's formation in 1961, and subsequently the AFSC's Research Technology Division (RTD) in 1963. The AFFDL and its sister laboratories were a product of General "Hap" Arnold's postwar drive to improve the scientific and technical capabilities of the USAAF (and subsequently the USAF), in order to reduce its reliance on the NACA. During the early 1960s, AFFDL engineer Alfred Draper began an effort to identify promising hypersonic configurations for both upper-atmospheric and exo-atmospheric applications, and by 1965 this work had coalesced into a series of three lifting body designs (labelled FDL-5, FDL-6, and FDL-7) that promised high hypersonic L/D and a cross range of up to 1,500 miles to either side of the orbital path. These new flat-bottomed, highly swept deltas would be able to reach almost any site within the North American continent during a return from orbit.

Having been configured purely for high performance during hypersonic flight, the FDL shapes would offer even less subsonic L/D than the Ames M2 shape—the poorest performing of the "plough horse" configurations. Although some of the new shapes were designed with deployable wings to generate lift at the slower speeds required for approach and landing, Draper and his colleagues believed that their FDL-7 shape could be modified to improve its subsonic L/D and handling, allowing it to land without the need for additional surfaces. Meanwhile, researchers at Langley had been thinking along similar lines to Draper's team and had come up with their own advanced lifting body configuration known as the Hyper III. To Dale Reed, these promising new shapes were "racehorses" when compared to the M2-F2, the M2-F3, the HL-10, and the X-24A, and the FRC engineer soon came up with a plan to get the Hyper III flying.

Since making the decision to move away from project management and back into engineering, Dale Reed had become increasingly interested in using radio-controlled models as a precursor to piloted flight tests. Having already pioneered this technique during his efforts to test the M2 shape back in the early 1960s, Reed's efforts now grew more sophisticated. Together with longtime collaborator Dick Eldredge, he constructed a large twin-engine mothership to carry his models to altitude, where they could be launched at the flick of a switch with one engineer taking control of the test vehicle while the other brought the launch aircraft home.

By following this low-cost approach, Reed and Eldredge were able to carry out a series of flight tests using the Langley Hyper III shape. The narrow delta-shaped lifting body was a far cry from the blunt half-cone M2 and featured two small winglike strakes at the rear of its wedge-shaped fuselage as well as two outward canted fins near the top of its steeply sloped sides. Like the FDL-7 design, the Hyper III had very low subsonic L/D, making controlled landings impractical without some form of deployable wing. Using their subscale model, Reed and Eldredge were able to assess various different types of rigid and flexible wings, eventually settling on a straight one-piece pivoting design housed in the upper fuselage. After proving that the Hyper III was stable and controllable during low-speed glide tests during 1969, Reed again approached Paul Bikle with a proposal for a full-scale test vehicle. In order to keep costs down and avoid additional complexity, the engineer suggested that the initial flights should be controlled from the ground, but this idea did not go down well with either the director or the FRC pilots' office. As a compromise, Bikle suggested that the vehicle be constructed in such a way that a cockpit could be added later once initial tests had proven the vehicle's basic airworthiness.

The Hyper III vehicle was built in the FRC shops on a low-priority basis by a volunteer workforce. Many FRC technicians and engineers were avid aircraft homebuilders, and the chance to work on the unusual lifting body project appealed to their sense of adventure. Constructed mainly from steel tubing covered with Dacron fabric, the lightweight test vehicle measured thirty-five feet long by twenty feet wide (at the widest point of the strakes). The aircraft's four control surfaces and pivoting wing were fashioned from aluminum, and parts for the vehicle's control system were salvaged from another lifting body—the air force's SV-5D PRIME subscale demonstrator. Reed had managed to secure the short-term loan of a navy Sikorsky SH-3 helicopter via Max Faget and planned to use this to lift the Hyper III to a launch altitude of some 10,000 ft. above the lake bed before releasing it into free flight, but he still needed a pilot to fly the lifting body. With the pilots' office showing little interest in remotely flying the vehicle, Reed once again turned to Milt Thompson.

Although he had not flown since 1966, Thompson was intrigued by Reed's remote-controlled concept and agreed to act as project pilot for the early tests. A "ground cockpit" featuring basic flight instruments and a control stick was constructed, allowing Thompson to sit out on the lake bed as he "flew" the Hyper III home. As the vehicle entered the final stage of its approach, the former research pilot would hand over control to engineer Dick Fischer, who would then use a standard radio-controlled box to visually flare and land the vehicle. With the flight plan finalized and all checks complete, the first flight of the Hyper III was set for 12 December 1969. Bruce Peterson, veteran of the horrific M2-F2 crash in 1967, would pilot the sH-3 launch helicopter for the flight. Although no longer an active research pilot, Peterson was still able to perform some flying duties in a support capacity and remained an active member of the Marine reserves. As he lifted off that morning, the former test pilot found it difficult to keep the Hyper III stable at the end of its four-hundred-foot-long cable. But by the time he reached the planned launch altitude, he had the lifting body pointing forward, and as he reached launch speed, the cargo hook was released, freeing the Hyper III.

While Peterson was climbing to the launch point, Milt Thompson had taken up position in his lake bed cockpit, taking time for a few last-minute cigarettes, but now as the glider fell free, he found himself concentrating every bit as intently as he would if he had been in the air. The Hyper III, with its pivoting wing fixed in the deployed position, launched on a northerly heading, continuing on this track for approximately three miles before Thompson performed a 180° turn back into the upwind approach leg. As the Hyper III flew onward, Thompson made a number of control inputs to assess the vehicle's stability, with constant radar updates from the FRC's control room offering assurance that he was still on course. Descending through 1,000 ft., the lifting body finally loomed into view through the morning's haze, and Thompson handed control to Fischer, allowing the engineer to bring the Hyper III in for a safe landing in front of the chariot-like ground cockpit.

Although he found the vehicle extremely sensitive to roll inputs, Milt Thompson had verified that the Hyper III shape was essentially stable, just as the earlier wind tunnel research at Langley had suggested. Unfortunately, the actual L/D turned out to be considerably poorer than predicted, and this, together with NASA Headquarters' refusal to allow Bikle to purchase a suitable tow aircraft, meant the Hyper III would never make another flight. Fortunately, all was not lost for the "racehorse" lifting bodies.

Having been impressed by the FRC's low-cost approach to testing its proof-of-concept lifting body vehicles, the AFFDL decided to follow a similar path in order to validate its own designs. Rather than build an entirely new vehicle, Alfred Draper began to investigate using one of the unflown Martin sV-5J jet-powered lifting bodies as the foundation for a new configuration. By gloving an extended fuselage and lateral strakes derived from the lab's FDL-7 shape onto the existing three-finned "flying potato" fuselage of the sv-5j, Draper came up with a hybrid double-delta shape known as the FDL-8x, which appeared to offer sufficient L/D to approach and land without the use of deployable wings. In 1969 the AFFDL produced a development plan for the new vehicle, but as his technical inquiries progressed, Draper began to have second thoughts about using an sv-5j.

Although Martin had produced two of the jet-powered lifting bodies, hoping that they would be procured by the ARPS as training aircraft, the resulting vehicles left much to be desired. With a single Pratt and Whitney J60-PW-1 engine fed by a small ventral intake, the sv-5j was woefully underpowered given the high drag and low lift offered by its portly fuselage. Although larger air intakes might have improved the situation, they would also have created additional drag—the last thing the vehicle needed. With Yeager and Martin both still keen to see the sv-5j fly, the company approached Milt Thompson, inquiring whether the former NASA pilot would consider performing a series of demonstration flights. After traveling to Baltimore to fly the company's simulator and discuss the aircraft's performance, Thompson was alarmed to discover just how marginal a flight in the sv-5j would be. Assuming that he was able to get the aircraft airborne, he would then have mere seconds to retract the gear, reducing the vehicle's drag enough to remain airborne. If he failed to achieve this, the fully fueled lifting body would settle back onto the runway with potentially disastrous consequences. After informing the project's managers that he regarded their aircraft as being too dangerous to fly, Thompson returned to the FRC, assuming that would be the end of the matter, but Martin was not about to give up just yet. Within days, a company representative was back in contact, asking the ex-pilot to name his price to fly the sv-5j. Having again indicated that he was not interested, Martin continued to pursue the matter until Thompson eventually quoted them a price of $25,000. For this, he intended to accelerate down the runway before hitting a carefully positioned two-by-four board that, he hoped, would send the vehicle airborne for a few brief seconds, after which he would decelerate, taxi back to the hangar, and ask for his money. To the NASA engineer's great relief, he never received a reply.

Although Draper intended to air-launch his FDL-8x from one of the AFFTC's NB-52 motherships, he still felt concerned that the sv-5j's low thrust would hamper the aircraft's potential research value. During a visit to Edwards, Draper concluded that the project would be better served by using the rocket-propelled x-24A rather than its jet-powered sibling as the base for the new vehicle, and having gained the support of both Fred DeMerritte and Paul Bikle from NASA and the AFFTC's Bob Hoey, the AFFDL only needed final approval from the AFSC to move its plans forward. Unlike previous research aircraft, the new FDL-8X was to be jointly funded by the air force and NASA, and this unusual arrangement caused some procrastination among AFSC managers, but in his usual dynamic manner, the FRC's John McTigue decided to break the deadlock by sending NASA's share of the funding ($550,000) straight to the AFFDL. On 21 April 1971 AFSC officially gave the program its go-ahead, and when the x-24A completed its final flight on 4 June, the aircraft's various subsystems were removed and placed into storage before the lifting body's airframe was transported to Martin's Denver plant, where the conversion work would take place. As air force and NASA officials met to sign a memorandum of understanding to cover the new program, the FDL-8X received a new name—the x-24B.

The x-24A's conversion followed the now-familiar Skunk Works template, with Martin assembling a small, highly skilled team, many of whom had helped build the original lifting body. Working closely with their air force and NASA counterparts, the team again used off-the-shelf items wherever possible, with surplus components from the x-15, the F-104, the T-38, and the F-106 all making an appearance. The new fuselage was gloved over the existing airframe, with the cockpit and the three vertical fins among the only recognizable elements of the x-24A left visible. When complete, the x-24B was over fourteen feet longer than its predecessor, with the new wing-like horizontal strakes adding ten feet to the lifting body's width. The rear fuselage was boat-tailed to help improve the vehicle's subsonic L/D, while the majority of the lower surface was now a flat double delta, with the long nose sloping up slightly to improve the vehicle's hypersonic trim (although the x-24B itself was never intended to reach these speeds).

Following its conversion, the x-24B was expected to have a launch weight of 13,800 lbs., by far the heaviest of the lifting bodies but still considerably lighter than some configurations of the x-15. With its long narrow nose, the new high-performance lifting body's center of gravity had shifted forward, meaning that the vehicle would need to be carried farther back beneath the NB-52's wing than the x-24A had been. Unfortunately, this meant that the

X-24B's pilot would be unable to eject from the vehicle while still attached to the launch aircraft, but this was deemed an acceptable risk by program managers. The increased length of the vehicle also meant that the nose gear would have to be moved forward, but with the main gear remaining in their original position, this looked likely to cause substantial loads as the nose slapped down in a similar fashion to the X-15's. A new, more robust nose gear, taken from the now-obsolete Grumman F11F-1F Super Tiger naval fighter was selected to deal with this. With the X-24B featuring two additional control surfaces in the form of elevons located to the rear of its horizontal strakes, the lifting body's stability and handling were expected to be excellent when compared to the "plough horses" that had preceded it.

As the conversion process continued in Denver, a highly experienced X-24B team was assembled at Edwards. Following on from his role on the X-24A, Johnny Armstrong remained as the AFFTC's program manager, while the FRC's Jack Kolf performed the same role for NASA. Bob Hoey would be in charge of flight planning. Having already flown the HL-10, the X-24A, and the M2-F3, John Manke was named as NASA project pilot, and he would be joined by air force test pilot Major Mike Love. Building on the work done for the X-24A, a new simulator was developed by the AFFTC's simulation lab. An attempt to provide the pilots with external visual cues using a light-table projection technique then under development at the lab proved less than successful, however, with John Manke repeatedly crashing the simulated aircraft as the projected horizon got out of sync with his instruments. As with the earlier lifting body and X-15 programs, the X-24B's pilots also flew numerous low L/D approaches in preparation for their research flights, with the F-104 remaining the aircraft of choice to simulate these rapid, high-drag descents. In 1972 the AFFTC began to phase out its F-104 chase aircraft, switching over to the T-38 and McDonnell Douglas's F-4 Phantom II instead. Although both of these aircraft were evaluated to determine how well they would perform low L/D approaches, only the T-38 proved suitable, and even then, a special waiver was required as the jet trainer needed to extend its gear above the 345 mph recommended limit to create the required drag. Fortunately, the FRC's fleet of F-104s were made available to AFFTC pilots for proficiency flights throughout the X-24B program.

On 24 October 1972 the X-24B arrived back at Edwards to begin the long process of having its subsystems and research instrumentation reinstalled before it underwent a thorough checkout procedure. Working under FRC ops engineer Norm DeMar, the X-24B crew pushed on through the later months of 1972 and into 1973, putting the X-24 through a rigorous set of ground tests, including high-speed taxi runs across the lake bed using the landing rockets and XLR11 (nicknamed Bonneville racer tests by the program team). Finally, by mid-July, the X-24B was ready to make its first glide flight.

The first attempt at a free flight took place on 24 July, but unfortunately, a gyro failure in the lifting body led to an abort and a trip home under the NB-52's wing for John Manke. The flight was rescheduled for 1 August, and on this occasion, everything went right, with the X-24B falling away from the launch aircraft at an altitude of 40,000 ft. In keeping with earlier lifting body debuts, Manke's main aim during this first flight was to check out how the new configuration handled during a series of standard test maneuvers, including the now-customary practice flare. Racing toward the lake bed with an airspeed of 460 mph, Manke set the gleaming silver-and-white X-24B up for a landing on Runway 18, making a perfect flare and settling the lifting body down onto the hard clay at 200 mph to end a four-minute, eleven-second flight. The X-24B flew well, exhibiting none of the first-flight instabilities experienced by the three original heavyweight lifting bodies.

With his first glide behind him, Manke made two more flights during August, using a higher launch altitude of 45,000 ft. to enable him to reach transonic speeds during his descents. On both occasions, the NASA pilot was also able to test the X-24B's performance with its dampers off, finding that the vehicle remained stable and easy to control under all conditions. During his third flight, Manke also tested the aircraft's fuel jettison system, but this seemingly routine action produced surprising results. Following an engine fire during one of the X-24A's powered flights, engineers had traced the cause back to jettisoned fuel that had recirculated into contact with the still-hot XLR11 due to the low-pressure zone around the lifting body's rear. In order to prevent a similar occurrence on the X-24B, the AFFDL had conducted a wind tunnel analysis to identify a new location for the jettison port and had settled on a position at the base of the right-hand fin just above the new elevon. When Manke opened the jettison valve in flight, the high-pressure flow of liquid began to interfere with the vehicle's aerodynamics, causing it to roll sharply to the right. As Manke fought to counteract the unexpected movement, he halted the jettison, at which point the lifting body returned to stable flight. Once again, a seemingly minor change turned out to have major implications for these unusual wingless shapes, and what had seemed a reasonable solution in the wind tunnel turned out to be anything but, during flight tests. The jettison port was subsequently moved to the rear of the fin and tested during the fourth glide on 18 September without problems. Manke now gave way to Mike Love, who made a single familiarization glide in the aircraft before powered flights began. Like his NASA colleague, Love was pleased with the x-24B's handling, which the pair felt was comparable to the F-104 or the T-38—high praise indeed.

With the x-24B's basic airworthiness established, it was time to begin the rocket-propelled envelope-expansion phase. John Manke's first powered flight attempt, on 31 October, was scrubbed after prelaunch checks revealed problems with the XLR11's igniters, and the bad luck continued thirteen days later when a second attempt was abandoned after the weather over Edwards deteriorated during the NB-52's climb to altitude. It was a case of third time lucky for Manke, though, as the next attempt, on 15 November, saw Manke use three of the XLR11's chambers to reach Mach 0.93 as the x-24B, sporting a smart new white-and-blue scheme, soared to 52,764 ft. Just over a month later, the NASA pilot pushed the lifting body to Mach 0.98 to test its transonic stability and control, while recording vibration and pressure data. Mike Love made the first x-24B flight of 1974, performing a second glide flight on 15 February, after which two project pilots began to alternate the envelope-expansion flights, with Manke taking the aircraft through Mach 1 for the first time, on 5 March, before Love made his first rocket-powered flight on 30 April.

Although it used the same XLR11 rocket engine as the other powered lifting bodies, the x-24B's additional weight looked likely to limit the aircraft's performance to a modest Mach 1.4. In order to counteract this, Thiokol's Reaction Motors Division and the AFFTC's rocket technicians proposed that the engine's four chambers should be switched to a higher pressure once the pilot confirmed they were all running correctly. By doing this, the engine's overall thrust could be increased from 8,600 lbf. to 9,800 lbf., giving the x-24B the potential to reach Mach 1.76. On 24 May, John Manke made the first flight using the increased chamber pressure, hitting the overdrive switch thirty seconds after ignition to reach Mach 1.14 (slightly slower than planned due to an ignition failure in one of the chambers). Over the next five months, the pilots took turns pushing the x-24b to ever-higher speeds and altitudes until Mike Love reached Mach 1.75 at 72,150 ft. on 25 October—the fastest the aircraft would ever travel and 0.15 of a Mach number faster than the x-24a's best mark.

Edwards often played host to vip visitors, and following one of his x-24b flights, John Manke was greeted by the famed French aquanaut Jacques Cousteau. Given Cousteau's general fascination with exploration, it was not surprising that he had an interest in activities out at Rogers Dry Lake, but as Manke was helped out of the x-24b's cockpit, he and the crew decided to make their esteemed guest feel more at home with a small preplanned gag. As the reception party walked out toward the angular lifting body, Manke—still wearing his white pressure suit—waddled out to meet them wearing a pair of black flippers over his flight boots! Following much laughter all round, the undersea explorer took great interest in hearing Manke's description of the flight before inspecting the aircraft's cockpit for himself.

By the end of 1974 Manke and Love had completed the x-24b's envelope-expansion phase, with the final two flights of the year marking the start of the dedicated research program. During a series of ten flights, the two pilots put the high-performance lifting body through a variety of aerodynamic maneuvers, testing its stability and control while investigating pressure, vibration, and boundary-layer noise in both supersonic and transonic flight. Toward the end of this phase, the x-24b was also used to test the silica-based thermal protection tiles for the upcoming space shuttle, and as the research flights drew to a close in June 1975, a new set of tests was planned in support of nasa's forthcoming space plane.

Throughout the envelope-expansion and research phases, Manke and Love both tested the x-24b's landing accuracy and, finding they were consistently able to bring the lifting body down within five hundred feet of a mark on Runway 18, both pilots felt confident that they would be able to land on Edwards afb's fifteen-thousand-foot main runway (Runway 4). Although the first shuttle orbiter was already under construction in nearby Palmdale (without landing engines), the x-24b team still believed that a timely demonstration of an unpowered, low l/d approach and landing using the lifting body would help build confidence in the shuttle's proposed approach and landing techniques. With the plan approved by both

FRC director Scherer and his AFFTC counterpart, former X-15 pilot Brigadier General Robert Rushworth, Manke and Love began an intense three-week period of training, including numerous high-drag approaches in F-104s and T-38s—Manke alone logged one hundred practice approaches in preparation for his first runway landing attempt, on 5 August. On that day, the X-24B launched from the NB-52 at an altitude of 45,000 ft., aiming for a target speed of Mach 1.5 at 71,000 ft. Unfortunately, the recalcitrant XLR11 again refused to perform as planned, and with only three chambers burning, Manke switched to his alternate flight plan. After reaching Mach 1.19 at 57,000 ft., the pilot put the aircraft through a planned series of data-gathering maneuvers before turning his attention to the all-important approach and landing.

The flight plan called for Manke to make a 165° turn into his final approach to Runway 4 while still at 24,000 ft. In order to provide an extra degree of safety, the flight planners had included a commit point some three miles before this turn, allowing Manke the option to carry on to a lake bed landing if he felt that a landing on the concrete runway could not be safely attempted. Upon reaching this point, Manke was reassured by Mike Love, who was watching the radar plots back in the FRC control room, "You can plan on landing on 04." With his turn made, Manke now had to aim for a single Joshua tree he had picked out on his approach path as the point at which he would begin his flare. The team had originally hoped to bring the X-24B down approximately 2,500 ft. from the start of the long concrete runway, but the experienced Rushworth vetoed this idea, insisting that they aim for the 5,000 ft. mark instead, instructing the pilots, "I will forgive you if you land long and end up rolling onto the lake bed, but landing short of the concrete runway will be unacceptable."

Although Manke later commented on the high workload during the final leg of his approach, the NASA pilot hit his flare point perfectly and brought the X-24B down to a pinpoint landing on the white line painted 5,000 ft. along the runway. In the control room a happy Mike Love indicated that Manke had just won a wager between the two men: "Yeah, it looks to me like you got that line for about a dollar's worth up here." Having successfully completed the X-24B's twenty-seventh flight, John Manke bowed out of the lifting body program, taking the opportunity of his post-flight debriefing to draw attention to the teamwork that had underpinned

32. The last of the rocket planes, the X-24B, with some of its pilots. Left to right: Einar Enevoldson, John Manke, Francis "Dick" Scobee, Tom McMurtry, Bill Dana, and Michael Love. Courtesy NASA.

the entire effort: "Any of you at the crew briefing yesterday are aware of the many people that were involved in something like this. That's more of a thrill I think than anything else around here, just watching everybody working and watching them put all the pieces together."

Mike Love matched Manke's feat by making the second pinpoint runway landing, on 20 August, following a flight to Mach 1.55 at 72,000 ft. Like his NASA counterpart, this marked the air force test pilot's final X-24B flight, and between them, the two pilots were happy to report that they had actually found landing on the concrete runway slightly easier than the usual lake bed landings, as the various runway markers and taxiways gave better visual cues than they were used to. With both Manke and Love having now left the lifting body program, it fell to FRC veteran Bill Dana to make the X-24B's final two powered flights. The first of these, on 9 September, saw Dana reach Mach 1.48 at 71,000 ft. before the former X-15 pilot made a standard lake bed landing. Bill Dana's second flight in the arrow-like lifting body took place on 23 September and marked the end of an era at Rogers Dry Lake.

Since Chalmers Goodlin had first lit the Bell XS-1's XLR11 engine on 9 December 1946, the skies above Rogers Dry Lake had borne witness to a steady stream of rocket research aircraft, as their brave pilots steadily extended the frontiers of flight. From Yeager's historic Mach 1 flight in October 1947 to Pete Knight's Mach 6.7 dash in the X-15A-2 some twenty years later, Bill Bridgeman's pioneering 1951 altitude record of 79,494 ft. in the Douglas Skyrocket to Joe Walker's trips beyond the Kármán line in X-15-3 during the summer of 1963, the rocket plane era had marked a golden age of aviation progress, moving flight from the lower reaches of the stratosphere all the way out into space. Now, the genial Dana was bringing the rocket plane era to a close, and to mark the occasion, the NASA pilot wore some of the most unusual pieces of flight equipment ever to grace the flight test facility.

Having flown the X-15, Dana was no stranger to the David Clark Company's A/P22S-2 pressure suit, and on joining the HL-10 program in 1969, he had been dispatched to the company's Worcester, Massachusetts, facility to be fitted for an improved version of the suit. Whereas the X-15 era A/P22S-2s had all used Scott Crossfield's preferred silver outer garment, the David Clark Company now used a white cover layer and was producing white flight boots to match. NASA had decided to retain the traditional black boots when ordering A/P22S-2s for its lifting body pilots, but when Dana went to test his new suit, he was horrified to find that he had been given white boots—not a color he deemed appropriate for a test pilot. Having complained to the David Clark Company technicians that he may as well wear pink boots, the company took Dana at his word, and when the finished suit was delivered, it came complete with two pairs of boots— one pair in "test-pilot" black and the other in garish pink with added daisy stickers. Taking the gesture in good humor, Dana decided to wear the pink boots on his next HL-10 flight, going so far as to stand on a table during the postflight debriefing to show them off to the assembled crew. Now, for his final lifting body flight, the NASA pilot decided to give the pink boots another airing.

Unfortunately, the flight itself became something of an anticlimax when one of the XLR11's chambers failed to light, forcing Dana to fly an alternate lower and slower profile. The lifting body program had perhaps been a step too far for the venerable rocket engines, some of which had been in the High Desert since the X-1 days. Following a safe lake bed landing,

Dana allowed the x-24B to roll to a halt as the four chase aircraft, two NASA F-104s and two air force T-38s, flew overhead in noisy salute. After posing for group photographs to mark the final rocket flight, the participants conducted their postflight duties before retiring to the Longhorn Bar near Lancaster. Milt Thompson would later note that the usual exuberance of a postflight party was somewhat subdued that night: "It was a wake, rather than a party, as more than an era was ending."

The x-24B did make six more glide flights before 1975 was out, allowing three new pilots to gain low L/D approach and landing experience. The FRC's Einar Enevoldson and Tom McMurtry shared these flights with air force captain Francis "Dick" Scobee. Scobee later went on to fly the shuttle, as pilot on STS-41-C in 1984, before tragically losing his life in January 1986, when the shuttle *Challenger* was lost during the launch of STS-51-L.

For a while at least, it looked like the x-24B might not mark the final chapter of the rocket plane story out at Edwards. Following the 1967 abandonment of plans to test the hypersonic research engine (HRE) using the x-15, advocates within NASA and the air force continued the push for a new aircraft to pick up the baton of hypersonic flight research. Considering the immense costs involved in the development and operation of such a vehicle, it made sense for the two organizations to pool their efforts into another joint flight research program. In July 1974, as John Manke and Mike Love continued their envelope-expansion flights in the x-24B, NASA and the air force began working together toward creating a successor, and once again the AFFDL's FDL-8 shape seemed to hold the most promise. With its large internal volume and now-proven stability across a range of flight conditions, the FDL-8 shape would be able to accommodate the rocket motors and associated tankage needed to take the vehicle from air-launch to a cruising speed of Mach 6 at an altitude between 80,000 to 100,000 ft. Here, the belly-mounted scramjet engine (fueled by a separate hydrogen tank) would take over, allowing the vehicle to cruise for a short period. By August 1975, mere weeks before Bill Dana's final rocket-powered flight in the x-24B, engineers from NASA Langley and the AFFDL presented a paper covering their initial concept for a joint hypersonic research aircraft—now known as the x-24C—to a meeting of the American Institute of Aeronautics and Astronautics, held in Los Angeles. With the preliminary concept definition now in place, an x-24c Joint Steering Committee was assembled before the end of 1975, and selected aerospace companies were soon approached to conduct trade studies for the new aircraft.

Kelly Johnson's Lockheed Skunk Works had not previously worked on a dedicated high-speed research aircraft program for the air force or NASA; indeed, Johnson himself had called the utility of such aircraft into question during the early development of the x-15. However, following their groundbreaking work on the a-12 family of triple-sonic reconnaissance aircraft, few could match the Skunk Works' abilities when it came to creating small numbers of bespoke aircraft able to fly in the most challenging of environments. A Lockheed team under the leadership of Harry Combs won a contract from Langley to examine the various technical and operational challenges posed by the x-24c. The base vehicle that had emerged from the earlier joint Langley-AFFDL studies bore a strong resemblance to the x-24b, with winglike strakes and a delta configuration. The x-24c design featured an x-15 style canopy and also carried that aircraft's xlr99 engine, but as Lockheed began refining the design, the configuration changed, with the strakes becoming larger and more pronounced upward-canted wing structures (although the aircraft was still regarded as a lifting body, since its broad fuselage would generate the majority of its lift). During the initial phases of the Lockheed study, a great deal of effort went into examining propulsion options, with the xlr99 or Rocketdyne's xlr105 being the preferred main engine options, with additional xlr11 or xl101 rockets being used during cruise (both the xlr105 and xlr101 were derived from the Atlas missile program). A large recessed bay in the belly of the aircraft would house research instrumentation and the scramjet units, with different propulsion modules fitted as required.

By 1975 work on the x-24c's design missions led to a requirement for a forty-second hypersonic cruise capability at 90,000 ft., meaning the research aircraft would be exposed to extremely high temperatures through the friction of air molecules as it streaked above the High Range. Lockheed examined two different approaches in order to cope with this extreme thermal environment: a heat-resistant hot structure and a heat-shielded conventional structure. The company proposed an approach using their own beryllium-aluminum alloy, Lockalloy, for the hot structure. Although the panels would need to be of a thick gauge and consequently relatively heavy, they would act as a heat sink during flight and need very little maintenance or refurbishment between flights. The heat-shielding approach would use a standard aluminum structure coated with either a spray-on ablative material, similar to that used on the x-15A-2, or heat-absorbing tiles formed from Lockheed's LI-900 silica-based material. Both of these materials would be lighter than Lockalloy, but both would require significant amounts of time and labor for refurbishment between flights, while they would also obscure the underlying structure, making inspections far more difficult. Interestingly, Lockheed discarded the LI-900 tiles as the highest-risk option, even though this approach had been adopted by Rockwell International for its shuttle orbiters.

As the Skunk Works' studies continued through 1976, NASA estimated that the program (now known as the National Hypersonic Flight Research Facility—NHFRF, pronounced as "nerf") would cost around $200 million, with flights due to commence in 1983. Echoing the x-15's final tally, the joint steering committee intended NHFRF to perform two hundred research flights over a ten-year period, offering the FRC and AFFTC a means to continue their pioneering flight research, while ushering in a new age of hypersonic lifting vehicles. Unfortunately, by 1977 the program costs were growing, as was the aircraft itself. By the end of Lockheed's third phase study, the x-24C had expanded into the far-larger L301 configuration, straining the B-52's carrying capabilities. Although the aircraft itself appeared technologically viable, NASA's interest in hypersonics had waned once more, and with the agency's budgets stretched thin by the space shuttle program, its support for NHFRF ended during September 1977. Unable to shoulder the financial burden alone, the air force also withdrew, and all official work on the x-24C NHFRF ground to a halt.

The failure of either the minishuttle or the x-24C to move beyond the drawing board came as a blow to the FRC. Bill Dana's final powered flight in the x-24B would remain as the last act in the nearly thirty-year story of rocket research aircraft that had played out on the shores of Rogers Dry Lake. In many ways, the lifting body programs marked a perfect denouement to this period of unparalleled progress. By taking a chance on Dale Reed's lightweight M2-F1, Paul Bikle had not only helped to secure the FRC's immediate future at a time when its continued existence beyond the x-15 program was being openly questioned; he had also set in motion a major new flight research effort that would have a significant influence on the nation's spacefaring future. What had started as a low-key, in-house project, conducted without the knowledge of NASA Headquarters, grew into a sustained research program involving numerous NASA centers, alongside the AFFDL and the AFFTC.

By flight-testing a number of promising lifting body configurations, the joint ranks of NASA and air force engineers and test pilots had demonstrated not only that these oddball, wingless aircraft could indeed be flown back to make a safe horizontal landing but also, more fundamentally, that flight research still had a vital role to play even in an age when the use of increasingly powerful digital computers in aerodynamic analysis was becoming more commonplace. In spite of their extensive wind tunnel testing and preflight simulation, the M2-F2, the HL-10, and the X-24A had all revealed flaws once they actually flew, flaws that were readily assessed by highly skilled pilots and analyzed by the engineers of the FRC and the AFFTC. The fact that all three lifting bodies were subsequently modified and returned to flight, carrying out highly successful research programs, stood as testament to Hugh Dryden's belief in flight-testing as a means to "separate the real from the imagined problems and to make known the overlooked and the unexpected problems." It was all the more fitting, then, that as the FRC looked toward its post–rocket plane future, it was dedicated in honor of the NACA's final director, becoming the Hugh L. Dryden Flight Research Center on 26 March 1976.

As the now-560-strong staff of NASA Dryden and their counterparts at the AFFTC moved on to their respective roles supporting the shuttle's forthcoming Approach and Landing Test (ALT) program and testing a new generation of combat aircraft designed with the hard-earned lessons of the Vietnam War in mind, the rocket test stands that had held the mighty power of the XLR11, the XLR25, and the XLR99 fell silent for good. The rocket planes themselves, or at least those that had survived their test careers, now became museum exhibits, static reminders of the imperative to fly higher and faster that had prevailed for so long and seen the first winged vehicles reach beyond the atmosphere and into space itself.

The end of the rocket plane era also coincided with the retirement of many of the stalwarts who had first traveled to Muroc AAF back in the late

1940s to work on the early transonic and supersonic research aircraft. Having served as deputy director to Walt Williams, Paul Bikle, and Lee Scherer, De Beeler, who also served as acting director following Bikle's retirement, left the FRC in 1974. Roxanah Yancey, the pioneering computer who had worked so tirelessly to ensure that the raw data gathered during research flights was reduced into a workable form for the center's engineers (before later graduating to become an engineer herself), had left one year earlier. As the ALT program got underway in 1977, NASA Dryden bid farewell to Jack Russell, the man whose long career in the High Desert had started as Bell Aircraft tested their XP-59 jet prototype in 1942.

Perhaps more than any other individual at Edwards, Russell's story was inextricably woven together with that of the rocket research aircraft. Having become an indispensable authority on the X-1 and its XLR11 engine, Russell had worked for Bell before transferring to the air force as a contractor when the government took control of the program in 1947. After moving to the NACA in 1950, he went on to work on the D-558-II, the advanced X-1s, the X-2, the X-15, and finally the lifting bodies. As his long and storied career drew to a close, Jack Russell received two honors. In recognition for his invaluable contributions to the success of the rocket research programs, he was presented with the NASA Exceptional Service Medal, but perhaps more touchingly, his many colleagues in the NASA rocket shop got together to present him with a more unusual souvenir of his time in the High Desert—a specially prepared and mounted XLR11 thrust chamber, including a chamber extension that had been used on Bill Dana's final rocket research flight in the X-24B.

Epilogue

The weird and wonderful rocket planes that flew over the High Desert between 1946 and 1975 were never built simply to break records, to go higher or faster for the sake of proving that this was possible. Although new records (albeit usually unofficial) were set with great regularity up until 1963, these accolades were a mere byproduct of the research aircraft's true purpose—the methodical collection of flight data as a means to verify or improve experimental models. This does not mean that speed and altitude records served no practical purpose; during the postwar period when both the navy and the newly formed air force were involved in high-speed flight research over the High Desert, the rocket planes offered a visible means to attract new recruits and Pentagon funding, while also demonstrating American engineering prowess to a world facing a new ideological conflict.

Just as Ezra Kotcher had surmised back in 1939, the rocket engine had provided a quick and efficient means of placing research vehicles into a desired high-speed flight condition. But as the frontiers of flight pushed ever outward into the thin air of the upper stratosphere and the high temperatures beyond Mach 2, the rocket became the only practical means available to propel these exotic aircraft.

At the sharp end of this high-stakes adventure sat the test pilots. Far from the gung ho heroes portrayed by Hollywood, these men were cautious and methodical, but that is not to say that they weren't possessed of great skill and courage. It took a special kind of bravery to flick a switch or advance a throttle, unleashing a barely contained explosion at their backs. In those moments, the pilot was truly alone, but they always flew with the knowledge that the ops engineer, the crew chief, and the dedicated cadre of skilled technicians responsible for maintaining those thoroughbred machines had toiled through the heat of the day and the chill of the night alike to ensure that their aircraft would not be found wanting.

For the most part, these pilots were not given to improvisation or to chasing demons and personal glory. Instead, they tended to be modest men who relied on their experience and the methodical preparation that preceded every flight. Their companions throughout this process were the flight planners—accomplished engineers who also spent countless hours interpreting the needs of aerodynamicists and researchers and distilling a cloud of desirable data points into a workable flight plan (complete with the oft called on contingencies, should things go awry). When the pilot finally brought their exhausted ship down onto the welcoming expanse of the lake bed, the freshly harvested data became the province of the computers— the dedicated, often-unheralded women who parsed filmstrips or telemetry readouts to provide the engineers with answers to their questions (or in many cases, posing whole new questions to be answered).

As the space race dawned in the late 1950s, it was hardly surprising that the newly formed NASA turned to the rocket plane programs at Edwards for operational knowledge and practical experience. Although America's early forays into human spaceflight all benefitted from work carried out in the High Desert, it would be the space shuttle that became the major beneficiary of the work done during the rocket plane years. When *Columbia* finally made its orbital debut in April 1981, it seemed perfectly natural that the orbiter's final destination at the end of a long lifting reentry would be the hard clay of Rogers Dry Lake. When the shuttle made a second test flight later that same year, its commander—former X-15 pilot Joe Engle—could not resist calling "Eddy Tower" during his final approach to the desert test site.

The flame of progress continued to burn at the Air Force Flight Test Center and the NASA Flight Research Center long after the last rocket engine had fallen silent and the dramas that had unfolded in the skies above were consigned to flight reports and research papers. There have been many false dawns during the intervening years, with the single-stage-to-orbit promise of the X-30 National Aero-Space Plane and the X-33 technology demonstrator proving as elusive as the mirages that shimmer above the sunbaked lake bed. For a brief period at the turn of the century, a new lifting body graced the skies above Edwards as the X-38, a prototype crew return vehicle (based on Hans Multhopp's sv-5 shape), conducted a series of recovery tests before succumbing to cancellation in 2002. More recently, the Sierra Nevada Corporation's Dream Chaser has revived hopes that an operational lifting body will finally fulfill the space station logistics role first envisioned for these wingless craft during the early 1960s.

Although time has taken its inevitable toll on the men and women who made the rocket plane programs a reality, those who remain maintain a deep pride in what was achieved during the three decades between 1946 and 1975. That heady mixture of camaraderie and competence fueled by a postwar can-do attitude and those hard-won victories and tragic losses marked by long days in the desert heat and long nights in the bars of Rosamond may lie in the past now, but the opportunities embodied by the seemingly endless runways of Rogers Dry Lake and the deep blue expanses of the High Desert's sky remain, awaiting the next generation of pioneers.

Sources

Books

Bridgeman, William, and Jacqueline Hazard. *The Lonely Sky*. London: Cassel, 1956.

Brzezinski, Matthew. *Red Moon Rising: Sputnik and the Rivalries That Ignited the Space Age*. London: Bloomsbury, 2007.

Crossfield, A. Scott, and Clay Blair Jr. *Always Another Dawn: The Story of a Rocket Test Pilot*. London: Hodder and Stoughton, 1961.

Davies, Peter E. *X-Planes 6: Bell x-2*. Oxford: Osprey Publishing, 2017.

Edwards AFB: Then and Now—A Pictorial Tour. Edwards AFB CA: Air Force Flight Test Centre History Office, 2001.

Evans, Michelle. *The x-15 Rocket Plane: Flying the First Wings into Space*. Lincoln: University of Nebraska Press, 2013.

Everest, Frank J., and John Guenther. *The Fastest Man Alive*. London: Cassel, 1958.

French, Francis, and Colin Burgess. *In the Shadow of the Moon: A Challenging Journey to Tranquility, 1965–1969*. Lincoln: University of Nebraska Press, 2007.

Godwin, Robert, ed. *Dyna-Soar: Hypersonic Strategic Weapons System*. Burlington ON: Apogee Books, 2003.

———. *x-15: The NASA Mission Reports*. Burlington ON: Apogee Books, 2000.

Goecke Powers, Sheryll. *Women in Flight Research at NASA Flight Research Center from 1946–1995*. Washington DC: NASA, 1997.

Gorn, Michael H. *Expanding the Envelope: Flight Research at NASA and NACA*. Lexington: University Press of Kentucky, 2001.

Gray, Mike. *Angle of Attack: Harrison Storms and the Race to the Moon*. New York: Norton, 1992.

Gunston, Bill. *Faster Than Sound: The Story of Supersonic Flight*. Yeovil, UK: Haynes, 2008.

———. *Jane's Aerospace Dictionary*. London: Jane's, 1980.

Hallion, Richard P. *Supersonic Flight—Breaking the Sound Barrier and Beyond: The Story of the Bell x-1 and the Douglas D-558*. New York: Macmillan, 1972.

———. *Test Pilots: The Frontiersmen of Flight*. Washington DC: Smithsonian Books, 1988.

Hallion, Richard P., and Michael H. Gorn. *On the Frontier: Experimental Flight at NASA Dryden*. Washington DC: Smithsonian Books, 2003.

Hansen, James R. *Engineer in Charge: A History of the Langley Aeronautical Laboratory, 1917–1958*. Washington DC: NASA, 1987.

———. *First Man: The Life of Neil Armstrong*. London: Simon and Schuster, 2005.

Hunley, J. D., ed. *Toward Mach 2: The Douglas D-558 Program*. Washington DC: NASA, 1999.

Jenkins, Dennis R. *Dressing for Altitude: U.S. Aviation Pressure Suits—Wiley Post to Space Shuttle*. Washington DC: NASA, 2012.

———. *Space Shuttle: The History of the National Space Transportation System—The First 100 Missions*. North Branch: Speciality Press, 2001.

———. *X-15: Extending the Frontiers of Flight*. Washington DC: NASA, 2007.

Jenkins, Dennis R., and Tony R. Landis. *Hypersonic: The Story of the North American X-15*. North Branch MN: Specialty Press, 2003.

———. *Valkyrie: North American's Mach 3 Superbomber*. North Branch MN: Specialty Press, 2004.

Libis, Scott. *Skystreak, Skyrocket, and Stiletto: Douglas High-Speed X-Planes*. North Branch MN: Specialty Press, 2005.

Matthews, Henry. *The Saga of the Bell X-2: First of the Spaceships—The Untold Story*. Beirut, Lebanon: HPM, 1999.

Miller, Jay. *The X-Planes—X-1 to X-45*. Hinckley OH: Midland, 2001.

Ottinger, Wayne C. *Four Winds Boomeranged: The Life Story of a Mechanical Engineer's Luck of the Draw*. Aletro, 2016.

Peebles, Curtis, ed. *The Spoken Word: Recollections of Dryden History—Beyond the Sky*. Washington DC: NASA, 2011.

———. *The Spoken Word: Recollections of Dryden History—The Early Years*. Washington DC: NASA, 2003.

Peebles, Curtis, and Richard P. Hallion. *Probing the Sky: Selected NACA Research Airplanes and Their Contributions to Flight*. Washington DC: NASA, 2014.

Reed, R. Dale, and Darlene Lister. *Wingless Flight: The Lifting Body Story*. Washington DC: NASA, 1997.

Rotundo, Louis. *Into the Unknown: The X-1 Story*. Shrewsbury, UK: Airlife, 1994.

Schultz, James. *Winds of Change: Expanding the Frontiers of Flight—Langley Research Center's 75 Years of Accomplishment, 1917–1992*. Washington DC: NASA, 1992.

Siddiqi, Asif A. *Sputnik and the Soviet Space Challenge*. Gainesville: University Press of Florida, 2003.

Smith, Melvyn. *Space Shuttle—US Winged Spacecraft: X-15 to Orbiter*. Yeovil, UK: Haynes, 1985.

Stoliker, Fred, Bob Hoey, and Johnny Armstrong, eds. *Flight Testing at Edwards: Flight Test Engineers' Stories, 1946–1975*. Lancaster CA: Flight Test Historical Foundation, 1996.

Sweetman, Bill. *High Speed Flight*. London: Jane's, 1983.

Swenson, Loyd S., Jr., James M. Grimwood, and Charles C. Alexander. *This New Ocean: A History of Project Mercury*. Washington DC: NASA, 1966.

Thompson, Milton O. *At the Edge of Space: The X-15 Flight Program*. Washington DC: Smithsonian Books, 1992.

Thompson, Milton O., and Curtis Peebles. *Flying without Wings: NASA Lifting Bodies and the Birth of the Space Shuttle*. Washington DC: Smithsonian Books, 1999.

Tregaskis, Richard. *X-15 Diary: The Story of America's First Space Ship*. Lincoln: University of Nebraska Press, 2004.

White, Robert M., and Jack L. Summers. *Higher and Faster: Memoir of a Pioneering Air Force Test Pilot*. Jefferson: McFarland, 2010.

Winter, Frank H. *America's First Rocket Company: Reaction Motors, Inc.* Reston VA: American Institute of Aeronautics and Astronautics, 2017.

Wolfe, Tom. *The Right Stuff*. London: Bantam Books, 1981.

Yeager, Chuck, Bob Cardenas, Bob Hoover, Jack Russell, and James Young. *The Quest for Mach One: A First-Person Account of Breaking the Sound Barrier*. London: Penguin, 1997.

Yeager, Chuck, and Leo Janos. *Yeager*. London: Pimlico, 1985.

Periodicals and Online Articles

Bergman, Jules V. "The Road to Beyond." *Air Force Magazine*, October 1960.

"Beyond Mach 1: The Remarkable Vehicles of Airborne Supersonic Research." *Flight*, 4 December 1953.

Blair, Clay, Jr. "The Last Flight of the X-2." *Air Force Magazine*, March 1957.

Boslaugh, David L. "First Hand: The Second Generation of X-1s." Chap. 7 in *The Experimental Research Airplanes and the Sound Barrier*. Engineering and Technology History Wiki. Last updated 15 January 2018. https://ethw.org/First-Hand:The_Second_Generation_X-1s _-_Chapter_7_of_The_Experimental_Research_Airplanes_and_the _Sound_Barrier.

Butz, J. S., Jr. "USAF Aerospace Research Pilot School: Toughest Flying School in the World." *Air Force Magazine*, September 1963.

Dupont, Ronald J., Jr. "Power for Progress: A Brief History of Reaction Motors Inc., 1941–1972." *Garden State Legacy*, no. 12 (June 2011).

Evans, Stanley H. "Tonic Sonics: Reflections on American Research, with Special Reference to the Douglas Skyrocket." *Flight*, 5 January 1950.

Kempel, Robert W. "Piloted Orbital Reentry Lifting Body Candidates Tested at Edwards Air Force Base." *American Aviation Historical Society Journal*, Fall 2016.

Kempel, Robert W., and Richard E. Day. "A Shadow over the Horizon—The Bell X-2." *American Aviation Historical Society Journal*, Spring 2003.

Leavitt, William. "Blueprint for Tomorrow's Space Crews." *Air Force Magazine*, May 1958.

"Nixon Unveils Rocket Ship Designed to Carry Spaceman." *Fort Lauderdale News*, 6 October 1958.

Peebles, Curtis. "Then and Now: Flight Research in the Second Half of the 20th Century." SAFE *Journal* 34, no. 1 (Fall 2006).

Powers, Tom. "He Flew Faster Than Sound." *Flying Magazine*, October 1951.

"Skyrocket: Turbo-Jet and Liquid Rocket Power in Douglas Research Aircraft." *Flight*, 5 February 1948.

SMA-ADMIN. "1958." Space Medicine Association. 31 July 2014. http://spacemedicineassociation.org/history1958/.

"Spaceflight: Lifting-Body Round-Up." *Flight International*, 9 July 1970.

"U.S. Military Rank Insignia." U.S. Department of Defense. https://www.defense.gov/Our-Story/Insignias/#enlisted-insignias.

Walker, Joseph A. "I Fly the X-15, Half Plane, Half Missile." *National Geographic Magazine*, September 1962.

"X-15 Pilots Receive Collier Trophy at White House." *New York Times*, 19 July 1962.

"X-15 Pilot Will Be Thoroughly Checked." *Washington Daily News*, 3 November 1958.

Interviews and Personal Communications

Armstrong, Neil A. NASA Oral History interview by Stephen E. Ambrose and Douglas Brinkley. 19 September 2001.

Bailey, Cylde, Richard Cox, Don Borchers, and Ralph Sparks. NASA Oral History interview by Michael Gorn. 30 March 1999.

Baker, Charlie. Email correspondence. December 2018.

———. Telephone interview. 17 December 2018.

Becker, John V. NASA Oral History interview by Rebecca Wright. 3 May 2008.

Butchart, Stanley P. NASA Oral History interview by Curt Asher. 15 September 1997.

Capasso, Vincent. Email correspondence. December 2018–May 2019.

Crossfield, Scott A. NASA Oral History interview by Peter Merlin. 3 February 1998.

———. Royal Aeronautical Society interview by Rodney Giesler. 8 May 1979.

Dana, William H. NASA Oral History interview by Peter Merlin. 14 November 1997.

Day, Richard E. NASA Oral History interview by J. D. Hunley. 1 May 1997.

Gelzer, Christian. Edwards CA. 10 July 2018.

———. Email correspondence. June 2018–August 2019.

———. Lancaster CA. 13 August 2019.

Griffith, John. NASA Oral History interview by Peter Merlin. 2 February 1998.

Hoey, Bob, and Johnny Armstrong. Email correspondence (via Dennis R. Jenkins). March 2019.

Jenkins, Dennis R. Email correspondence. September 2018–September 2019.

Kempel, Robert W. Email correspondence. December 2018–August 2019.

———. Lancaster CA. 13 August 2019.

———. Telephone interview. 19 December 2018.

Knight, William J. Royal Aeronautical Society interview by Rodney Giesler. 10 May 1979.

McTigue, John G. NASA Oral History interview by Rebecca Wright. 29 September 2005.

Merlin, Peter W. Email correspondence. May–August 2019.

———. Palmdale CA. 13 August 2019.

———. Telephone interview. 19 May 2019.

Moore, Tony. Email correspondence. April–August 2019.

———. Palmdale CA. 13 August 2019.

Ottinger, Wayne C. Email correspondence. October–December 2018.

———. Telephone interview. 27 November 2018.

Perry, John. Interview by Michelle Evans. August 2018.

Pyle, Jon. Email correspondence. December 2018–January 2019.

———. Telephone interview. 7 January 2019.

Reedy, Jerry. NASA Oral History interview by Curtis Peebles. 4 April 2001.

Stoddard, David. Email correspondence. October 2018–August 2019.

———. Lancaster CA. 14 August 2019.

———. Telephone interview. 15 November 2018.

Welsh, George B. Email correspondence. August 2018–January 2019.

Whiteside, Walter W. NASA Oral History interview by J. D. Hunley. 30 July 1997.

Williams, Walter C. NASA Oral History interview by John Terreo. 6 August 1993.

Yeager, Chuck. Royal Aeronautical Society interview by Rodney Giesler. 17 May 1979.

Other Sources

Centennial Symposium: X-1 Panel. Society of Experimental Test Pilots. Video. 2003. Wayne C. Ottinger Collection.

Centennial Symposium: X-15 and Lifting Bodies Panel. Society of Experimental Test Pilots. Video. 2003. Wayne C. Ottinger Collection.

Clarke, R. W., and Victor J. Handmacher. *Evaluation Report of North American Aviation Proposal: Saturn/X-15 Vehicle.* Los Angeles CA: Air Force Space Systems Division, 1960.

Combs, Henry G., et al. *Configuration Development Study of the X-24C Hypersonic Research Airplane: Executive Summary.* NASA CR 145274. October 1977.

Day, Richard E. *Coupling Dynamics in Aircraft: A Historical Perspective.* NASA Special Publication 532. Edwards CA: NASA, 1997.

Day, Richard E., and Donald Reisert. *Flight Behavior of the X-2 Research Airplane to a Mach Number of 3.20 and a Geometric Altitude of 126,200 Feet.* NASA Technical Memorandum X-137. Edwards CA: NASA, 1959.

Dennehy, Cornelius J., Jeb S. Orr, Immanuel Barshi, and Irving C. Statler. *A Comprehensive Analysis of the X-15 Flight 3-65 Accident.* NASA TM-2014-218538. Hampton VA: Langley Research Center, October 2014.

Draper, Alfred C., and Thomas S. Sieron. *Evolution and Development of Hypersonic Configurations, 1958–1990.* Baltimore MD: Air Force Systems Command, 1991.

Flight Test Results Pertaining to the Space Shuttlecraft: A Symposium Held at Flight Research Center Edwards, California. NASA TM X-2101. Washington DC: NASA, October 1970.

Geiger, Clarence J. *History of the X-20A Dyna-Soar: Volume I.* Historical Publications Series. Dayton OH: Aeronautical Systems Division Information Office, 1963.

Hallion, Richard P., ed. *The Hypersonic Revolution: Case Studies in the History of Hypersonic Technology—Volume I: From Max Valier to Project PRIME (1924–1967).* Washington DC: Air Force History and Museums Program, 1998.

———. *The Hypersonic Revolution: Eight Case Studies in the History of Hypersonic Technology—Volume II: From Scramjet to the National Aero-Space Plane.* Washington DC: Air Force History and Museums Program, 1987.

History of the Air Force Flight Test Center: Edwards Air Force Base, California. Baltimore MD: Air Force Systems Command, June 1961.

Hoey, Robert G. *Testing Lifting Bodies at Edwards.* Palmdale CA: PAT Projects, September 1994.

Houchin, Roy F. *The Rise and Fall of Dyna-Soar: A History of Air Force Hypersonic R&D, 1944–1963.* Auburn AL: Auburn University, August 1995.

Kempel, Robert W., Weneth D. Painter, and Milton O. Thompson. *Developing and Flight Testing the HL-10 Lifting Body: A Precursor to the Space Shuttle.* NASA Reference Publication 1332. Washington DC: NASA, April 1994.

NACA Conference on High-Speed Aerodynamics: A Compilation of the Papers Presented. NACA Technical Memorandum x-67369. Moffit Field CA: Ames Aeronautical Laboratory, 1958.

NASA. "Flying without Wings—NASA (AFRC/DFRC) Documentary about Lifting Bodies, by John McTigue." By Robert Hecker and David Bowen. NASA HQ-145. John J. Hennessy Motion Pictures, 1969. Uploaded by MDx Media, YouTube video, 14:27. https://www.youtube.com/watch?v=ez2clDxiDv8.

Parker Temple, L. *x-15B: The Spaceplane That Almost Was.* Paper. AIAA 57th International Astronautical Congress. October 2006.

Proceedings of the x-15 First Flight 30th Anniversary Celebration. NASA Conference Publication 3105. Edwards CA: NASA Ames Research Center, Dryden Flight Research Facility, 1991.

Rainey, Robert W., and Charles L. Ladson. *HL-10 Historical Review.* Hampton VA: NASA Langley Research Center, July 1969.

Research Airplane Committee Report on Conference on the Progress of the x-15 Project: A Compilation of Papers Presented. Hampton VA: NACA Langley Aeronautical Laboratory, 1956.

Rohrer, Robert A. *W. N. Y. Launches the Space Age: The x-1A Story.* Video. 1997. Dave Stoddard Collection.

Stillwell, Wendell H. *x-15 Research Results, with a Selected Bibliography.* Washington DC: NASA Scientific and Technical Information Division, 1965.

System Failure Case Studies: Vicious Cycle. Cleveland OH: NASA Safety Centre, March 2011.

Thompson, Wayne, and Bernard C. Nalty. *Within Limits: The U.S. Air Force and the Korean War.* Washington DC: Air Force History and Museums Program, 1996.

Toward the Unknown. Toluca Productions/Warner Brothers. Motion Picture. 1956.

USAF and Bell Aircraft. "Flight to the Future." 1956. Excerpted from the DVD *Mach 2: D-558 and x-2.* Uploaded by rocket.aero, YouTube video, 22:04. https://www.youtube.com/watch?v=IsPFTKyojmU.

x-15 Panel. Spacefest VIII. Video. 2017. Wayne C. Ottinger Collection.

x-15: Research at the Edge of Space. NASA Publication EP-9. Washington DC: NASA, 1963.

Zalovcik, John A. *A Radar Method of Calibrating Airspeed Installations on Airplanes in Maneuvers at High Altitudes and at Transonic and Supersonic Speeds.* NACA Technical Report 985. Washington DC: NACA, 1950.

Index

316; RMI motors in museum at, 281; USAAF Materiel Command at, 14; X-15 office at, 170, 180
Wright-Patterson Air Force Base (AFB), 49, 72, 128, 203
Wright R-3350 engine, 71, 141–42
WS-110A (North American Aviation), 170, 172
Wyld, James Hart, 21–22

X-1 (Bell Aircraft): air-launch of, 70; canopy on, 49, 58, 82, 90; emergency in, at low altitude, 73; and first Mach 1 test flight, 44–45, 308; funding for, 165; ground takeoff by, 60; impact of, xi; metal fatigue in nitrogen tanks of, 145; partial pressure suit for, 49; Pete Everest's high-altitude flights in, 48–49; Scott Crossfield as test pilot for, 90; slotted-throat test section verified by, 157; to Smithsonian Institution, 50; for transonic and supersonic research, 108; visual tracking of, 57; at Wright-Patterson AFB, 127. See also advanced X-1 (Bell Aircraft); XS-1 (Bell Aircraft)
X-1-3 Queenie (Bell Aircraft), 82–83, 85–86, 118, 123, 145, 185. See also XS-1 number 3 (Bell Aircraft)
X-1E "Little Joe," 145–48, 204, 221, 265, 281
X-2 (Bell Aircraft): accident investigation of, 118, 136–38; air-launch of, 112; airspeed achieved by, 130, 134–36, 201–2; altitude achieved by, 133, 201–2; canard surfaces on nose of, 119; canopy for, 112–13, 115–16, 133; and captive fuel jettison test in B-50 carrier, 118, 122; center of gravity of, 122; cockpit for, 111–12, 114; computer-based flight simulation for, 126–27; control stick for, 113–15; demonstration phase for, 122–32; detachable nose section on, 111–12, 114, 136–38, 173; electric-powered control system for, 113–15; emergency oxygen bottle on, 132; envelope-expansion phase for, 131–34; explosion during fueling tests of, 96, 108, 118, 123; fuel system for, 112, 129, 132; funding for, 165; fuselage on, 111–12; g-forces on, 129; glide tests for, 94, 115–17, 120–24; handover of, to NACA, 131, 133–34; high-altitude instability of, 142; horizontal stabilizers on, 112; hydraulic control system for, 120; images of, 116, 131; instrument panel for, 114, 127; Iven Kincheloe as test pilot

for, 128, 131–33; landing gear on, 112, 116–17, 120–23, 128; low-speed stall tests on, 116; maximum fuel load of, 112; Mel Apt as test pilot for, 127, 134–38; metal crash bar in cockpit of, 123; performance predictions for, 70; Pete Everest as pilot for, 94, 96, 117, 120–30, 134; powered flight tests of, 124–30; preflight checklist for, 124; research instrumentation in, 112, 120; Ronald-Bel Stifler on Mel Apt's death in, 138; Scott Crossfield as project pilot for, 119; Sergeant solid-fueled rockets for, 160; skin of, 112, 114, 119, 129–30; Skip Ziegler as test pilot for, 94, 96, 115–17; speed-expansion flights of, 126–27; stability testing of, at high speed, 126, 202; tail-high attitude of, 112, 116, 123; tail on, 112, 208; thrust misalignments on, 134, 208; tube bundles for storage of nitrogen replaced on, 96; Ulmer leather gaskets used on, 108, 123; USAF investment in, 124; weight of, when fully fueled, 125; wings on, 70, 111; XLR11 motor for, 118; XLR25 engine in, 112–18, 123, 125–26, 128–29, 132, 134. See also Model 52 (Bell Aircraft); MX-743 (Bell Aircraft); XS-2 (Bell Aircraft)
X-3 (Douglas Aircraft), 79, 167
X-4 (Northrop), 89–90, 106, 145
X-5 variable geometry test bed, 84
X-15 (North American Aviation): advanced digital flight data system for, 229; aeromedical data from, 211; air force investment in, 184; air-launch carrier for, studies of, 176–77; airspeed achieved by, with interim XLR11 motors, 192–93, 201–2, 205; airspeed achieved by, with XLR99 engine, 196, 205, 208–10, 214, 217, 225, 250; altitude achieved by, with interim XLR11 motors, 192–93, 201–2, 205; altitude achieved by, with XLR99 engine, 196, 205, 208–9, 213, 217–18, 220–21, 233–34, 241, 247, 250, 252; ammonia tanks on, 168, 195–96, 217, 241; approach speed for, 194; APU on, 186, 188–89, 219, 243–44; ballistic control jets on, 206, 243, 247; battery on, 243; boost guidance display on, 249; cameras on, 248; canopy on, 209–10; cockpit of, 179–80, 188, 195, 232; computerized attitude display in, 247–49; contract for, 124, 170; Dale

In the Outward Odyssey: A People's History of Spaceflight series

Into That Silent Sea: Trailblazers of the Space Era, 1961–1965
Francis French and Colin Burgess
Foreword by Paul Haney

In the Shadow of the Moon: A Challenging Journey to Tranquility, 1965–1969
Francis French and Colin Burgess
Foreword by Walter Cunningham

To a Distant Day: The Rocket Pioneers
Chris Gainor
Foreword by Alfred Worden

Homesteading Space: The Skylab Story
David Hitt, Owen Garriott, and Joe Kerwin
Foreword by Homer Hickam

Ambassadors from Earth: Pioneering Explorations with Unmanned Spacecraft
Jay Gallentine

Footprints in the Dust: The Epic Voyages of Apollo, 1969–1975
Edited by Colin Burgess
Foreword by Richard F. Gordon

Realizing Tomorrow: The Path to Private Spaceflight
Chris Dubbs and Emeline Paat-Dahlstrom
Foreword by Charles D. Walker

The X-15 Rocket Plane: Flying the First Wings into Space
Michelle Evans
Foreword by Joe H. Engle

Wheels Stop: The Tragedies and Triumphs of the Space Shuttle Program, 1986–2011
Rick Houston
Foreword by Jerry Ross

Bold They Rise
David Hitt and Heather R. Smith
Foreword by Bob Crippen

Go, Flight! The Unsung Heroes of Mission Control, 1965–1992
Rick Houston and Milt Heflin
Foreword by John Aaron

Infinity Beckoned: Adventuring Through the Inner Solar System, 1969–1989
Jay Gallentine
Foreword by Bobak Ferdowsi

Fallen Astronauts: Heroes Who Died Reaching for the Moon, Revised Edition
Colin Burgess and Kate Doolan with Bert Vis
Foreword by Eugene A. Cernan

Apollo Pilot: The Memoir of Astronaut Donn Eisele
Donn Eisele
Edited and with a foreword by Francis French
Afterword by Susie Eisele Black

Outposts on the Frontier: A Fifty-Year History of Space Stations
Jay Chladek
Foreword by Clayton C. Anderson

Come Fly with Us: NASA's Payload Specialist Program
Melvin Croft and John Youskauskas
Foreword by Don Thomas

Shattered Dreams: The Lost and Canceled Space Missions
Colin Burgess
Foreword by Don Thomas

*The Ultimate Engineer: The Remarkable Life
of NASA's Visionary Leader George M. Low*
Richard Jurek
Foreword by Gerald D. Griffin

Beyond Blue Skies: The Rocket Plane Programs That Led to the Space Age
Chris Petty
Foreword by Dennis R. Jenkins

To order or obtain more information on these or other University of Nebraska Press titles, visit nebraskapress.unl.edu.